Year 5A

A Guide to Teaching for Mastery

Series Editor: Tony Staneff

Contents

Introduction

Foreword by the series editor and author, Tony Staneff

For far too long in the UK, maths has been feared by learners – and by many teachers, too. As a result, most learners consistently underachieve. More crucially, negative beliefs about ability, aptitude and the nature of maths are entrenched in children's thinking from an early age.

Yet, as someone who has loved maths all my life, I've always believed that every child has the capacity to succeed in maths. I've also had the great pleasure of leading teams and departments who share that belief and passion. Teaching for mastery, as practised in China and other South-East Asian jurisdictions since the 1980s, has confirmed my conviction that maths really is for everyone and not just those who have a special talent. In recent years, my team and I at Trinity Academy, Halifax, have had the privilege of researching with and working alongside some of the finest mastery practitioners from the UK and beyond, whose impact on learners' confidence, achievement and attitude is an inspiration.

The mastery approach recognises the value of developing the power to think rather than just do. It also recognises the value of making a coherent journey in which whole-class groups tackle concepts in very small steps, one by one. You cannot build securely on loose foundations – and it is just the same with maths: by creating a solid foundation of deep understanding, our children's skills and confidence will be strong and secure. What's more, the mindset of learner and teacher alike is fundamental: everyone can do maths … EVERYONE CAN!

I am proud to have been part of the extensive team responsible for turning the best of the world's practice, research, insights, and shared experiences into *Power Maths*, a unique teaching and learning resource developed especially for UK classrooms. *Power Maths* embodies our vision to help and support primary maths teachers to transform every child's mathematical and personal development. 'Everyone can!' has become our mantra and our passion, and we hope it will be yours, too.

Now, explore and enjoy all the resources you need to teach for mastery, and please get back to us with your *Power Maths* experiences and stories!

What is *Power Maths*?

Created especially for UK primary schools, and aligned with the new National Curriculum, *Power Maths* is a whole-class, textbook-based mastery resource that empowers every child to understand and succeed. *Power Maths* rejects the notion that some people simply 'can't do' maths. Instead, it develops growth mindsets and encourages hard work, practice and a willingness to see mistakes as learning tools.

Best practice consistently shows that mastery of small, cumulative steps builds a solid foundation of deep mathematical understanding. *Power Maths* combines interactive teaching tools, high-quality textbooks and continuing professional development (CPD) to help you equip children with a deep and long lasting understanding. Based on extensive evidence, and developed in partnership with practising teachers, *Power Maths* ensures that it meets the needs of children in the UK.

Power Maths and Mastery

Power Maths makes mastery practical and achievable by providing the structures, pathways, content, tools and support you need to make it happen in your classroom.

To develop mastery in maths children need to be enabled to acquire a deep understanding of maths concepts, structures and procedures, step by step. Complex mathematical concepts are built on simpler conceptual components and when children understand every step in the learning sequence, maths becomes transparent and makes logical sense. Interactive lessons establish deep understanding in small steps, as well as effortless fluency in key facts such as tables and number bonds. The whole class works on the same content and no child is left behind.

Power Maths

- Builds every concept in small, progressive steps.
- Is built with interactive, whole-class teaching in mind.
- Provides the tools you need to develop growth mindsets.
- Helps you check understanding and ensure that every child is keeping up.
- Establishes core elements such as intelligent practice and reflection.

The *Power Maths* approach

Everyone can!

Founded on the conviction that every child can achieve, *Power Maths* enables children to build number fluency, confidence and understanding, step by step.

Child-centred learning

Children master concepts one step at a time in lessons that embrace a Concrete-Pictorial-Abstract (C-P-A) approach, avoid overload, build on prior learning and help them see patterns and connections. Same-day intervention ensures sustained progress.

Continuing professional development

Embedded teacher support and development offer every teacher the opportunity to continually improve their subject knowledge and manage whole-class teaching for mastery.

Whole-class teaching

An interactive, whole-class teaching model encourages thinking and precise mathematical language and allows children to deepen their understanding as far as they can.

Introduction to the author team

Power Maths arises from the work of maths mastery experts who are committed to proving that, given the right mastery mindset and approach, **everyone can do maths**. Based on robust research and best practice from around the world, *Power Maths* was developed in partnership with a group of UK teachers to make sure that it not only meets our children's wide-ranging needs but also aligns with the National Curriculum in England.

Tony Staneff, Series Editor and author

Vice Principal at Trinity Academy, Halifax, Tony also leads a team of mastery experts who help schools across the UK to develop teaching for mastery via nationally recognised CPD courses, problem-solving and reasoning resources, schemes of work, assessment materials and other tools.

A team of experienced authors, including:

 **Josh Lury** – a specialist maths teacher, author and maths consultant with a passion for innovative and effective maths education

Trinity Academy, Halifax (Michael Gosling CEO, Tony Staneff, Emily Fox, Kate Henshall, Rebecca Holland, Stephanie Kirk, Stephen Monaghan, Beth Smith and Rachel Webster)

David Board, Belle Cottingham, Jonathan East, Tim Handley, Derek Huby, Neil Jarrett, Timothy Weal, Paul Wrangles – skilled maths teachers and mastery experts

Cherri Moseley – a maths author, former teacher and professional development provider

Professors Liu Jian and Zhang Dan, Series Consultants and authors, and their team of mastery expert authors:

 **Wei Huinv, Huang Lihua, Zhu Dejiang, Zhu Yuhong, Hou Huiying, Yin Lili, Zhang Jing, Zhou Da and Liu Qimeng**

Used by over 20 million children, Professor Liu Jian's textbook programme is one of the most popular in China. He and his author team are highly experienced in intelligent practice and in embedding key maths concepts using a C-P-A approach.

A group of 15 teachers and maths co-ordinators

We have consulted our teacher group throughout the development of *Power Maths* to ensure we are meeting their real needs in the classroom.

Your *Power Maths* resources

To help you teach for mastery, *Power Maths* comprises a variety of high-quality resources.

Pupil Textbooks

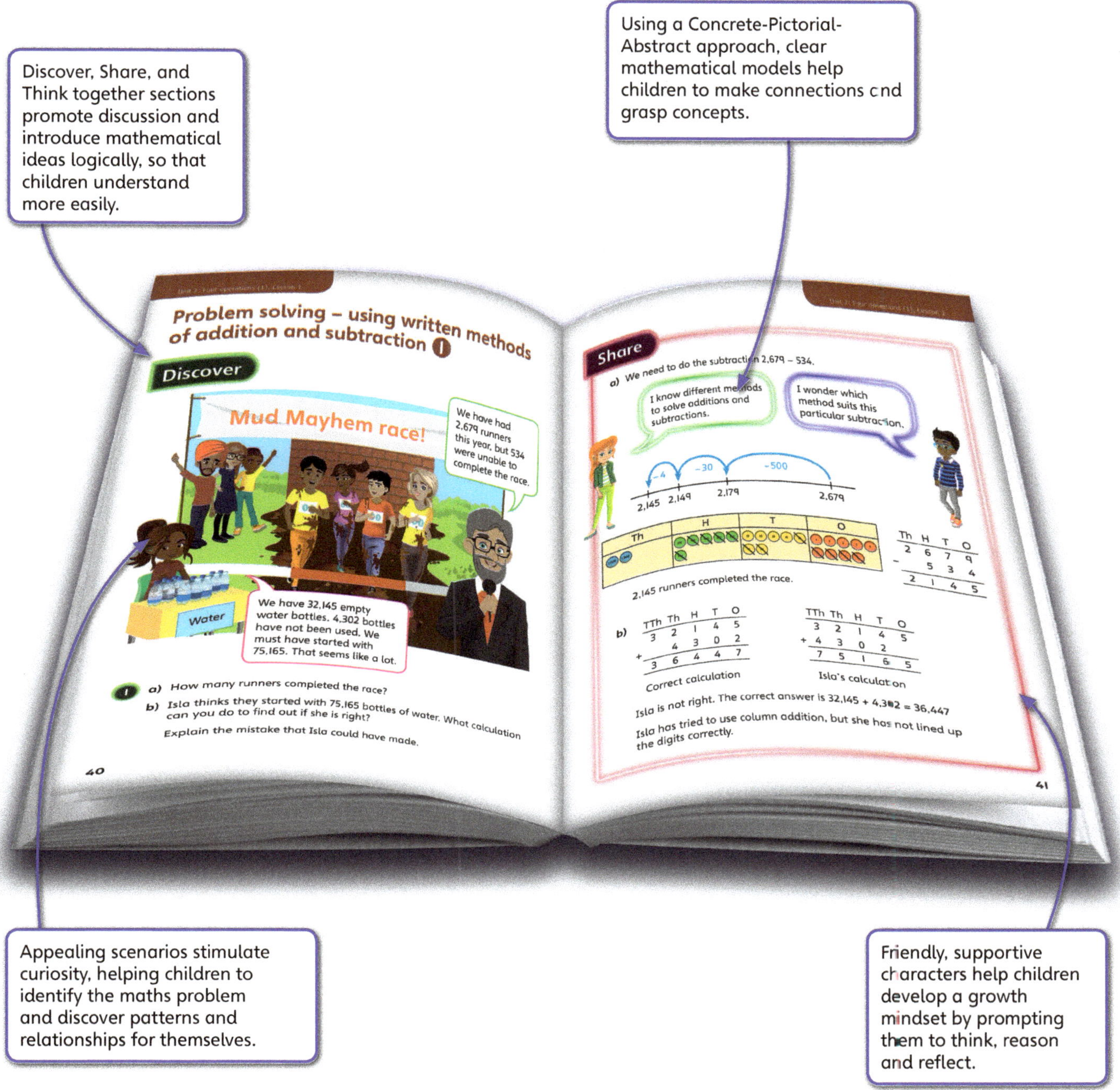

The coherent *Power Maths* lesson structure carries through into the vibrant, high-quality textbooks. Setting out the core learning objectives for each class, the lesson structure follows a carefully mapped journey through the curriculum and supports children on their journey to deeper understanding.

Pupil Practice Books

The Practice Books offer just the right amount of intelligent practice for children to complete independently in the final section of each lesson.

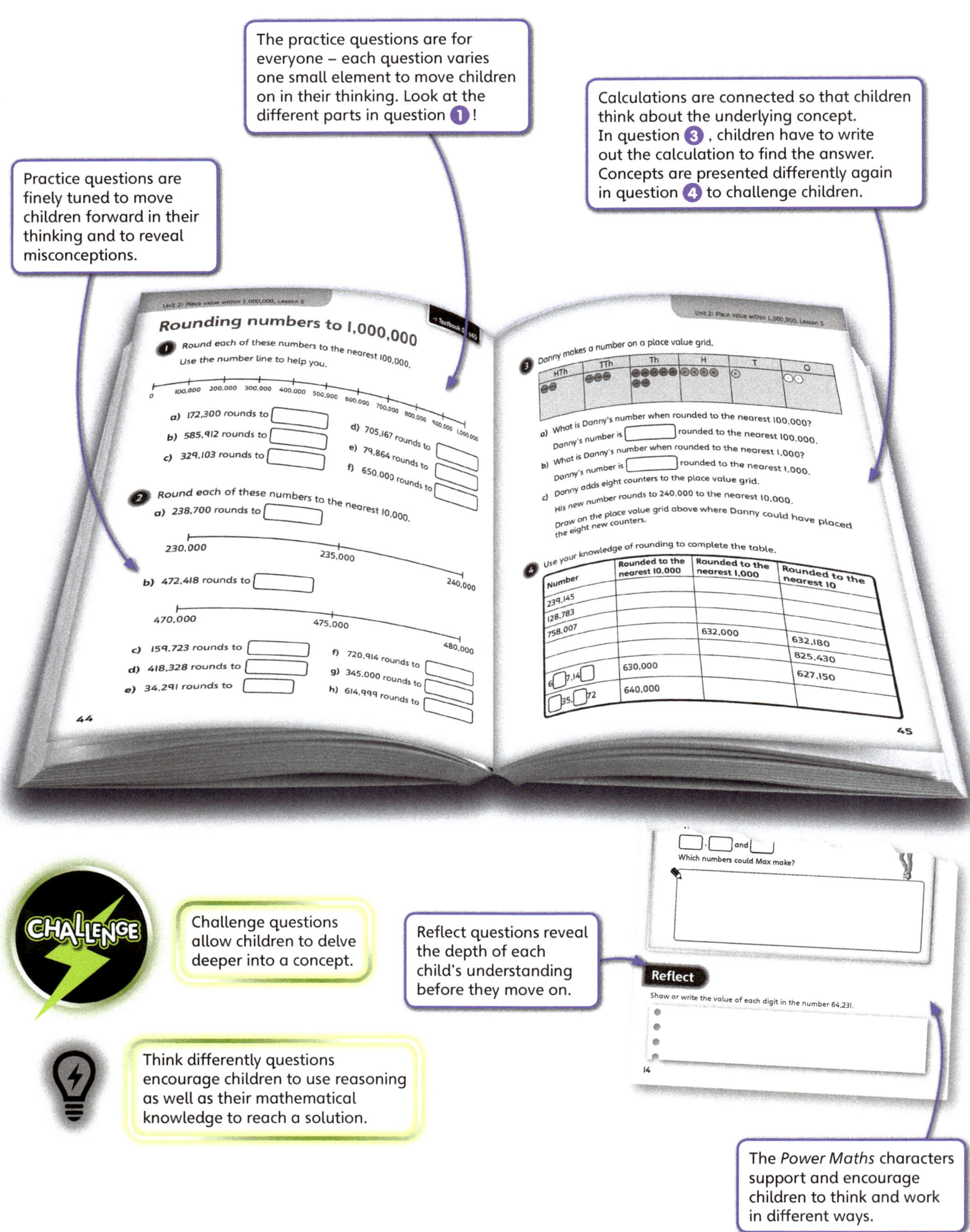

Online subscriptions

The online subscription will give you access to additional resources.

eTextbooks

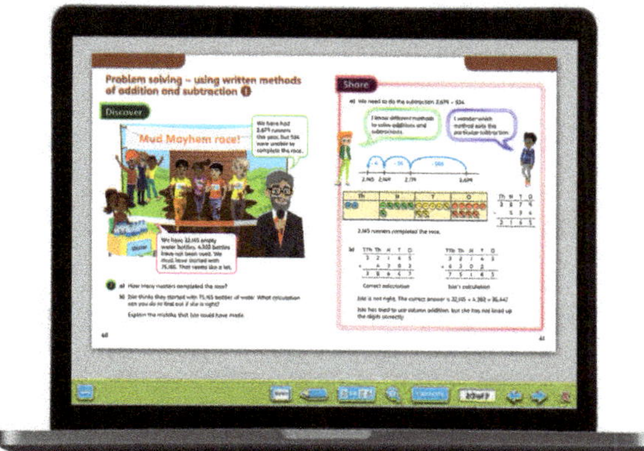

Digital versions of *Power Maths* Textbooks allow class groups to share and discuss questions, solutions and strategies. They allow you to project key structures and representations at the front of the class, to ensure all children are focusing on the same concept.

Teaching tools

Here you will find interactive vers ons of key *Power Maths* structures and representations.

Power Ups

Use this series of daily activities to promote and check number fluency.

Online versions of Teacher Guide pages

PDF pages give support at both unit and lesson levels. You will also find help with key strategies and templates for tracking progress.

Unit videos

Watch the professional development videos at the start of each unit to help you teach with confidence. The videos explore common misconceptions in the unit, and include intervention suggestions as well as suggestions on what to look out for when assessing mastery in your children.

End of unit Strengthen and Deepen materials

Each Strengthen activity at the end of every unit addresses a key misconception and can be used to support children who need it. The Deepen activities are designed to be 'Low Threshold High Ceiling' and will challenge those children who can understand more deeply. These resources will help you ensure that every child understands and will help you keep the class moving forward together. These printable activities provide an optional resource bank for use after the assessment stage.

Underpinning all of these resources, *Power Maths* is infused throughout with continual professional development, supporting you at every step.

The *Power Maths* teaching model

At the heart of *Power Maths* is a clearly structured teaching and learning process that helps you make certain that every child masters each maths concept securely and deeply. For each year group, the curriculum is broken down into core concepts, taught in units. A unit divides into smaller learning steps – lessons. Step by step, strong foundations of cumulative knowledge and understanding are built.

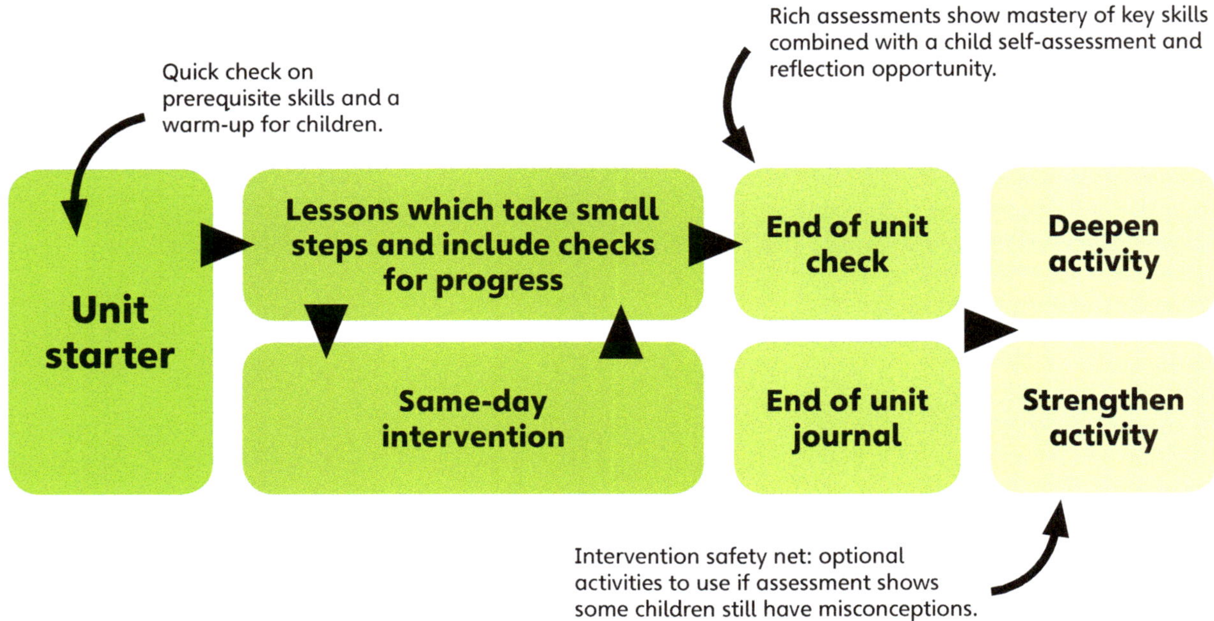

Unit starter

Each unit begins with a unit starter, which introduces the learning context along with key mathematical vocabulary, structures and representations.

- The Textbooks include a check on readiness and a warm-up task for children to complete.

- Your Teacher Guide gives support right from the start on important structures and representations, mathematical language, common misconceptions and intervention strategies.

- Unit-specific videos develop your subject knowledge and insights so you feel confident and fully equipped to teach each new unit. These are available via the online subscription.

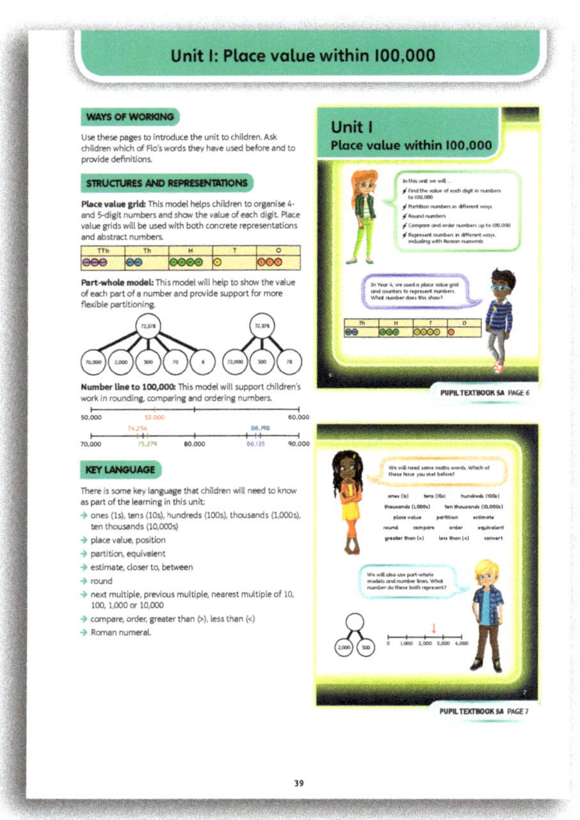

Lesson

Once a unit has been introduced, it is time to start teaching the series of lessons.

- Each lesson is scaffolded with Textbook and Practice Book activities and always begins with a Power Up activity (available via online subscription).
- *Power Maths* identifies lesson by lesson what concepts are to be taught.
- Your Teacher Guide offers lots of support for you to get the most from every child in every lesson. As well as highlighting key points, tricky areas and how to handle them, you will also find question prompts to check on understanding and clarification on why particular activities and questions are used.

Same-day intervention

Same-day interventions are vital in order to keep the class progressing together. Therefore, *Power Maths* provides plenty of support throughout the journey.

- Intervention is focused on keeping up now, not catching up later, so interventions should happen as soon as they are needed.
- Practice questions are designed to bring misconceptions to the surface, allowing you to identify these easily as you circulate during independent practice time.
- Child-friendly assessment questions in the Teacher Guide help you identify easily which children need to strengthen their understanding.

End of unit check and journal

At the end of a unit, summative assessment tasks reveal essential information on each child's understanding. An End of unit check in the Pupil Textbook lets you see which children have mastered the key concepts, which children have not and where their misconceptions lie. The Practice Book includes an End of unit journal in which children can reflect on what they have learnt. Each unit also offers Strengthen and Deepen activities, available via the online subscription.

The Teacher Guide offers support with handling misconceptions.

The End of unit check presents six to nine multiple-choice questions. These questions are designed to reveal misconceptions and help you target areas that need strengthening.

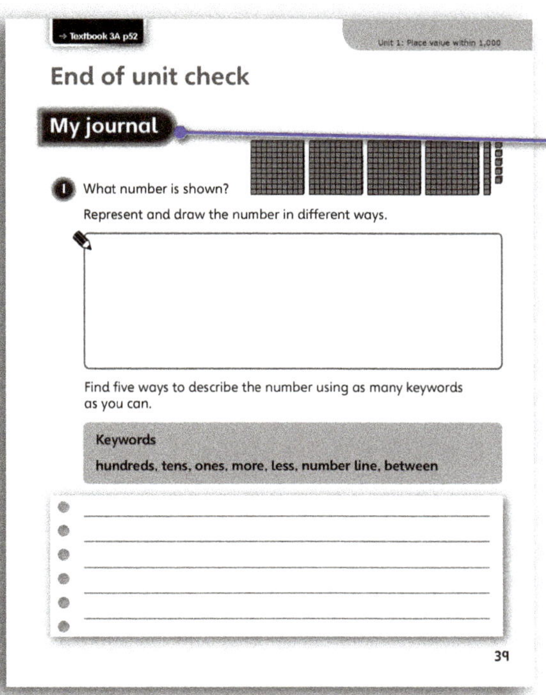

The End of unit journal is an opportunity for children to test out their learning and reflect on how they feel about it. Tackling the 'journal' problem reveals whether a child understands the concept deeply enough to move on to the next unit.

In KS2, the End of unit assessment will also include at least one SATs-style question.

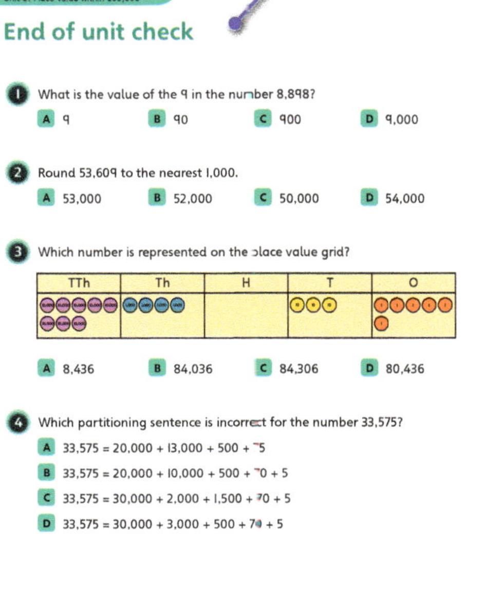

The *Power Maths* lesson sequence

At the heart of *Power Maths* is a unique lesson sequence designed to empower children to understand core concepts and grow in confidence. Embracing the National Centre for Excellence in the Teaching of Mathematics' (NCETM's) definition of mastery, the sequence guides and shapes every *Power Maths* lesson you teach.

Flexibility is built into the *Power Maths* programme so there is no one-to-one mapping of lessons and concepts meaning you can pace your teaching according to your class. While some children will need to spend longer on a particular concept (through interventions or additional lessons), others will reach deeper levels of understanding. However, it is important that the class moves forward together through the termly schedules.

Power Up ⏱ 5 minutes

Each lesson begins with a Power Up activity (available via the online subscription) which supports fluency in key number facts.

The whole-class approach depends on fluency, so the Power Up is a powerful and essential activity.

TOP TIP

If the class is struggling with the task, revisit it later and check understanding.

Power Ups reinforce key skills such as times-tables, number bonds and working with place value.

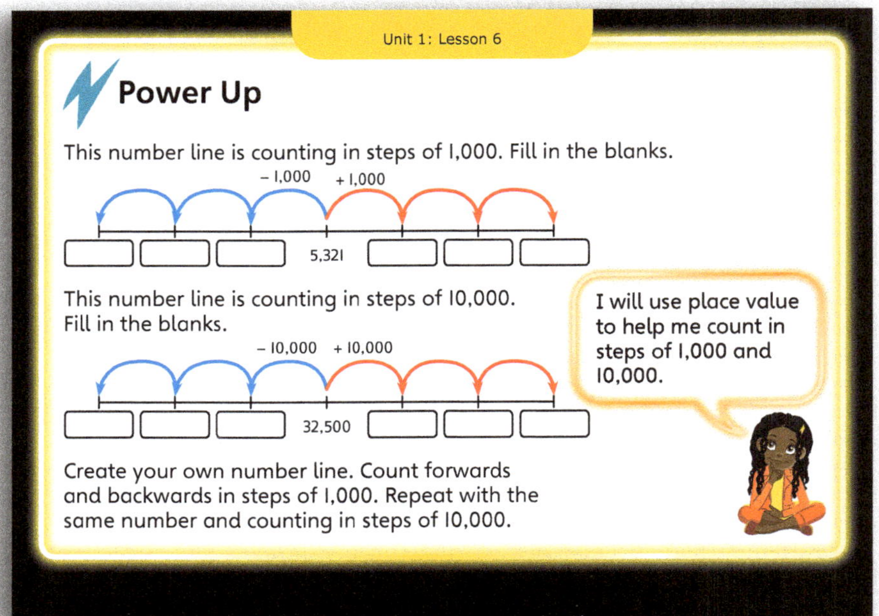

Discover ⏱ 10 minutes

A practical, real-life problem arouses curiosity. Children find the maths through story-telling.

A real-life scenario is provided for the Discover section but feel free to build upon these with your own examples that are more relevant to your class.

TOP TIP

Discover works best when run at tables, in pairs with concrete objects.

Question ❶ a) tackles the key concept and question ❶ b) digs a little deeper. Children have time to explore, play and discuss possible strategies.

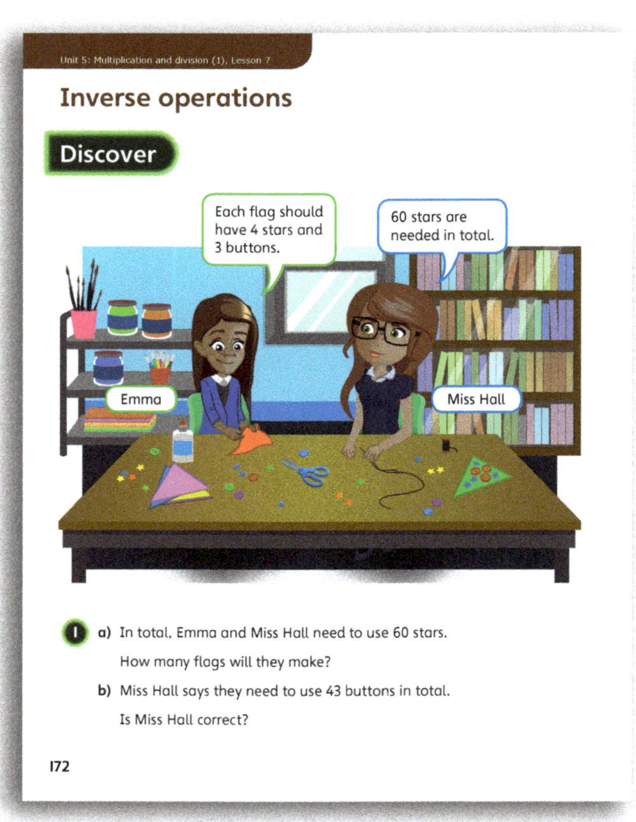

Share ⏱ 10 minutes

Teacher-led, this interactive section follows the Discover activity and highlights the variety of methods that can be used to solve a single problem.

TOP TIP

Ask children to discuss their methods. Pairs sharing a textbook is a great format for this!

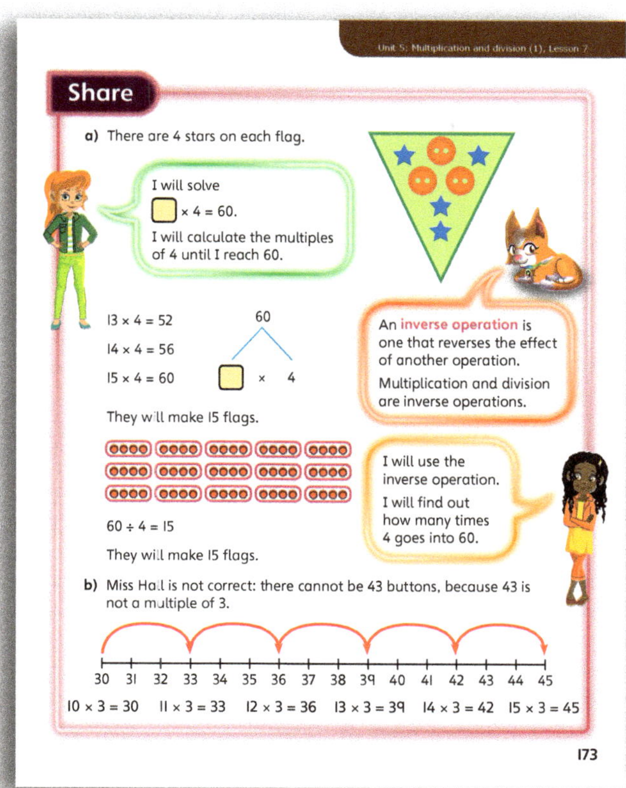

Your Teacher Guide gives target questions for children. The online toolkit provides interactive structures and representations to link concrete and pictorial to abstract concepts.

TOP TIP

Bring children to the front to share and celebrate their solutions and strategies.

Think together

⏱ 10 minutes

Children work in groups on the carpet or at tables, using their textbooks or eBooks.

TOP TIP

Make sure children have mini whiteboards or pads to write on if they are not at their tables.

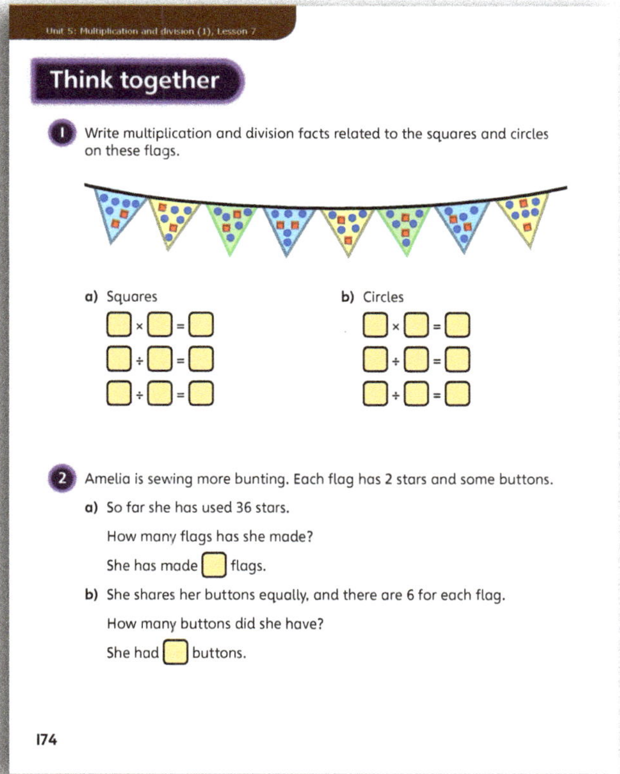

Using the Teacher Guide, model question ❶ for your class.

Question ❷ is less structured. Children will need to think together in their groups, then discuss their methods and solutions as a class.

In questions ❸ and ❹ children try working out the answer independently. The openness of the challenge question helps to check depth of understanding.

Using their Practice Books, children work independently while you circulate and check on progress.

Questions follow small steps of progression to deepen learning.

TOP TIP
Some children could work separately with a teacher or assistant.

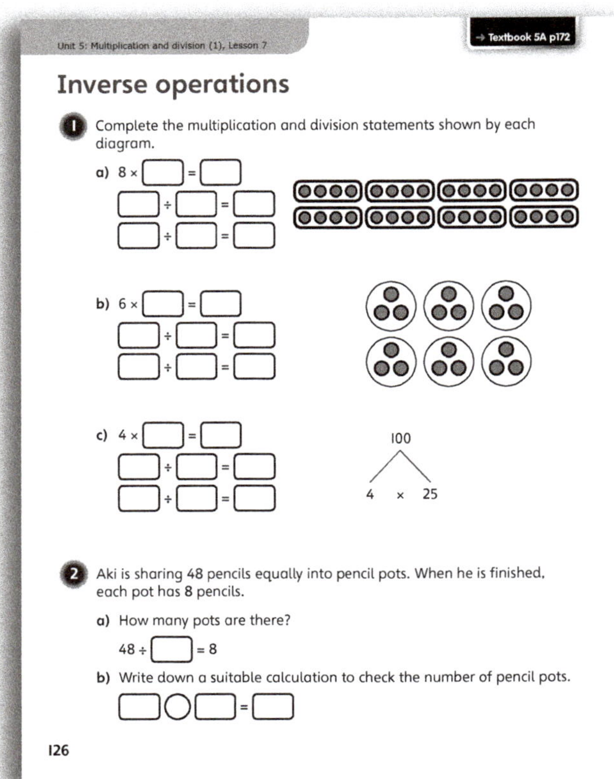

Are some children struggling? If so, work with them as a group, using mathematical structures and representations to support understanding as necessary.

There are no set routines: for real understanding, children need to think about the problem in different ways.

'Spot the mistake' questions are great for checking misconceptions.

The Reflect section is your opportunity to check how deeply children understand the target concept.

The Practice Books use various approaches to check that children have fully understood each concept.

Looking like they understand is not enough! It is essential that children can show they have grasped the concept.

Using the *Power Maths* Teacher Guide

Think of your Teacher Guides as *Power Maths* handbooks that will guide, support and inspire your day-to-day teaching. Clear and concise, and illustrated with helpful examples, your Teacher Guides will help you make the best possible use of every individual lesson. They also provide wrap-around professional development, enhancing your own subject knowledge and helping you to grow in confidence about moving your children forward together.

There is a Teacher Guide per year group for every term with unit and lesson level guidance and support.

Tips and advice on key elements such as C-P-A approaches, misconceptions, language, modelling growth mindsets and same-day intervention.

Annotations for every Pupil Textbook and Practice Book page, providing prompts for key questions to ask to expose understanding and explanations as to why key questions have been chosen.

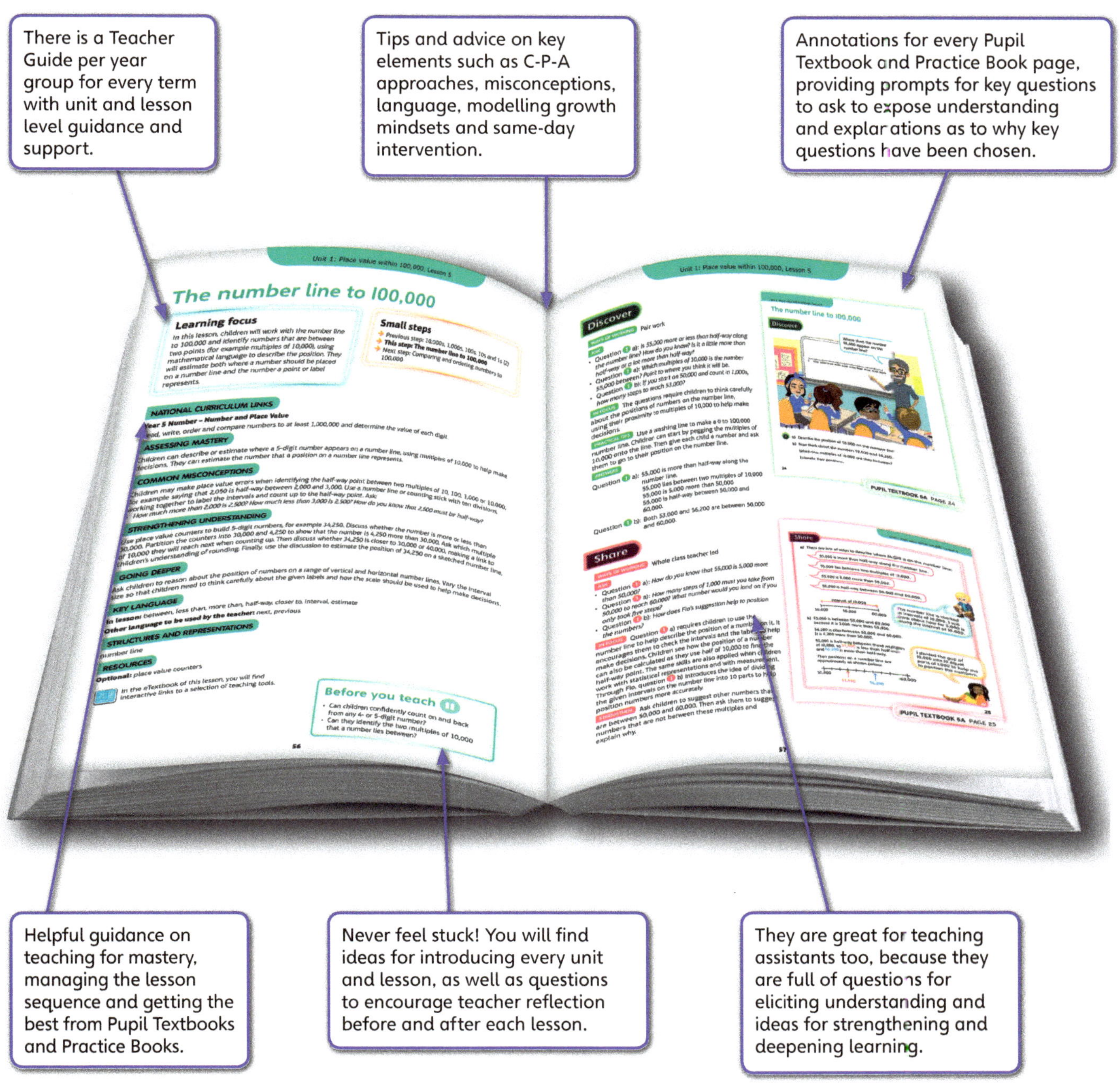

Helpful guidance on teaching for mastery, managing the lesson sequence and getting the best from Pupil Textbooks and Practice Books.

Never feel stuck! You will find ideas for introducing every unit and lesson, as well as questions to encourage teacher reflection before and after each lesson.

They are great for teaching assistants too, because they are full of questions for eliciting understanding and ideas for strengthening and deepening learning.

At the end of each unit, your Teacher Guide helps you identify who has fully grasped the concept, who has not and how to move every child forward. This is covered later in the Assessment strategies section.

Power Maths Year 5, yearly overview

Textbook	Strand	Unit		Number of Lessons
Textbook A / Practice Book A (Term 1)	Number – number and place value	1	Place value within 100,000	8
	Number – number and place value	2	Place value within 1,000,000	8
	Number – addition and subtraction	3	Addition and subtraction	10
	Statistics	4	Graphs and tables	5
	Number – multiplication and division	5	Multiplication and division (1)	10
	Measurement	6	Measure – area and perimeter	7
Textbook B / Practice Book B (Term 2)	Number – multiplication and division	7	Multiplication and division (2)	11
	Number – fractions (including decimals and percentages)	8	Fractions (1)	8
	Number – fractions (including decimals and percentages)	9	Fractions (2)	12
	Number – fractions (including decimals and percentages)	10	Fractions (3)	7
	Number – fractions (including decimals and percentages)	11	Decimals and percentages	12
Textbook C / Practice Book C (Term 3)	Number – fractions (including decimals and percentages)	12	Decimals	15
	Geometry – properties of shapes	13	Geometry – properties of shapes (1)	7
	Geometry – properties of shapes	14	Geometry – properties of shapes (2)	5
	Geometry – position and direction	15	Geometry – position and direction	4
	Measurement	16	Measure – converting units	10
	Measurement	17	Measure – volume and capacity	4

Power Maths Year 5, Textbook 5A (Term I) Overview

Strand 1	Strand 2	Unit		Lesson number	Lesson title	NC Objective 1	NC Objective 2	NC Objective 3
Number – number and place value		Unit 1	Place value within 100,000	1	Numbers to 10,000	Read, write, order and compare numbers to at least 1,000,000 and determine the value of each digit	Count forwards or backwards in steps of powers of 10 for any given number up to 1,000,000	
Number – number and place value		Unit 1	Place value within 100,000	2	Rounding to the nearest 10, 100 and 1,000	Round any number up to 1,000,000 to the nearest 10, 100, 1,000, 10,000 and 100,000		
Number – number and place value		Unit 1	Place value within 100,000	3	10,000s, 1,000s, 100s, 10s and 1s (1)	Read, write, order and compare numbers to at least 1,000,000 and determine the value of each digit		
Number – number and place value		Unit 1	Place value within 100,000	4	10,000s, 1,000s, 100s, 10s and 1s (2)	Solve number problems and practical problems that involve all of the above		
Number – number and place value		Unit 1	Place value within 100,000	5	The number line to 100,000	Read, write, order and compare numbers to at least 1,000,000 and determine the value of each digit		
Number – number and place value		Unit 1	Place value within 100,000	6	Comparing and ordering numbers to 100,000	Read, write, order and compare numbers to at least 1,000,000 and determine the value of each digit		

Strand 1	Strand 2	Unit		Lesson number	Lesson title	NC Objective 1	NC Objective 2	NC Objective 3
Number – number and place value		Unit 1	Place value within 100,000	7	Rounding numbers within 100,000	Round any number up to 1,000,000 to the nearest 10, 100, 1,000, 10,000 and 100,000		
Number – number and place value		Unit 1	Place value within 100,000	8	Roman numerals to 10,000	Read roman numerals to 1,000 (m) and recognise years written in roman numerals		
Number – number and place value		Unit 2	Place value within 1,000,000	1	100,000s 10,000s, 1,000s, 100s, 10s and 1s (1)	Read, write, order and compare numbers to at least 1,000,000 and determine the value of each digit		
Number – number and place value		Unit 2	Place value within 1,000,000	2	100,000s 10,000s, 1,000s, 100s, 10s and 1s (2)	Solve number problems and practical problems that involve all of the above		
Number – number and place value		Unit 2	Place value within 1,000,000	3	Number line to 1,000,000	Read, write, order and compare numbers to at least 1,000,000 and determine the value of each digit		
Number – number and place value		Unit 2	Place value within 1,000,000	4	Comparing and ordering numbers to 1,000,000	Read, write, order and compare numbers to at least 1,000,000 and determine the value of each digit		
Number – number and place value		Unit 2	Place value within 1,000,000	5	Rounding numbers to a 1,000,000	Round any number up to 1,000,000 to the nearest 10, 100, 1,000, 10,000 and 100,000		
Number – number and place value		Unit 2	Place value within 1,000,000	6	Negative numbers	Interpret negative numbers in context, count forwards and backwards with positive and negative whole numbers, including through zero		
Number – number and place value		Unit 2	Place value within 1,000,000	7	Counting in 10s, 100s, 1,000s, 10,000s	Count forwards or backwards in steps of powers of 10 for any given number up to 1,000,000		
Number – number and place value		Unit 2	Place value within 1,000,000	8	Number sequences	Solve number problems and practical problems that involve all of the above		
Number – addition and subtraction		Unit 3	Addition and subtraction	1	Adding whole numbers with more than 4 digits (1)	Add and subtract whole numbers with more than 4 digits, including using formal written methods (columnar addition and subtraction)		
Number – addition and subtraction		Unit 3	Addition and subtraction	2	Adding whole numbers with more than 4 digits (2)	Add and subtract whole numbers with more than 4 digits, including using formal written methods (columnar addition and subtraction)		
Number – addition and subtraction		Unit 3	Addition and subtraction	3	Subtracting whole numbers with more than 4 digits (1)	Add and subtract whole numbers with more than 4 digits, including using formal written methods (columnar addition and subtraction)		
Number – addition and subtraction		Unit 3	Addition and subtraction	4	Subtracting whole numbers with more than 4 digits (2)	Add and subtract whole numbers with more than 4 digits, including using formal written methods (columnar addition and subtraction)		
Number – addition and subtraction		Unit 3	Addition and subtraction	5	Using rounding to estimate and check answers	Use rounding to check answers to calculations and determine, in the context of a problem, levels of accuracy		
Number – addition and subtraction		Unit 3	Addition and subtraction	6	Mental addition and subtraction (1)	Add and subtract numbers mentally with increasingly large numbers		
Number – addition and subtraction		Unit 3	Addition and subtraction	7	Mental addition and subtraction (2)	Add and subtract numbers mentally with increasingly large numbers	Solve addition and subtraction multi-step problems in contexts, deciding which operations and methods to use and why	

Strand 1	Strand 2	Unit		Lesson number	Lesson title	NC Objective 1	NC Objective 2	NC Objective 3
Number – addition and subtraction		Unit 3	Addition and subtraction	8	Using inverse operations	Estimate and use inverse operations to check answers to a calculation		
Number – addition and subtraction		Unit 3	Addition and subtraction	9	Problem solving – addition and subtraction (1)	Solve addition and subtraction multi-step problems in contexts, deciding which operations and methods to use and why		
Number – addition and subtraction		Unit 3	Addition and subtraction	10	Problem solving – addition and subtraction (2)	Solve addition and subtraction multi-step problems in contexts, deciding which operations and methods to use and why		
Statistics		Unit 4	Graphs and tables	1	Interpreting tables	Complete, read and interpret information in tables, including timetables		
Statistics		Unit 4	Graphs and tables	2	Two-way tables	Complete, read and interpret information in tables, including timetables		
Statistics		Unit 4	Graphs and tables	3	Interpreting line graphs (1)	Solve comparison, sum and difference problems using information presented in a line graph		
Statistics		Unit 4	Graphs and tables	4	Interpreting line graphs (2)	Solve comparison, sum and difference problems using information presented in a line graph		
Statistics		Unit 4	Graphs and tables	5	Drawing line graphs	Solve comparison, sum and difference problems using information presented in a line graph		
Number – multiplication and division		Unit 5	Multiplication and division (1)	1	Multiples	Identify multiples and factors, including finding all factor pairs of a number, and common factors of two numbers	Solve problems involving multiplication and division including using their knowledge of factors and multiples, squares and cubes	
Number – multiplication and division		Unit 5	Multiplication and division (1)	2	Factors	Identify multiples and factors, including finding all factor pairs of a number, and common factors of two numbers		
Number – multiplication and division		Unit 5	Multiplication and division (1)	3	Prime numbers	Know and use the vocabulary of prime numbers, prime factors and composite (non-prime) numbers	Establish whether a number up to 100 is prime and recall prime numbers up to 19	
Number – multiplication and division		Unit 5	Multiplication and division (1)	4	Using factors	Solve problems involving multiplication and division including using their knowledge of factors and multiples, squares and cubes		
Number – multiplication and division		Unit 5	Multiplication and division (1)	5	Squares	Recognise and use square numbers and cube numbers, and the notation for squared (2) and cubed (3)	Solve problems involving multiplication and division including using their knowledge of factors and multiples, squares and cubes	
Number – multiplication and division		Unit 5	Multiplication and division (1)	6	Cubes	Recognise and use square numbers and cube numbers, and the notation for squared (2) and cubed (3)	Identify multiples and factors, including finding all factor pairs of a number, and common factors of two numbers	Solve problems involving multiplication and division including using their knowledge of factors and multiples, squares and cubes
Number – multiplication and division		Unit 5	Multiplication and division (1)	7	Inverse operations	Solve problems involving multiplication and division, including scaling by simple fractions and problems involving simple rates		

Strand 1	Strand 2	Unit		Lesson number	Lesson title	NC Objective 1	NC Objective 2	NC Objective 3
Number – multiplication and division		Unit 5	Multiplication and division (1)	8	Multiplying whole numbers by 10, 100 and 1,000	Multiply and divide whole numbers and those involving decimals by 10, 100 and 1,000		
Number – multiplication and division		Unit 5	Multiplication and division (1)	9	Dividing whole numbers by 10, 100 and 1,000	Multiply and divide whole numbers and those involving decimals by 10, 100 and 1,000	Solve problems involving multiplication and division, including scaling by simple fractions and problems involving simple rates	
Number – multiplication and division		Unit 5	Multiplication and division (1)	10	Multiplying and dividing by multiples of 10, 100 and 1,000	Multiply and divide whole numbers and those involving decimals by 10, 100 and 1,000		
Measurement		Unit 6	Measure – area and perimeter	1	Measuring perimeter	Measure and calculate the perimeter of composite rectilinear shapes in centimetres and metres		
Measurement		Unit 6	Measure – area and perimeter	2	Calculating perimeter (1)	Measure and calculate the perimeter of composite rectilinear shapes in centimetres and metres		
Measurement		Unit 6	Measure – area and perimeter	3	Calculating perimeter (2)	Measure and calculate the perimeter of composite rectilinear shapes in centimetres and metres		
Measurement		Unit 6	Measure – area and perimeter	4	Calculating area (1)	Calculate and compare the area of rectangles (including squares), and including using standard units, square centimetres (cm^2) and square metres (m^2) and estimate the area of irregular shapes		
Measurement		Unit 6	Measure – area and perimeter	5	Calculating area (2)	Calculate and compare the area of rectangles (including squares), and including using standard units, square centimetres (cm^2) and square metres (m^2) and estimate the area of irregular shapes		
Measurement		Unit 6	Measure – area and perimeter	6	Comparing area	Calculate and compare the area of rectangles (including squares), and including using standard units, square centimetres (cm^2) and square metres (m^2) and estimate the area of irregular shapes		
Measurement		Unit 6	Measure – area and perimeter	7	Estimating area	Calculate and compare the area of rectangles (including squares), and including using standard units, square centimetres (cm^2) and square metres (m^2) and estimate the area of irregular shapes		

Mindset: an introduction

Global research and best practice deliver the same message: learning is greatly affected by what learners perceive they can or cannot do. What is more, it is also shaped by what their parents, carers and teachers perceive they can do. Mindset – the thinking that determines our beliefs and behaviours – therefore has a fundamental impact on teaching and learning.

Everyone can!

Power Maths and mastery methods focus on the distinction between 'fixed' and 'growth' mindsets (Dweck, 2007).[1] Those with a fixed mindset believe that their basic qualities (for example, intelligence, talent and ability to learn) are pre-wired or fixed: 'If you have a talent for maths, you will succeed at it. If not, too bad!' By contrast, those with a growth mindset believe that hard work, effort and commitment drive success and that 'smart' is not something you are or are not, but something you become. In short, everyone can do maths!

Key mindset strategies

A growth mindset needs to be actively nurtured and developed. *Power Maths* offers some key strategies for fostering healthy growth mindsets in your classroom.

It is okay to get it wrong

Mistakes are valuable opportunities to re-think and understand more deeply. Learning is richer when children and teachers alike focus on spotting and sharing mistakes as well as solutions.

Praise hard work

Praise is a great motivator, and by focusing on praising effort and learning rather than success, children will be more willing to try harder, take risks and persist for longer.

Mind your language!

The language we use around learners has a profound effect on their mindsets. Make a habit of using growth phrases, such as, 'Everyone can!', 'Mistakes can help you learn' and 'Just try for a little longer'. The king of them all is one little word, 'yet …
I cannot solve this … yet!'
Encourage parents and carers to use the right language too.

Build in opportunities for success

The step-by-small-step approach enables children to enjoy the experience of success. In addition, avoid ability grouping and encourage every child to answer questions and explain or demonstrate their methods to others.

[1]Dweck, C (2007) *The New Psychology of Success*, Ballantine Books: New York

The *Power Maths* characters

The *Power Maths* characters model the traits of growth mindset learners and encourage resilience by prompting and questioning children as they work. Appearing frequently in the Textbooks and Practice Books, they are your allies in teaching and discussion, helping to model methods, alternatives and misconceptions, and to pose questions. They encourage and support your children, too: they are all hardworking, enthusiastic and unafraid of making and talking about mistakes.

Meet the team!

Creative Flo is open-minded and sometimes indecisive. She likes to think differently and come up with a variety of methods or ideas.

Determined Dexter is resolute, resilient and systematic. He concentrates hard, always tries his best and he'll never give up – even though he doesn't always choose the most efficient methods!

Curious Ash is eager, interested and inquisitive, and he loves solving puzzles and problems. Ash asks lots of questions but sometimes gets distracted.

Sparks the Cat

Brave Astrid is confident, willing to take risks and unafraid of failure. She is never scared to jump straight into a problem or question, and although she often makes simple mistakes she is happy to talk them through with others.

Mathematical language

Traditionally, we in the UK have tended to try simplifying mathematical language to make it easier for young children to understand. By contrast, evidence and experience show that by diluting the correct language, we actually mask concepts and meanings for children. We then wonder why they are confused by new and different terminology later down the line! *Power Maths* is not afraid of 'hard' words and avoids placing any barriers between children and their understanding of mathematical concepts. As a result, we need to be planned, precise and thorough in building every child's understanding of the language of maths. Throughout the Teacher Guides you will find support and guidance on how to deliver this, as well as individual explanations throughout the Pupil Textbooks.

Use the following key strategies to build children's mathematical vocabulary, understanding and confidence.

Precise and consistent

Everyone in the classroom should use the correct mathematical terms in full, every time. For example, refer to 'equal parts', not 'parts'. Used consistently, precise maths language will be a familiar and non-threatening part of children's everyday experience.

Full sentences

Teachers and children alike need to use full sentences to explain or respond. When children use complete sentences, it both reveals their understanding and embeds their knowledge.

Stem sentences

These important sentences help children express mathematical concepts accurately, and are used throughout the *Power Maths* books. Encourage children to repeat them frequently, whether working independently or with others. Examples of stem sentences are:

'4 is a part, 5 is a part, 9 is the whole.'

'There are … groups. There are … in each group.'

Key vocabulary

The unit starters highlight essential vocabulary for every lesson. In the Pupil Textbooks, characters flag new terminology and the Teacher Guide lists important mathematical language for every unit and lesson. New terms are never introduced without a clear explanation.

Mathematical signs

Mathematical signs are used early on so that children quickly become familiar with them and their meaning. Often, the *Power Maths* characters will highlight the connection between language and particular signs.

The role of talk and discussion

When children learn to talk purposefully together about maths, barriers of fear and anxiety are broken down and they grow in confidence, skills and understanding. Building a healthy culture of 'maths talk' empowers their learning from day one.

Explanation and discussion are integral to the *Power Maths* structure, so by simply following the books your lessons will stimulate structured talk. The following key 'maths talk' strategies will help you strengthen that culture and ensure that every child is included.

Sentences, not words

Encourage children to use full sentences when reasoning, explaining or discussing maths. This helps both speaker and listeners to clarify their own understanding. It also reveals whether or not the speaker truly understands, enabling you to address misconceptions as they arise.

Working together

Working with others in pairs, groups or as a whole class is a great way to support maths talk and discussion. Use different group structures to add variety and challenge. For example, children could take timed turns for talking, work independently alongside a 'discussion buddy', or perhaps play different *Power Maths* character roles within their group.

Think first – then talk

Provide clear opportunities within each lesson for children to think and reflect, so that their talk is purposeful, relevant and focused.

Give every child a voice

Where the 'hands up' model allows only the more confident child to shine, *Power Maths* involves everyone. Make sure that no child dominates and that even the shyest child is encouraged to contribute – and is praised when they do.

Assessment strategies

Teaching for mastery demands that you are confident about what each child knows and where their misconceptions lie: therefore, practical and effective assessment is vitally important.

Formative assessment within lessons

The Think together section will often reveal any confusions or insecurities: try ironing these out by doing the first Think together question as a class. For children who continue to struggle, you or your teaching assistant should provide support and enable them to move on.

Performance in Practice can be very revealing: check Practice Books and listen out both during and after practice to identify misconceptions.

The Reflect section is designed to check on the all-important depth of understanding. Be sure to review how children performed in this final stage before you teach the next lesson.

End of unit check – Textbook

Each unit concludes with a summative check to help you assess quickly and clearly each child's understanding, fluency, reasoning and problem-solving skills. In KS2 this check also contains a SATs-style question to help children become familiar with answering this type of question.

In KS2 we would suggest the End of unit check is completed independently in children's exercise books, but you can adapt this to suit the needs of your class.

End of unit check – Practice Book

The Practice Book contains further opportunities for assessment, and can be completed by children independently whilst you are carrying out diagnostic assessment with small groups. Your Teacher Guide will advise you on what to do if children struggle to articulate an explanation – or perhaps encourage you to write down something they have explained well. It will also offer insights into children's answers and their implications for the next learning steps. It is split into three main sections, outlined below.

My journal

My journal is designed to allow children to show their depth of understanding of the unit. It can also serve as a way of checking that children have grasped key mathematical vocabulary. Children should have some time to think about how they want to answer the question, and you could ask them to talk to a partner about their ideas. Then children should write their answer in their Practice Book.

Power check

The Power check allows children to self-assess their level of confidence on the topic by colouring in different smiley faces. You may want to introduce the faces as follows:

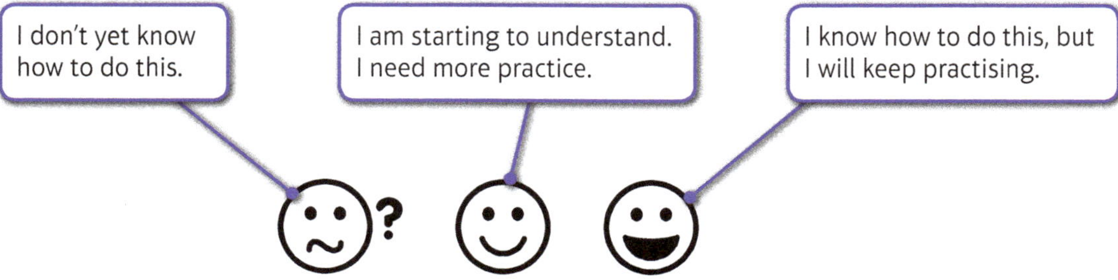

Power play or Power puzzle

Each unit ends with either a Power play or a Power puzzle. This is an activity, puzzle or game that allows children to use their new knowledge in a fun, informal way. In Key Stage 2 we have also included a deeper level to each game to help challenge those children who have grasped a concept quickly.

How to use diagnostic questions

The diagnostic questions provided in *Power Maths* Textbooks are carefully structured to identify both understanding and misconceptions (if children answer in a particular way, you will know why). The simple procedure below may be helpful:

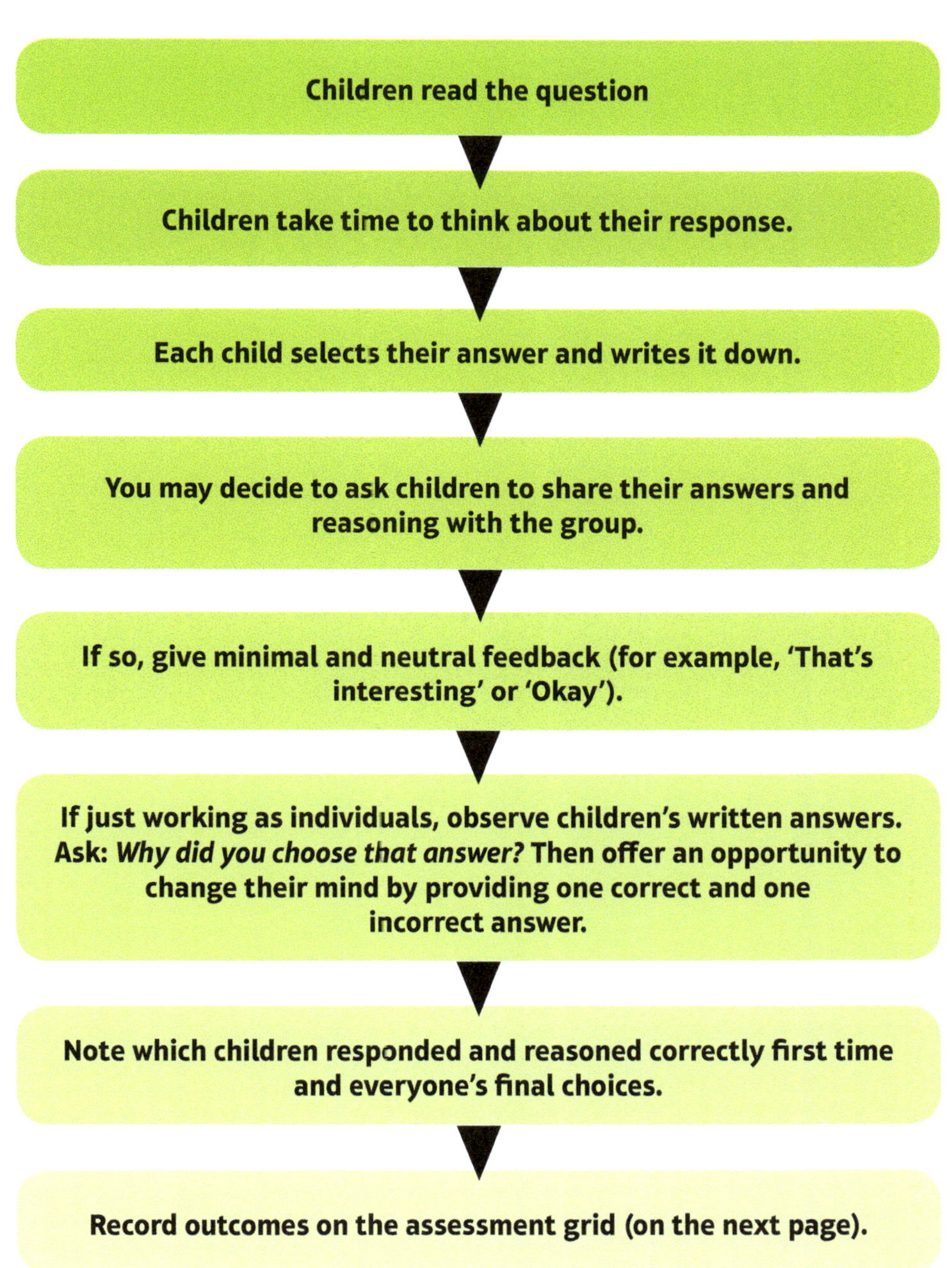

Power Maths unit assessment grid

Year ___   **Unit** ___ ____

Record only as much information as you judge appropriate for your assessment of each child's mastery of the unit and any steps needed for intervention.

Name	Diagnostic questions	SATs-style question	My journal	Power check	Power play/puzzle	Mastery	Intervention/ Strengthen

Keeping the class together

Traditionally, children who learn quickly have been accelerated through the curriculum. As a consequence, their learning may be superficial and will lack the many benefits of enabling children to learn with and from each other.

By contrast, *Power Maths'* mastery approach values real understanding and richer, deeper learning above speed. It sees all children learning the same concept in small, cumulative steps, each finding and mastering challenge at their own level. Remember that when you teach for mastery, EVERYONE can do maths! Those who grasp a concept easily have time to explore and understand that concept at a deeper level. The whole class therefore moves through the curriculum at broadly the same pace via individual learning journeys.

For some teachers, the idea that a whole class can move forward together is revolutionary and challenging. However, the evidence of global good practice clearly shows that this approach drives engagement, confidence, motivation and success for all learners, and not just the high flyers. The strategies below will help you keep your class together on their maths journey.

Mix it up

Do not stick to set groups at each table. Every child should be working on the same concept, and mixing up the groupings widens children's opportunities for exploring, discussing and sharing their understanding with others.

Recycling questions

Reuse the Pupil Textbook and Practice Book questions with concrete materials to allow children to explore concepts and relationships and deepen their understanding. This strategy is especially useful for reinforcing learning in same-day interventions.

Strengthen at every opportunity

The next lesson in a *Power Maths* sequence always revises and builds on the previous step to help embed learning. These activities provide golden opportunities for individual children to strengthen their learning with the support of teaching assistants.

Prepare to be surprised!

Children may grasp a concept quickly or more slowly. The 'fast graspers' won't always be the same individuals, nor does the speed at which a child understands a concept predict their success in maths. Are they struggling or just working more slowly?

Depth and breadth

Just as prescribed in the National Curriculum, the goal of *Power Maths* is never to accelerate through a topic but rather to gain a clear, deep and broad understanding.

"Pupils who grasp concepts rapidly should be challenged through being offered rich and sophisticated problems before any acceleration through new content. Those who are not sufficiently fluent with earlier material should consolidate their understanding, including through additional practice, before moving on."

National Curriculum: Mathematics programmes of study: KS1 & 2, 2013

The lesson sequence offers many opportunities for you to deepen and broaden children's learning, some of which are suggested below.

Discover

As well as using the questions in the Teacher Guide, check that children are really delving into why something is true. It is not enough to simply recite facts, such as '6 + 3 = 9'. They need to be able to see why, explain it, and to demonstrate the solution in several ways.

Share

Make sure that every child is given chances to offer answers and expand their knowledge and not just those with the greatest confidence.

Think together

Encourage children to think about how they found the solution and explain it to their partner. Be sure to make concrete materials available on group tables throughout the lesson to support and reinforce learning.

Practice

Avoid any temptation to select questions according to your assessment of ability: practice questions are presented in a logical sequence and it is important that each child works through every question.

Reflect

Open-ended questions allow children to deepen their understanding as far as they can by discovering new ways of finding answers. For example, *Give me another way of working out how high the wall is … And another way?*

Online materials

For each unit you will find additional strengthening activities to support those children who need it and to deepen the understanding of those who need the additional challenge.

Same-day intervention

Since maths competence depends on mastering concepts one-by-one in a logical progression, it is important that no gaps in understanding are ever left unfilled. Same-day interventions – either within or after a lesson – are a crucial safety net for any child who has not fully made the small step covered that day. In other words, intervention is always about keeping up, not catching up, so that every child has the skills and understanding they need to tackle the next lesson. That means presenting the same problems used in the lesson, with a variety of concrete materials to help children model their solutions.

We offer two intervention strategies below, but you should feel free to choose others if they work better for your class.

Within-lesson intervention

The Think together activity will reveal those who are struggling, so when it is time for Practice, bring these children together to work with you on the first Practice questions. Observe these children carefully, ask questions, encourage them to use concrete models and check that they reach and can demonstrate their understanding.

After-lesson intervention

You might like to use Think together before an assembly, giving you or teaching assistants time to recap and expand with slow graspers during assembly time. Teaching assistants could also work with strugglers at other convenient points in the school day.

The role of practice

Practice plays a pivotal role in the *Power Maths* approach. It takes place in class groups, smaller groups, pairs and independently, so that children always have the opportunities for thinking as well as the models and support they need to practise meaningfully and with understanding.

Intelligent practice

In *Power Maths*, practice never equates to the simple repetition of a process. Instead we embrace the concept of intelligent practice, in which all children become fluent in maths through varied, frequent and thoughtful practice that deepens and embeds conceptual understanding in a logical, planned sequence. To see the difference, take a look at the following examples.

Traditional practice

- Repetition can be rote – no need for a child to think hard about what they are doing.

- Praise may be misplaced.

- Does this prove understanding?

Intelligent practice

- Varied methods – concrete, pictorial and abstract.

- Calculations expressed in different ways, requiring thought and understanding.

- Constructive feedback.

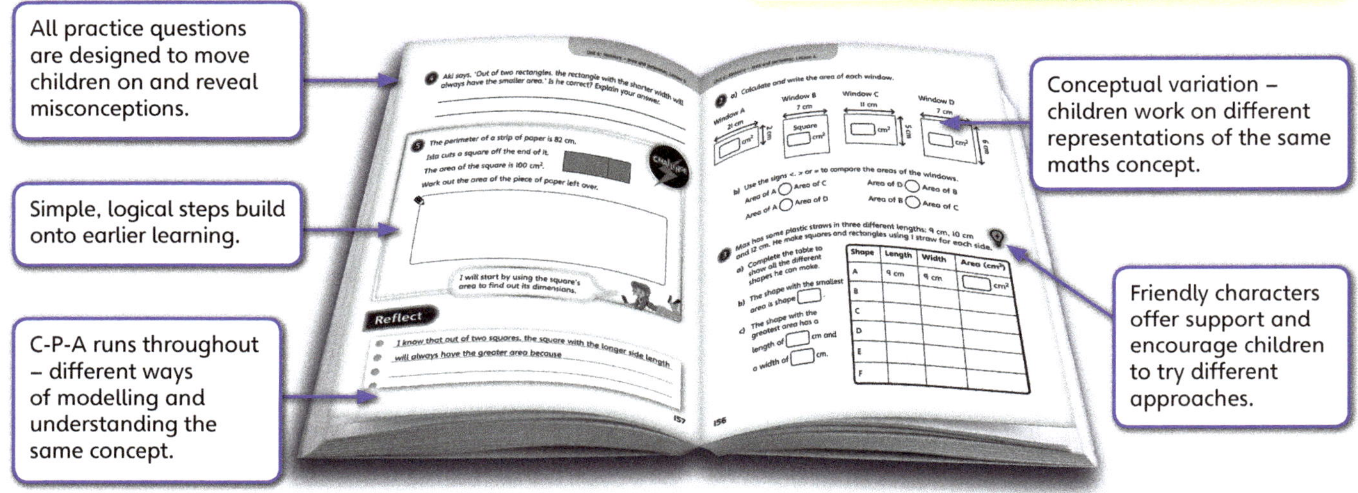

A carefully designed progression

The Practice Books provide just the right amount of intelligent practice for children to complete independently in the final sections of each lesson. It is really important that all children are exposed to the Practice questions, and that children are not directed to complete different sections. That is because each question is different and has been designed to challenge children to think about the maths they are doing. The questions become more challenging so children grasping concepts more quickly will start to slow down as they progress. Meanwhile, you have the chance to circulate and spot any misconceptions before they become barriers to further learning.

Homework and the role of carers

While *Power Maths* does not prescribe any particular homework structure, we acknowledge the potential value of practice at home. For example, practising fluency in key facts, such as number bonds and times-tables, is an ideal homework task, and carers could work through uncompleted Practice Book questions with children at either primary stage.

However, it is important to recognise that many parents and carers may themselves lack confidence in maths, and few, if any, will be familiar with mastery methods. A Parents' and Carers' Evening that helps them understand the basics of mindsets, mastery and mathematical language is a great way to ensure that children benefit from their homework. It could be a fun opportunity for children to teach their families that everyone can do maths!

Structures and representations

Unlike most other subjects, maths comprises a wide array of abstract concepts – and that is why children and adults so often find it difficult. By taking a Concrete-Pictorial-Abstract (C-P-A) approach, *Power Maths* allows children to tackle concepts in a tangible and more comfortable way.

Non-linear stages

Concrete

Replacing the traditional approach of a teacher working through a problem in front of the class, the concrete stage introduces real objects that children can use to 'do' the maths – any familiar object that a child can manipulate and move to help bring the maths to life. It is important to appreciate, however, that children must always understand the link between models and the objects they represent. For example, children need to first understand that three cakes could be represented by three pretend cakes, and then by three counters or bricks. Frequent practice helps consolidate this essential insight. Although they can be used at any time, good concrete models are an essential first step in understanding.

Pictorial

This stage uses pictorial representations of objects to let children 'see' what particular maths problems look like. It helps them make connections between the concrete and pictorial representations and the abstract maths concept. Children can also create or view a pictorial representation together, enabling discussion and comparisons. The *Power Maths* teaching tools are fantastic for this learning stage, and bar modelling is invaluable for problem solving throughout the primary curriculum.

Abstract

Our ultimate goal is for children to understand abstract mathematical concepts, signs and notation and, of course, some children will reach this stage far more quickly than others. To work with abstract concepts, a child needs to be comfortable with the meaning of, and relationships between, concrete, pictorial and abstract models and representations. The C-P-A approach is not linear, and children may need different types of models at different times. However, when a child demonstrates with concrete models and pictorial representations that they have grasped a concept, we can be confident that they are ready to explore or model it with abstract signs such as numbers and notation.

Use at any time and with any age to support understanding.

Practical aspects of *Power Maths*

One of the key underlying elements of *Power Maths* is its practical approach, allowing you to make maths real and relevant to your children, no matter their age.

Manipulatives are essential resources for both key stages and *Power Maths* encourages teachers to use these at every opportunity, and to continue the Concrete-Pictorial-Abstract approach right through to Year 6.

The Textbooks and Teacher Guides include lots of opportunities for teaching in a practical way to show children what maths means in real life.

Discover and Share

The Discover and Share sections of the Textbook give you scope to turn a real-life scenario into a practical and hands-on section of the lesson. Use these sections as inspiration to get active in the classroom. Where appropriate, use the Discover contexts as a springboard for your own examples that have particular resonance for your children – and allow them to get their hands dirty trying out the mathematics for themselves.

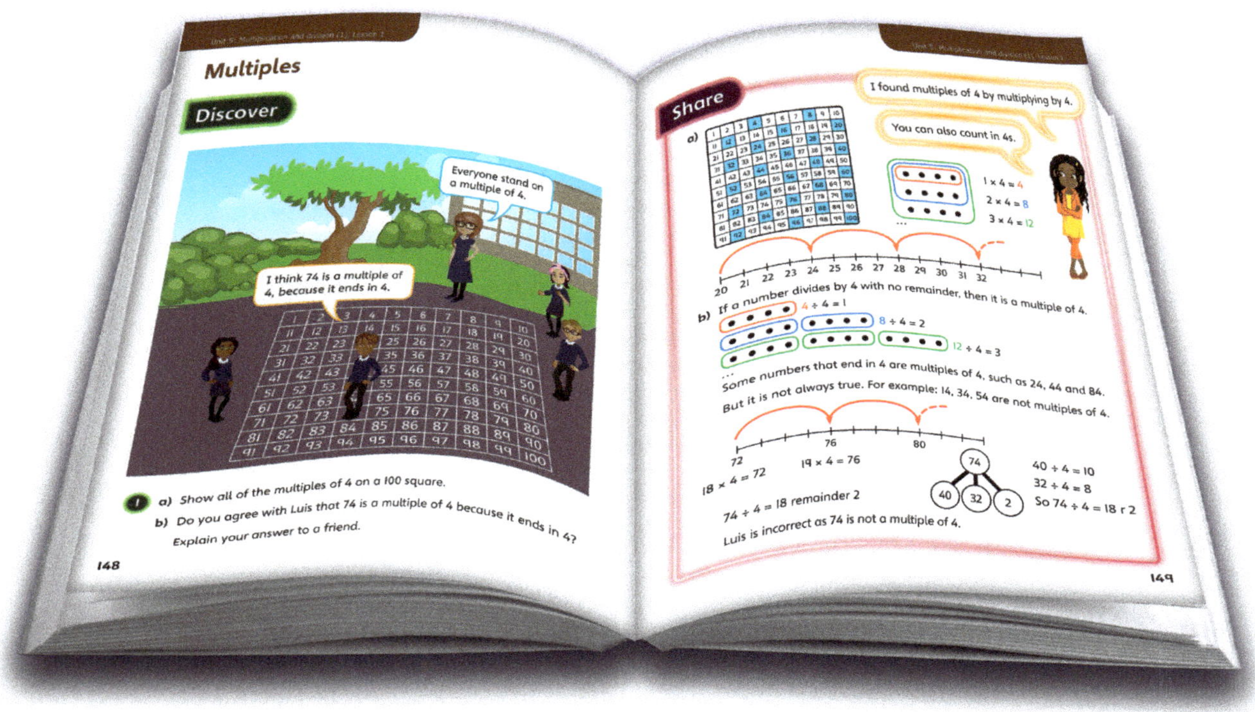

Unit videos

Every unit has a video which incorporates real-life classroom sequences.

These videos show you how the reasoning behind mathematics can be carried out in a practical manner by showing real children using various concrete and pictorial methods to come to the solution. You can see how using these practical models, such as part-whole and bar models, helps them to find and articulate their answer.

Mastery tips

Mastery Experts give anecdotal advice on where they have used hands-on and real-life elements to inspire their children.

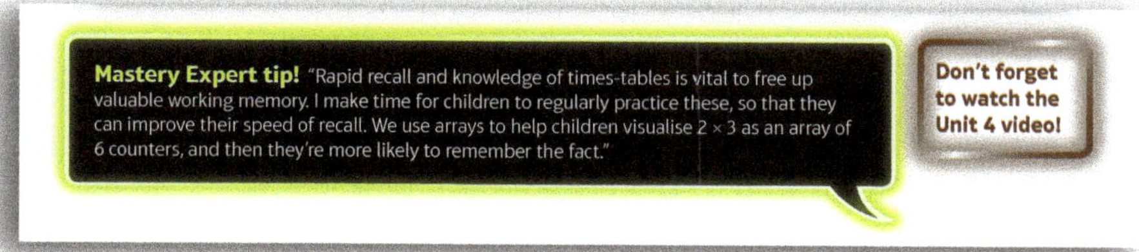

Concrete-Pictorial-Abstract (C-P-A) approach

Each Share section uses various methods to explain an answer, helping children to access abstract concepts by using concrete tools, such as counters. Remember this isn't a linear process, so even children who appear confident using the more abstract method can deepen their knowledge by exploring the concrete representations. Encourage children to use all three methods to really solidify their understanding of a concept.

Pictorial representation – drawing the problem in a logical way that helps children visualise the maths

Concrete representation – using manipulatives to represent the problem. Encourage children to physically use resources to explore the maths.

Abstract representation – using words and calculations to represent the problem.

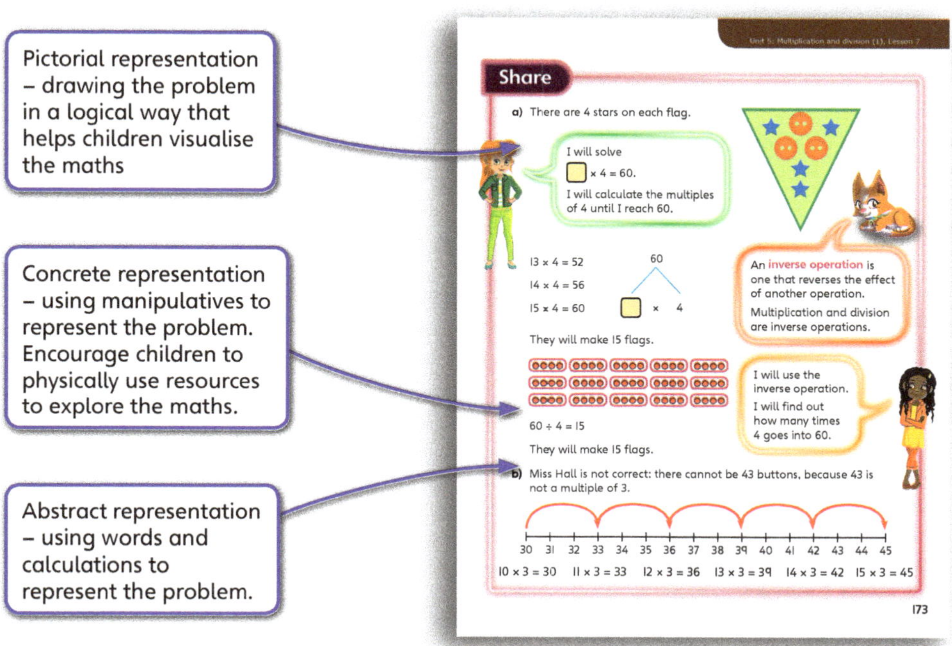

Practical tips

Every lesson suggests how to draw out the practical side of the Discover context.

You'll find these in the Discover section of the Teacher Guide for each lesson.

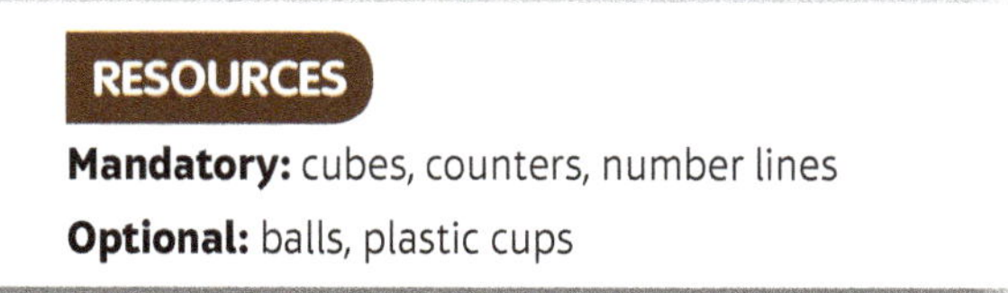

Resources

Every lesson lists the practical resources you will need or might want to use. There is also a summary of all of the resources used throughout the term on page 34 to help you be prepared.

RESOURCES

Mandatory: cubes, counters, number lines
Optional: balls, plastic cups

List of practical resources

Year 5A Mandatory resources

Resource	Lesson
100 square	**Unit 5** lessons 1, 3
2d squares and rectangles	**Unit 6** lessons 2, 3
Base 10 equipment	**Unit 5** lessons 8, 9, 10
Squared paper (cm)	**Unit 4** lesson 5 **Unit 5** lesson 2 **Unit 6** lessons 4, 6
Counters	**Unit 3** lessons 3, 4 **Unit 5** lessons 2, 3, 4, 7
Cubes	**Unit 5** lessons 6, 7
Digit cards	**Unit 2** lesson 5 **Unit 3** lesson 4
Metre rulers	**Unit 6** lesson 4
Place value counters	**Unit 2** lessons 1, 2, 4, 7 **Unit 3** lessons 1, 2 **Unit 5** lessons 9, 10
Rulers/measuring tapes (cm)	**Unit 6** lessons 1, 2, 3

Year 5A Optional resources

Resource	Lesson
2D shapes in 2 colours	**Unit 4** lesson 2
Base 10 equipment	**Unit 1** lessons 1, 2, 3, 4 **Unit 2** lessons 1, 7 **Unit 5** lessons 6, 4
Bead strings	**Unit 2** lesson 3
Calculators	**Unit 1** lesson 4
Chessboard	**Unit 5** lesson 5
Counters	**Unit 4** lesson 2 **Unit 5** lessons 1, 5 **Unit 6** lessons 5, 6, 7
Cubes	**Unit 2** lesson 8 **Unit 5** lessons 2, 5
Digit cards	**Unit 2** lessons 3, 4, 6
Examples of maps of the local area	**Unit 6** lesson 4
Multiplication squares	**Unit 5** lesson 10
Newspapers	**Unit 6** lesson 4
Number lines	**Unit 2** lesson 8 **Unit 4** lesson 3
Paper squares	**Unit 6** lesson 6
Paper straws (or similar)	**Unit 6** lessons 2, 3
Place value cards	**Unit 2** lessons 1, 2, 4
Place value counters	**Unit 1** lessons 1, 2, 3, 4, 5, 6, 7 **Unit 2** lesson 5 **Unit 3** lessons 6, 7, 8
Place value grid	**Unit 1** lessons 1, 4 **Unit 2** lesson 7 **Unit 5** lessons 8, 9
Play money	**Unit 2** lesson 2

Resource	Lesson
Prepared grid for question 4 practice; survey data in tables	**Unit 4** lesson 5
Printed number lines	**Unit 2** lesson 8 **Unit 4** lesson 3
Printed place value grids	**Unit 1** lessons 1, 4 **Unit 2** lesson 7 **Unit 5** lessons 8, 9
Rectilinear shapes to cut out	**Unit 6** lesson 3
Rulers	**Unit 4** lesson 3 **Unit 6** lesson 6
Sets of cards showing: i, i, i, v, x, x, x, l, c, set of matching cards for 4, 9, 40, 90, 400, 900 in numbers and roman numerals	**Unit 1** lesson 8
Squared paper (cm)	**Unit 2** lesson 8 **Unit 5** lesson 5 **Unit 6** lessons 2, 5
Stopwatch or egg timer	**Unit 3** lesson 7
Straws	**Unit 6** lessons 2, 3, 6
String	**Unit 2** lesson 3 **Unit 6** lessons 2, 6
Thermometers	**Unit 2** lesson 6 **Unit 4** lesson 3
Transparent rulers	**Unit 4** lesson 3
Transparent squared overlay	**Unit 6** lesson 7
Trundle wheel	**Unit 6** lesson 1
Weather maps or forecasts	**Unit 4** lesson 3

Variation helps visualisation

Children find it much easier to visualise and grasp concepts if they see them presented in a number of ways, so be prepared to offer and encourage many different representations.

For example, the number six could be represented in various ways:

Getting started with *Power Maths*

As you prepare to put *Power Maths* into action, you might find the tips and advice below helpful.

STEP 1: Train up!

A practical, up-front, full-day professional development course will give you and your team a brilliant head-start as you begin your *Power Maths* journey. You will learn more about the ethos, how it works and why.

STEP 2: Check out the progression

Take a look at the yearly and termly overviews. Next take a look at the unit overview for the unit you are about to teach in your Teacher Guide, remembering that you can match your lessons and pacing to your class.

STEP 3: Explore the context

Take a little time to look at the context for this unit: what are the implications for the unit ahead? (Think about key language, common misunderstandings and intervention strategies, for example.) If you have the online subscription, don't forget to watch the corresponding unit video.

STEP 4: Prepare for your first lesson

Familiarise yourself with the objectives, essential questions to ask and the resources you will need. The Teacher Guide offers tips, ideas and guidance on individual lessons to help you anticipate children's misconceptions and challenge those who are ready to think more deeply.

STEP 5: Teach and reflect

Deliver your lesson – and enjoy!

Afterwards, reflect on how it went … Did you cover all five stages?
Does the lesson need more time? How could you improve it?
What percentage of your class do you think mastered the concept?
How can you help those that didn't?

Unit I
Place value within 100,000

Don't forget to watch the Unit 1 video!

WHY THIS UNIT IS IMPORTANT

This unit builds on Year 4 work on numbers within 10,000 to further develop children's sense of number, working with numbers within 100,000. It is essential for children to have a secure understanding of place value in order to estimate, make decisions when calculating and working with measurements. Knowledge of the number line and identifying or estimating points on it are key to work on scales in measurement and statistics.

WHERE THIS UNIT FITS

→ **Unit 1: Place value within 100,000**

→ Unit 2: Place value within 1,000,000

The unit builds on children's work from Year 4 on 4-digit numbers. Many of the models and images used previously will be further extended to include 5-digit numbers so that children can flexibly work with all numbers to 100,000. This unit provides the foundation for working with numbers up to 1,000,000 and develops fluency with place value to support calculating during the year.

Before they start this unit, it is expected that children:
- know that 4-digit whole numbers are made up of 1,000s, 100s, 10s and 1s, and can represent 3- and 4-digit numbers in different ways (for example, part-whole model, place value grid)
- understand that all numbers have a position on the number line and can explain when a number is closer or further away from a significant boundary (for example, that 1,315 is closer to 1,000 than to 2,000)
- can explain or show why one number is larger than another and apply this understanding to order a set of 3- or 4-digit numbers.

ASSESSING MASTERY

Children know the value of each digit in numbers up to 100,000 and can represent them in different ways. They can identify the two multiples of 10, 100, 1,000 or 10,000 that a number lies between and apply this understanding to rounding. They can flexibly partition numbers, appreciating that the combined parts must still be equivalent to the whole. Also, they can apply their knowledge of place value and the number line to compare and order 4- and 5-digit numbers.

COMMON MISCONCEPTIONS	STRENGTHENING UNDERSTANDING	GOING DEEPER
Children may think that parts shown in different ways are not equivalent to the whole because they do not look the same, for example: $3,465 = 3,000 + 400 + 60 + 5$ $3,465 = 2,000 + 1,400 + 65$	Build numbers with place value counters, then move them to partition differently. Children should find the value of each part before recombining them to confirm they are still equivalent to the whole.	Explore more flexible partitioning into two, three, four, five or six parts. Children should be prepared to prove that their combined value of parts each time is equivalent to the whole. Ask children to compile a set of 'Helpful tips' that can be used by others to compare numbers. Encourage them to think about what mistakes others may make.
Children may simply compare first digits without checking place value, assuming that, for example, $736 > 1,400$ because the first digit is 7.	Build pairs of numbers with a varying number of digits using base 10 equipment. Focus on the first digit of each number, checking their values to help make decisions.	

Unit I: Place value within 100,000

WAYS OF WORKING

Use these pages to introduce the unit to children. Ask children which of Flo's words they have used before and to provide definitions.

STRUCTURES AND REPRESENTATIONS

Place value grid: This model helps children to organise 4- and 5-digit numbers and show the value of each digit. Place value grids will be used with both concrete representations and abstract numbers.

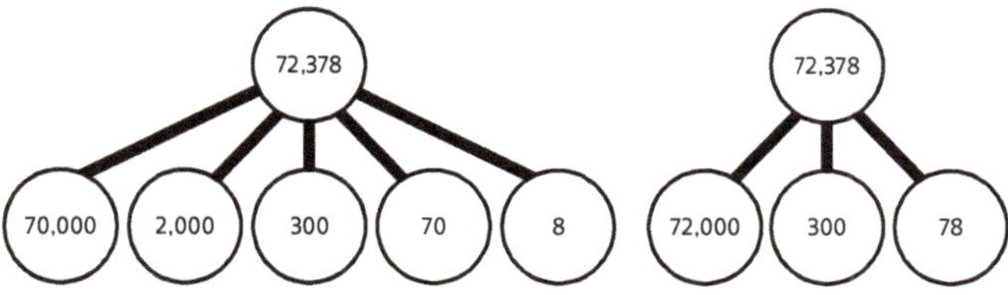

Part-whole model: This model will help to show the value of each part of a number and provide support for more flexible partitioning.

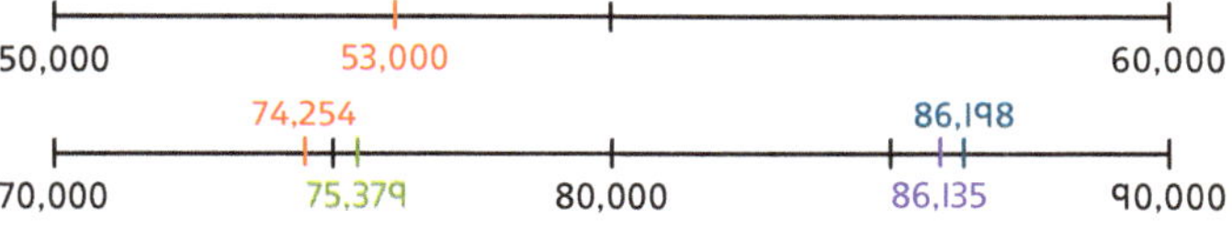

Number line to 100,000: This model will support children's work in rounding, comparing and ordering numbers.

KEY LANGUAGE

There is some key language that children will need to know as part of the learning in this unit:

→ ones (1s), tens (10s), hundreds (100s), thousands (1,000s), ten thousands (10,000s)
→ place value, position
→ partition, equivalent
→ estimate, closer to, between
→ round
→ next multiple, previous multiple, nearest multiple of 10, 100, 1,000 or 10,000
→ compare, order, greater than (>), less than (<)
→ Roman numeral

PUPIL TEXTBOOK 5A PAGE 6

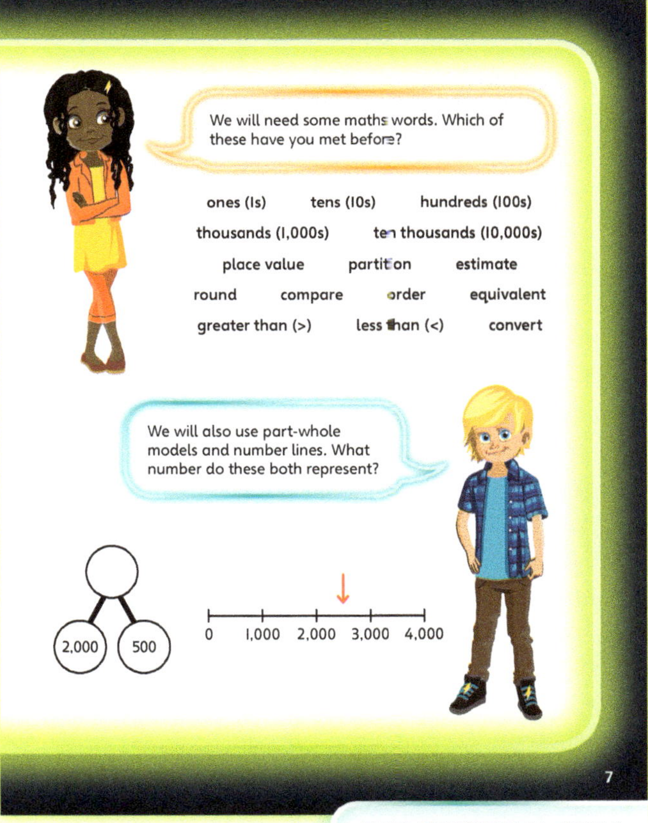

PUPIL TEXTBOOK 5A PAGE 7

Numbers to 10,000

Learning focus

In this lesson, children will revisit numbers to 10,000 with a focus on place value and count in 1,000s from different numbers. They will build numbers and break them down using what they know about the value of each digit and knowledge of zero as a place holder.

Small steps

→ **This step: Numbers to 10,000**
→ Next step: Rounding to the nearest 10, 100 and 1,000

NATIONAL CURRICULUM LINKS

Year 5 Number – Number and Place Value

- Read, write, order and compare numbers to at least 1,000,000 and determine the value of each digit.
- Count forwards or backwards in steps of powers of 10 for any given number up to 1,000,000.

ASSESSING MASTERY

Children can say or write the value of each digit in numbers up to 10,000, representing them in different ways using manipulables, and can partition or build numbers using knowledge of 1,000s, 100s, 10s and 1s (explaining the role of zero as a place holder). Children can explain that when they count on or back in steps of 1,000, only the 1,000s value will change.

COMMON MISCONCEPTIONS

Children often forget to use zero to show when a place has no value; for example, recording 'three thousand and forty-five' as 345, forgetting to use 0 to show there are no 100s in the hundreds position. Ask:
- *What is the same about the numbers 345 and 3,045? What is different?*

STRENGTHENING UNDERSTANDING

Give children pairs of 3- and 4-digit numbers to build with place value counters. Each pair should include at least two digits that are the same but have different values, for example 4,521 and 5,362. Ask them to explain what the value of digit 5 is in each number and then repeat this for digit 2.

GOING DEEPER

Ask children to represent 4-digit numbers in different ways; for example in numerals, words, partitioning, manipulables, money and so on. Ensure that some of the numbers contain one or more zeros, for example 3,501, 6,020 and 4,500.

KEY LANGUAGE

In lesson: place value, digits, thousands (1,000s), hundreds (100s), tens (10s), ones (1s), zero (0), count on, count back

Other language to be used by the teacher: partition

STRUCTURES AND REPRESENTATIONS

place value grid

RESOURCES

Optional: place value counters, base 10 equipment

 In the eTextbook of this lesson, you will find interactive links to a selection of teaching tools.

Before you teach 

- Can children confidently explain the value of each digit in 3-digit numbers?
- Can they use place value to help them count on and back from any 4-digit number (such as 4,218, 4,219, 4,220)?

Discover

 Pair work

ASK

- Question ❶ a): *What are the scores on the ducks that Lee has already hooked? What do you notice about them?*
- Question ❶ a): *How many 1,000s can you see? How many 100s?*
- Question ❶ b): *Read out Olivia's total score. How many 100s are in the 100s position?*
- Question ❶ b): *What is this question asking you to find? What should you do first?*

IN FOCUS Question ❶ b) requires children to first build the number that is shown by the six given ducks, reasoning about the number of 1,000s, 100s, 10s and 1s that can be seen. They need to recognise that the missing scores must change the value of the number to read 3,122.

PRACTICAL TIPS Put some place value counters into a bag to represent the ducks. Play a game with children, where they need to take a number of counters from the bag and work out the total. Provide a large place value grid for children to place their counters on.

ANSWERS

Question ❶ a): Lee's total score is 2,402.

Question ❶ b): The missing scores on the two ducks are 1,000 and 10.

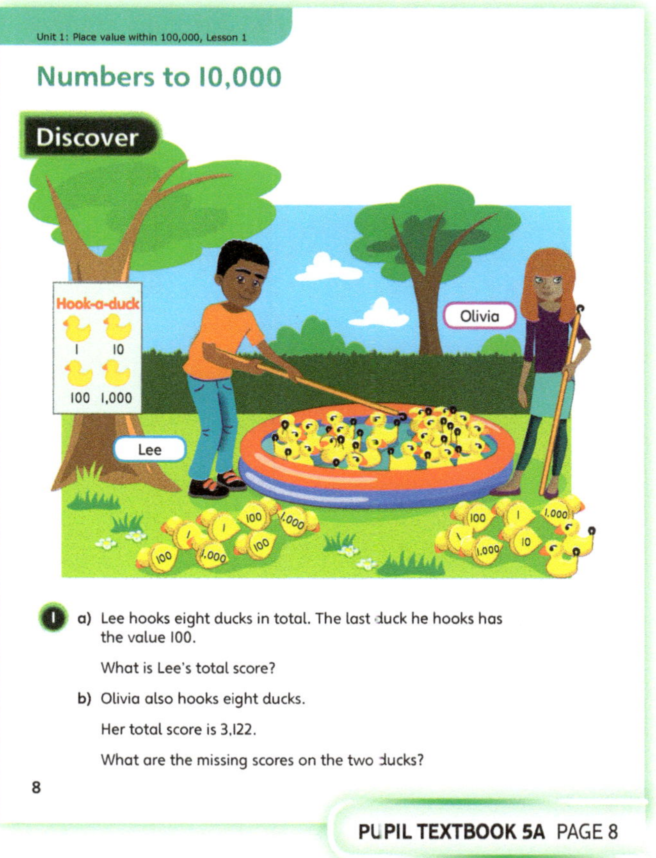

PUPIL TEXTBOOK 5A PAGE 8

Share

 Whole class teacher led

ASK

- Question ❶ a): *Do the place value counters on the grid match Lee's total of 2,402? How do you know?*
- Question ❶ a): *The number 2,402 has four digits but it is only shown partitioned into the three parts 2,000 + 400 + 2. Explain why.*
- Question ❶ b): *What digit changes when you count on 1,000? Why do the other digits not change?*
- Question ❶ b): *Why can the missing scores not be 100 and 1?*

IN FOCUS Question ❶ a) uses place value counters and a place value grid to represent the scores on the ducks. It focuses on the use of zero as a place holder, ensuring that children recognise that the number being represented is 2,402, not 242. Partitioning of 2,402 using addition is also introduced to show the value of each digit as 2,000 + 400 + 2.

STRENGTHEN Ask children to read the numbers aloud together to identify the value of each digit. Record the numbers in words so that children see how the words reflect the numerals. In question ❶ b), children could represent Olivia's total score on another place value grid and compare this with the image in the textbook, explaining what is the same and what is different about them. Look together at counting on 1,000 from 2,112, identifying the digits that change and the ones that stay the same.

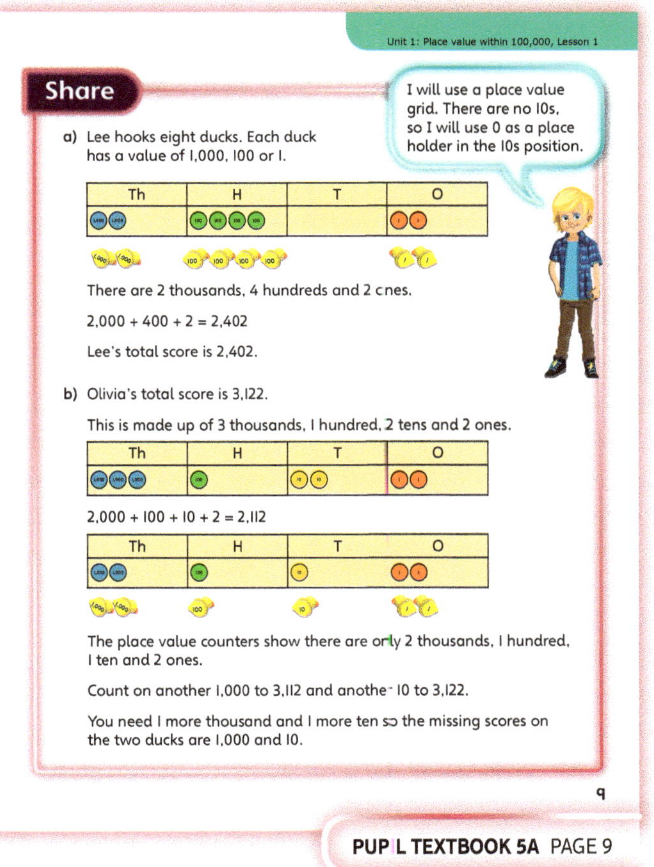

PUPIL TEXTBOOK 5A PAGE 9

Think together

WAYS OF WORKING Whole class teacher led (I do, We do, You do)

ASK

- Question **1** a): *How do you know that one of the digits in Bella's score must be a zero?*
- Question **1** b): *Which digits will change and which will stay the same when you count on 2,000? Why?*
- Question **2** : *What do you know so far about Zac's score? Which digits need to change so the score is 5,212?*
- Question **4** c): *Which position has the greatest or smallest value? Which digit should you use here to make the largest or smallest possible number?*

IN FOCUS In question **1**, children need to use zero as a place holder, recognising Bella's score as 4,043 and not 443.

Question **3** gives the opportunity to address any misconceptions with place value as children build 4-digit numbers. It also provides an example of the benefit of using a systematic approach to seek solutions.

STRENGTHEN In question **2**, encourage children to show Zac's total score using place value counters on a large place value grid. Use the grid to explain the difference between the values of the digit 2 in 5,212.

DEEPEN Ask children to think about how they can change the numbers that Aki could not make in question **3** a) into a number that he could make. Is it possible to just change one digit each time? Can two digits be changed each time? Encourage them to explain how the value of the digit is changing each time; for example in 2,351 they could change the digit 5 to a 4 so the value of the number is ten less.

ASSESSMENT CHECKPOINT Use question **1** to assess whether children can build a 4-digit number and count on in 1,000s. Use question **2** to assess whether they recognise how the digits (and the associated values) need to change to reach a specified number.

ANSWERS

Question **1** a): There are 4 thousands, 4 tens and 3 ones.

$$4,000 + 40 + 3 = 4,043$$

Bella's score is 4,043

Question **1** b): Bella's new score is 6,043.

Question **2** : The missing scores are 1,000, 100 and 10.

Question **3** a): 7,210, 4,222 and 5,302

Question **3** b): Other total scores include: 1,432, 8,200, 7,111, and so on.

Question **3** c): The highest possible score is 9,100.

Question **3** d): The lowest possible score is 1,000.

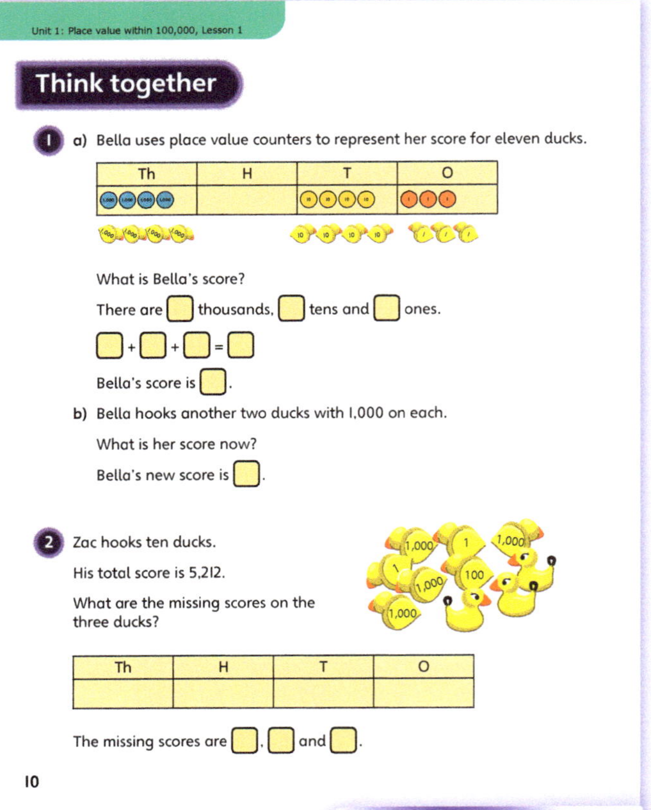

PUPIL TEXTBOOK 5A PAGE 10

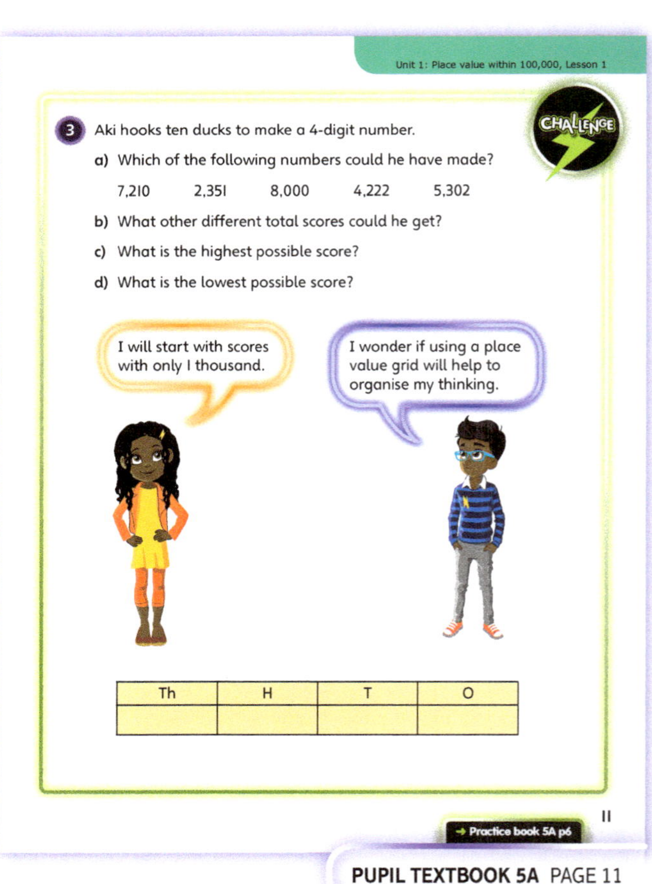

PUPIL TEXTBOOK 5A PAGE 11

Practice

WAYS OF WORKING Independent thinking

IN FOCUS Question ❶ covers the same content as the textbook, but using different manipulables. In the second example, the number of 10s is split and should be re-combined to show 4 tens.

In question ❷, children must first recognise what counters should be used to represent the number 5,632 and then decide what additional counters must be added to show this. They are required to show the value of each digit in an abstract way using partitioning.

STRENGTHEN In question ❶, give children both base 10 equipment and place value counters to build the numbers in a place value grid. Emphasise that the 1,000 place value counter and the base 10 equipment 1,000 cube represent the same number. Children may also find it useful to represent each of the prices in question ❹ using place value counters and then count back in steps of 1,000 by removing a counter at a time from the 1,000s position, saying the number after removing each counter.

DEEPEN Encourage children to explain their thinking in question ❺. They should first talk about the column that has the largest value and the one that has the smallest value, identifying these as 1,000s and 1s respectively. Ask them to make up a set of clues for two 4-digit numbers using four different digits for a partner to solve.

ASSESSMENT CHECKPOINT Use question ❸ to assess whether children can identify different representations of the same number shown using manipulables, words or in an abstract way. Encourage children to explain how 'four thousand, two hundred and twenty-five' should be changed so it does represent the same number.

Use question ❹ to assess whether children can count on and back in 1,000s, recognising which digit will need to change. In part b), £2,575 will result in a 3-digit price as the 2 thousands are removed.

ANSWERS Answers for the **Practice** part of the lesson appear in the separate **Practice and Reflect answer guide**.

Reflect

WAYS OF WORKING Independent thinking

IN FOCUS Children need to partition a 4-digit number with no accompanying representation. Each digit has a value so there are no zeros to deal with. Encourage them to read the number first and then say the value of each digit. In doing so, they will find the value of the digits easier to identify.

ASSESSMENT CHECKPOINT Look for children who are confident explaining the value of each digit and how the digits change or remain the same as they count back 3,000. They should recognise that only the 1,000s digit changes and that the other digits remain unchanged.

ANSWERS Answers for the **Reflect** part of the lesson appear in the separate **Practice and Reflect answer guide**.

After the lesson ⏸

- Are children confident explaining the value of each digit in numbers up to 10,000?
- Can they represent numbers with manipulables, in words or using partitioning?
- Can they explain why 3,545 will not be in a court of 1,000 starting from 2,544?

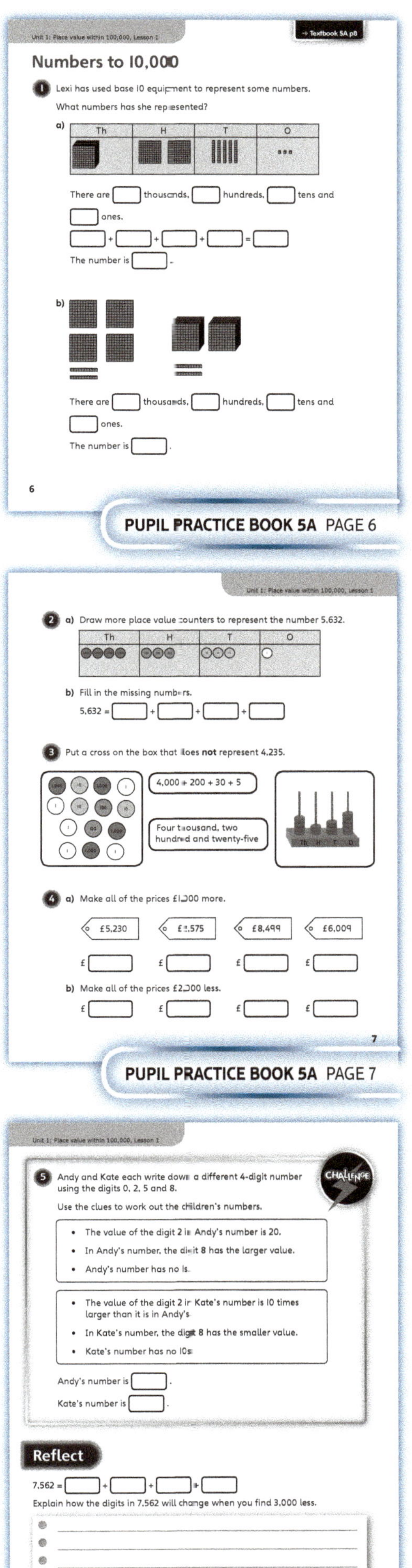

43

Rounding to the nearest 10, 100 and 1,000

Learning focus

In this lesson, children will round numbers up to 10,000 to the nearest 10, 100 or 1,000, and learn that it is important to know which two multiples of 10, 100 or 1,000 a number lies between, so they can make appropriate decisions.

Also, they will identify the digit in a number that should be looked at to help them with rounding.

Small steps

→ Previous step: Numbers to 10,000
→ **This step: Rounding to the nearest 10, 100 and 1,000**
→ Next step: 10,000s, 1,000s, 100s, 10s and 1s (1)

NATIONAL CURRICULUM LINKS

Year 5 Number – Number and Place Value

Round any number up to 1,000,000 to the nearest 10, 100, 1,000, 10,000 and 100,000.

ASSESSING MASTERY

Children can identify the two multiples of 10, 100 or 1,000 that a number lies between and can explain which it is closer to. They can apply this understanding to round numbers to the nearest 10, 100 or 1,000.

COMMON MISCONCEPTIONS

When rounding to the nearest multiple (such as 100) children state the previous multiple incorrectly. For example, they may say that the previous multiple of 100 before 432 is 300, not 400, because they see the pattern 300, 400, 500. Encourage children to talk about numbers in other ways; for example, 432 is 32 more than 400 rather than only about how many 100s, 10s and 1s it has. Ask:
• *How many more than 3,000 is 3,450? Which two multiples of 1,000 does it lie between?*

STRENGTHENING UNDERSTANDING

Each time children are rounding a number, encourage them to explain which digit determines how the number should be rounded. Use number lines to support their thinking.

GOING DEEPER

When children have rounded a number, ask them to suggest a range of other numbers that round to the same multiple of 10, 100 and 1,000 and to explain why another example does not.

KEY LANGUAGE

In lesson: round, multiple, digit, thousands (1,000s), hundreds (100s), tens (10s), ones (1s), closer to, nearest, next, previous

STRUCTURES AND REPRESENTATIONS

number line

RESOURCES

Optional: base 10 equipment, place value counters

 In the eTextbook of this lesson, you will find interactive links to a selection of teaching tools.

Before you teach 

• Can children confidently identify the next and previous multiple of 10 or 100 for 2- and 3-digit numbers?
• Can they remember any of the rules for rounding to the nearest 10 or 100?

Discover

 Pair work

ASK

- Question **1** a): *What does it mean to round a number to the nearest 10? What rules for rounding can you remember?*
- Question **1** a): *How many 10s can you see in 3,425? How many 100s? How does this help?*
- Question **1** b): *Why can't Mo's score be 5,900 or 8,200? Which multiple of 1,000 are these numbers closest to?*
- Question **1** b): *Can you explain whether Mo's score can be less than 6,500?*

IN FOCUS Question **1** b) requires children to start to identify possible ranges of numbers that round to a given multiple, 1,000 in this case. Children are more likely to think about the numbers that round up but often forget those that round down. This question requires rounding up and down.

PRACTICAL TIPS Encourage children to sketch their own number lines as needed. It is also helpful to describe the numbers in a way that identifies a previous multiple, for example 3,425 is 5 more than 3,420 and 25 more than 3,400.

ANSWERS

Question **1** a): 3,425 rounds up to 3,430 to the nearest 10.

3,425 rounds down to 3,400 to the nearest 100.

Question **1** b): Mo's score could be any number between 6,500 and 7,499.

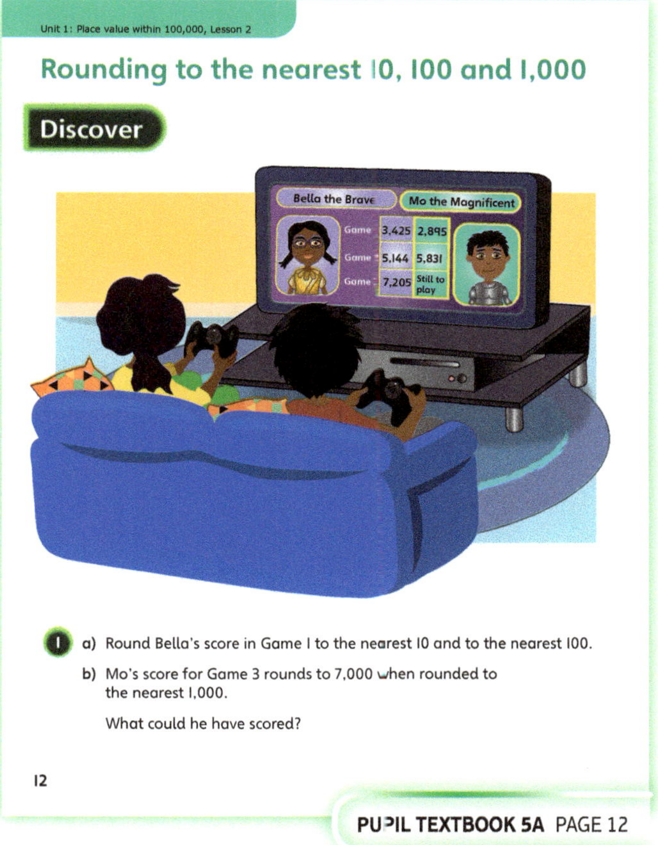

PUPIL TEXTBOOK 5A PAGE 12

Share

 Whole class teacher led

ASK

- Question **1** a): *Why does Dexter want to find the multiples of 10 and 100 that the number lies between?*
- Question **1** a): *Which digit will tell you whether 3,425 is closer to 3,420 or 3,430? Why?*
- Question **1** a): *What do you know about the digit 5 when rounding? Should you round up or down?*
- Question **1** b): *Why does Flo want to think about the largest and smallest numbers that round to 7,000?*

IN FOCUS Question **1** a) uses number lines to show which multiple of 10 or 100 the number is closer to. Sparks reminds children that when rounding the digit 5, which is midway between the two multiples, they need to round up.

STRENGTHEN Use computerised versions of a number line that allow children to clearly see how much closer a number is to one multiple than the other.

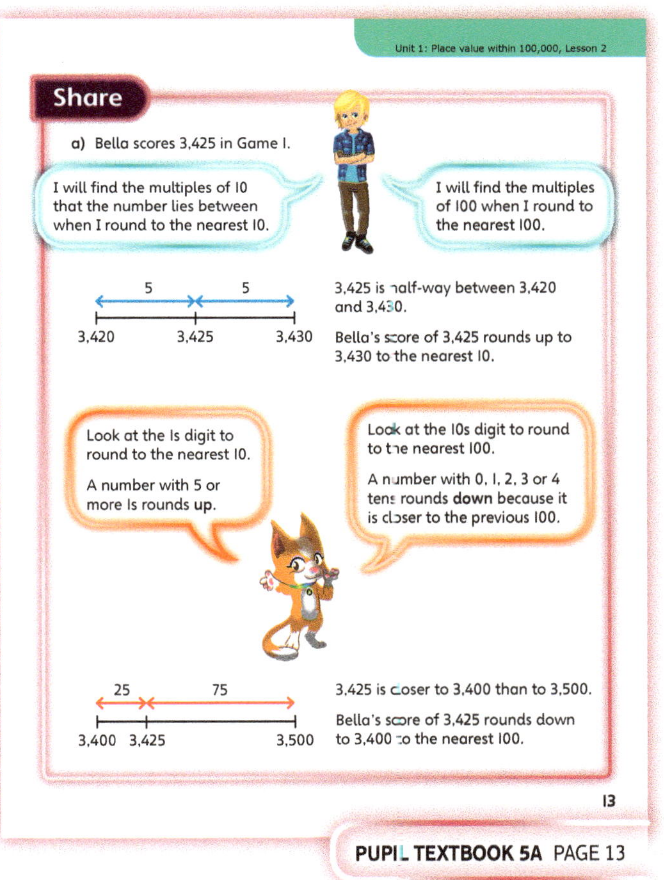

PUPIL TEXTBOOK 5A PAGE 13

Think together

WAYS OF WORKING Whole class teacher led (I do, We do, You do)

ASK

- Question **1** b): *Can you sketch a number line to show which two multiples of 100 Bella's score lies between? Which multiple should you write 5,144 closer to?*
- Question **2** a): *What is the best way to count the parts?*
- Question **2** b): *What digit must you look at?*
- Question **2** b): *What is the next multiple of 10 after 2,895? 2,900 has two 0s, is it still a multiple of 10?*
- Question **3** : *What do you know about numbers that round to 5,000 to the nearest 1,000? What could the 100s digit be? What could it not it be?*
- Question **3** : *How does Mo's statement change what you know about the 100s digit?*

IN FOCUS In question **2**, the number has been chosen so that the number rounded to the nearest 10 is also a multiple of 100. Children should recognise and explain why and when this may be the case, perhaps by suggesting other examples. This provides the opportunity to address the misconception that multiples of 10 only end in one 0, not recognising that multiples of 100 or 1,000 are also multiples of 10.

Question **3** uses the context of missing digits to provide the opportunity to address any misconceptions about rounding. The number has been chosen to include one digit that rounds down and the digit 5, which rounds up.

STRENGTHEN In question **3**, encourage children to use Bella's statement to find the 1,000s digit. Discuss why they can ignore Mo's statement when finding the 1,000s digit. When they have determined that the first digit must be 5, ask them which digit they need to use to decide whether to round up or down to the nearest 10. This will guide them towards the 10s digit being 7.

DEEPEN Ask children to design a poster or make up a rhyme that would help others remember the rules for rounding.

ASSESSMENT CHECKPOINT Use questions **1** and **2** to assess children's fluency with rounding. Use question **3** to check children's reasoning about ranges of numbers that will round to a given multiple.

ANSWERS

Question **1** a): 5,144 is between the numbers 5,100 and 5,200.

It is closer to 5,100 because it has only 4 tens.

5,144 rounds to 5,100 to the nearest 100.

Question **1** b): 5,144 is between 5,000 and 6,000.

It is closer to 5,000 because it has 1 in the 100s position.

5,144 rounds to 5,000 to the nearest 1,000.

Question **2** a): The base 10 equipment represents Mo's score of 2,895.

Question **2** b): The score is closer to 2,900.

2,895 rounds to 2,900 to the nearest 10.

Question **3** : Bella's score is 5,375.

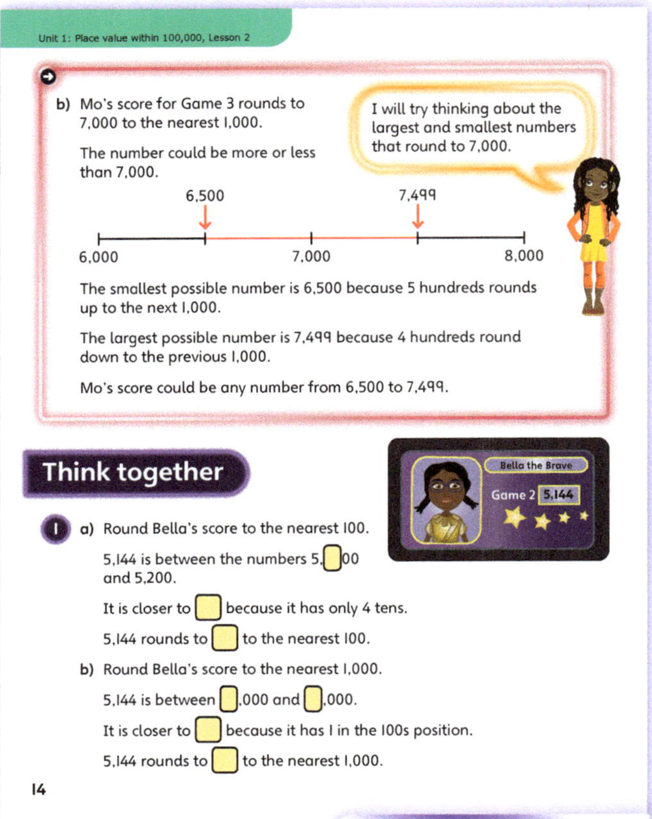

PUPIL TEXTBOOK 5A PAGE 14

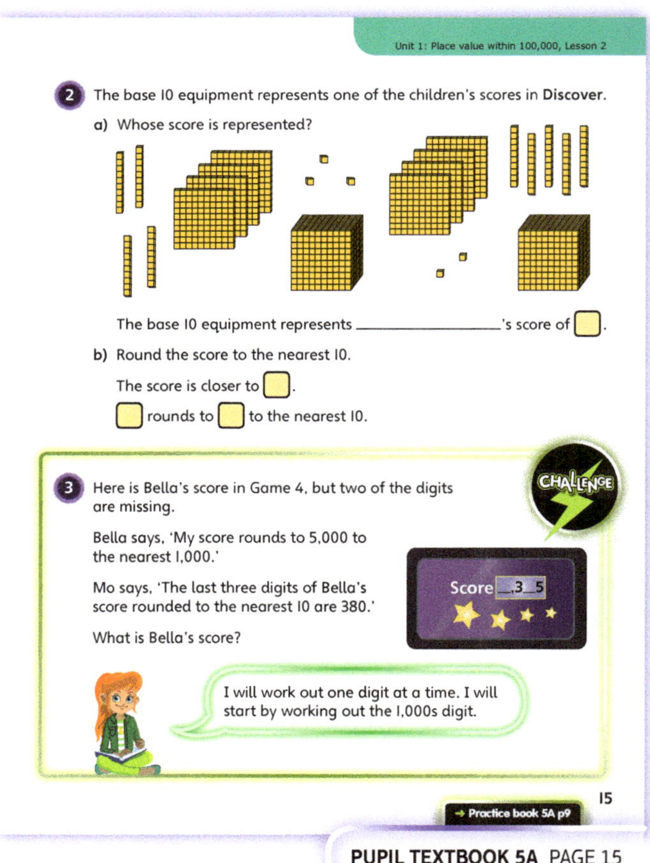

PUPIL TEXTBOOK 5A PAGE 15

Practice

WAYS OF WORKING Independent thinking

IN FOCUS In question **5**, children reason about the value of missing digits, making decisions based on the way each number has been rounded and what they know about rules for rounding.

Question **7** requires children to reason that the subtracted 10 must cross the 50s boundary. They need to recognise that 2,650 is the lowest possible number and 2,659 is the highest. They should be prepared to explain why, for example, 2,660 would not fit the criteria.

STRENGTHEN Encourage children to sketch number lines to help make decisions about rounding and to identify the next and previous multiple. For question **3** b), they could start by guessing digits and rounding them to check if the number fits the criteria or how the digit must be amended. Encourage them to develop their findings into a strategy for identifying possible numbers.

DEEPEN When children have completed question **7**, ask them to explain how the possible solutions would change if 20 was the number subtracted in the problem rather than 10. Then extend it to 25 or 30.

THINK DIFFERENTLY In question **6**, children need to consider two multiples when creating a new number. Encourage children to think about all the columns within the place value grid and what effect adding a counter to it would have on the overall number.

ASSESSMENT CHECKPOINT Use questions **1** and **2** to assess whether children can round numbers to the nearest 10, 100 or 1,000. Use question **5** to check that they know which digit they must check when rounding to a given multiple. Question **6** will demonstrate children's confidence in applying rounding in the context of money.

ANSWERS Answers for the **Practice** part of the lesson appear in the separate **Practice and Reflect answer guide**.

Reflect

WAYS OF WORKING Pair work

IN FOCUS Children need to think about what is the same and what is different about rounding to 10, 100 and 1,000. For example, digits 5, 6, 7, 8 and 9 round up and 0, 1, 2, 3, 4 round down in all cases and it is important to look at which pair of multiples a number lies between. However, the multiples they lie between will be different, depending on whether they are 10s, 100s or 1,000s.

ASSESSMENT CHECKPOINT Check children recognise that the digit they need to look at for rounding to 10, 100 and 1,000 is different each time.

ANSWERS Answers for the **Reflect** part of the lesson appear in the separate **Practice and Reflect answer guide**.

After the lesson ▮▮

- Can children say the two multiples of 1,000 that a number lies between?
- Can children confidently explain the rules for rounding to the nearest 10, 100 and 1,000? Can they explain which digit will help them?
- Can children reason about numbers that do or do not round to a given multiple?

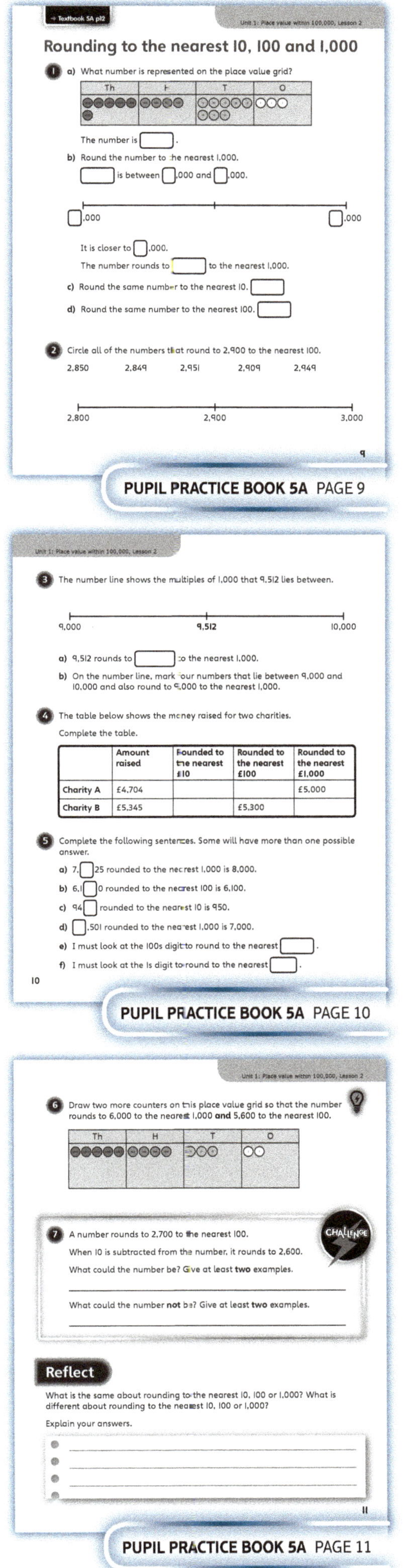

10,000s, 1,000s, 100s, 10s and 1s ①

Learning focus

In this lesson, children will work with numbers to 100,000, focusing on the position and value of each digit. They will represent numbers in different ways and break them down, explaining the value of each part.

Small steps

→ Previous step: Rounding to the nearest 10, 100 and 1,000
→ **This step: 10,000s, 1,000s, 100s, 10s and 1s (1)**
→ Next step: 10,000s, 1,000s, 100s, 10s and 1s (2)

NATIONAL CURRICULUM LINKS

Year 5 Number – Number and Place Value

Read, write, order and compare numbers to at least 1,000,000 and determine the value of each digit.

ASSESSING MASTERY

Children can say or write the value of each digit in numbers up to 100,000 and represent them in different ways. They can partition or build numbers using knowledge of 10,000s, 1,000s, 100s, 10s and 1s, explaining the role of 0 as a place holder.

COMMON MISCONCEPTIONS

When a number includes the same digit more than once (such as 45,435), children may not recognise that it is the position of the digit that is important. In this example, count across the columns, establishing that each column is 10 times larger than the previous one. Ask:

- *In the number 45,435, how many times smaller is the value of the digit 4 in the 100s column than in the 10,000s column? Counting across from left to right: 10 times smaller, 100 times smaller.*

STRENGTHENING UNDERSTANDING

Use place value counters on a part-whole model so that children can clearly see the value of each digit. Ask them first to write and represent the whole and then to break it up into parts to identify the value of each.

GOING DEEPER

Ask children to make up some clues about a 5-digit number for others to solve. First give some examples, for instance take the number 25,324: *The digit with the smallest value is 4. The digit with the largest value is 2. This digit has a value that is 1,000 times larger than the 10s digit. One of the digits has the value 300. The remaining digit has the value 5,000. What is the number?*

KEY LANGUAGE

In lesson: digit, place value, position, ten thousands (10,000s), thousands (1,000s), hundreds (100s), tens (10s), ones (1s), zero (0), multiple

Other language to be used by the teacher: partition

STRUCTURES AND REPRESENTATIONS

place value grid, part-whole model, number line

RESOURCES

Optional: place value counters, base 10 equipment

 In the eTextbook of this lesson, you will find interactive links to a selection of teaching tools.

Before you teach 

- Can children confidently explain the value of each digit in a 4-digit number?
- Can they explain the difference in value between a digit that appears more than once in a number, for example 4,245?

Discover

WAYS OF WORKING Pair work

ASK

- Question **1** a): *Where can you see how many passengers flew with the airline? How many digits are in the number? What is the value of the digit 2?*
- Question **1** b): *How many bags are there? What labels can you use to show the value of each position in the number?*
- Question **1** b): *How many parts do you think you will need for a part-whole model?*

IN FOCUS The airline figures introduce children to 5-digit numbers, and question **1** b) encourages them to use what they have learnt about 4-digit numbers to help partition a 5-digit number and write it in words.

PRACTICAL TIPS Ask children to read out the numbers on the poster that they are confident with. Talk together about the larger numbers, agreeing that the first column on the left is ten times larger than the 1,000s column to the right and has the value 10,000. Represent the difference using base 10 equipment or place value counters.

ANSWERS

Question **1** a): The digit 7 represents 70,000 and 70.

Question **1** b):

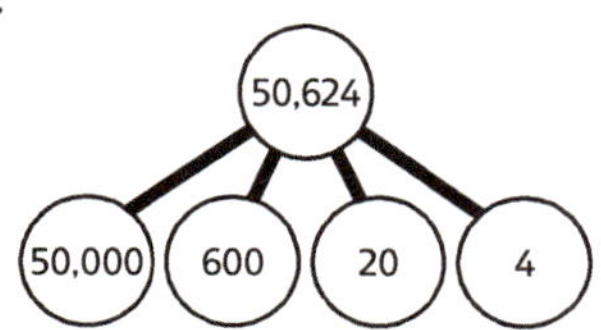

Fifty thousand, six hundred and twenty-four

Share

WAYS OF WORKING Whole class teacher led

ASK

- Question **1** a): *Which parts tell you about the values of the digit 7? Which columns are they in?*
- Question **1** b): *How does using the part-whole model help you to say and write the number? Would you need to put any counters in the thousands column to represent this number on a place value grid?*

IN FOCUS Question **1** a) uses a repeated digit to make children think about the place value of a digit rather than just the digit itself. The number chosen for question **1** b) has the digit 0 in one position, so that children need to understand that the part-whole model for the 5-digit number has only four parts. They should use the part-whole model to help say and write the number in words.

STRENGTHEN Encourage children to read the numbers in full, for example 72,378 as seventy-two thousand, three hundred and seventy-eight and not simply as digits 7-2-3-7-8, so that the place value is made explicit.

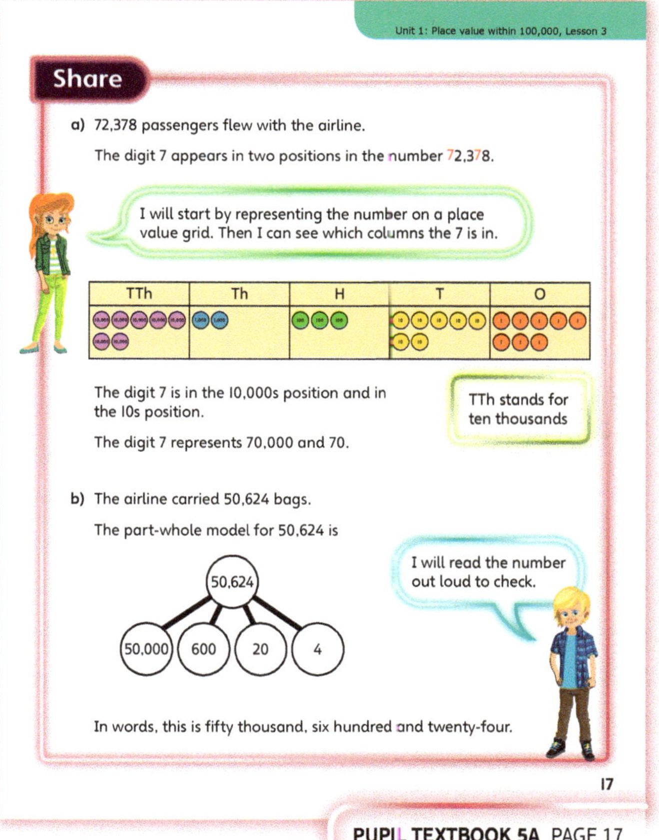

Think together

WAYS OF WORKING Whole class teacher led (I do, We do, You do)

ASK

- Question **1** a): *The digit 2 is repeated in this number, so which digit has the larger value? How do you know?*
- Question **2** : *Why is the tens column empty? How would you show this number on a part-whole model?*
- Question **3** : *Why do you think Astrid thought that?*
- Question **3** : *What is the next 10,000 after 60,000? Is 79,300 higher or lower?*

IN FOCUS The number in question **1** has been chosen to have two pairs of repeating digits, so that children need to reason about their values.

Question **3** introduces a number line with multiples of 10,000, so children need to think about the position of the number on a number line rather than the value of each digit. This question also addresses the misconception of trying to identify the previous multiple of 10,000 by using the 10,000s digit not the 1,000s one.

STRENGTHEN To strengthen understanding in question **1** b), suggest that children also represent the number on a part-whole model to match the counters and place value grid. They should use this to help say the number aloud and write it in words. They could also write the value of each part in words on separate cards and then place the cards together. Discuss why we say 'thirteen thousand' rather than 'ten thousand, three thousand'.

DEEPEN Give each child a set of blank part-whole models with a different number of parts (two parts, three parts, four parts and five parts). Ask them to write a 5-digit number for each of the wholes where the value of each digit is shown as one of the parts. For example, children may choose 20,040 for a model with two parts, labelling one part as 20,000 and the other as 40.

ASSESSMENT CHECKPOINT Use question **1** to assess whether children understand that the position of a digit is important to determine its value and that digits in a number may repeat, but they will have different values. Use question **2** to check that children can represent a 5-digit number in different ways and can explain why each representation is correct.

ANSWERS

Question **1** a): The missing numbers are 3,000, 200 and 20.

Question **1** b): Thirteen thousand, two hundred and eight (spoken and written).

The value of the digit 3 is 3,000.

Question **2** a): 45,206 represented in different ways, such as with place value counters, in words, on a part-whole model and so on.

Question **2** b): The number is 206 more than 45,000.

Question **3** a): The number lies between 70,000 and 80,000.

Question **3** b): Any numbers between 70,000 and 80,000, with explanations of the values of their digits.

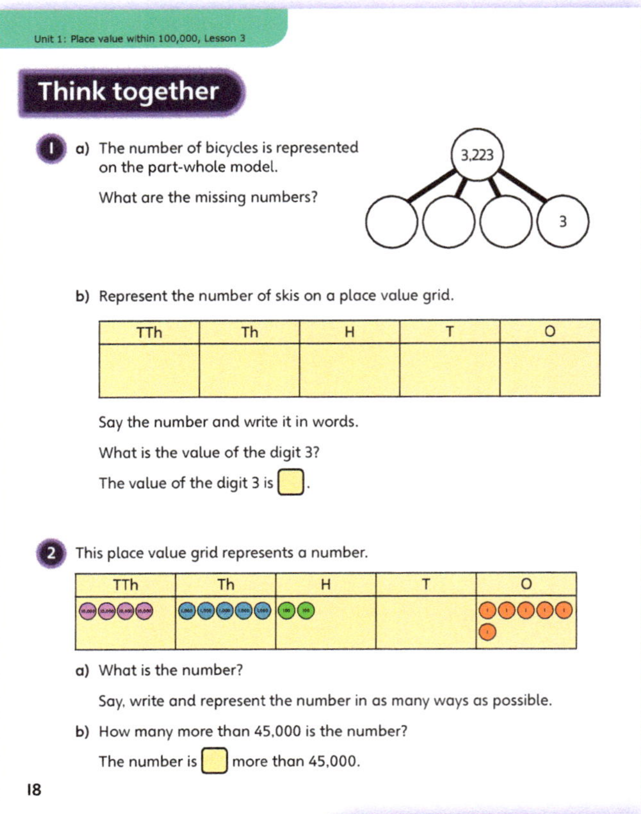

PUPIL TEXTBOOK 5A PAGE 18

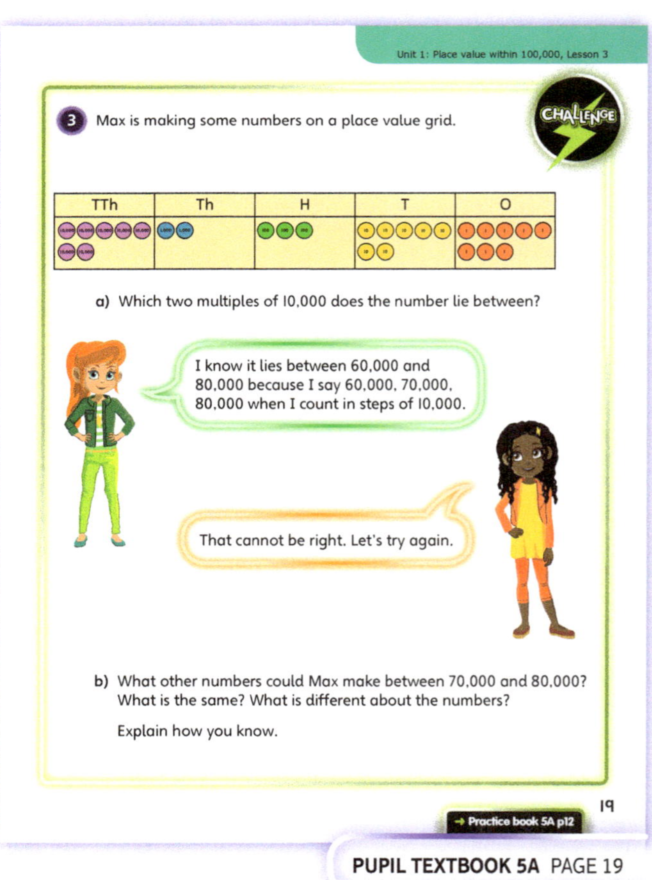

PUPIL TEXTBOOK 5A PAGE 19

Practice

WAYS OF WORKING Independent thinking

IN FOCUS Question **1** b) emphasises the fact that a digit's value depends on its position by swapping two of the digits and asking children to write the new number. Question **2** also emphasises this link by requiring them to match each digit in a number with its value.

Question **6** demands flexible reasoning as children have to find the missing digits that will allow numbers to be made to suit all of the given criteria.

STRENGTHEN Encourage children to represent the numbers in question **2** on a place value grid or with counters to help them make decisions about the value of the digit 4 each time.

DEEPEN When children have completed question **6**, ask them to identify two digits that Max could not have used. Then challenge them to change the clues so the solution uses these two digits.

ASSESSMENT CHECKPOINT Use questions **1** and **2** to assess whether children can confidently recognise and explain the value of each digit in numbers up to 100,000. Use questions **3** and **4** to assess whether children understand how 5-digit numbers can be represented in place value grids and part-whole models. Use question **5** to assess whether they understand what happens to a number when you add or subtract a multiple of 100, 1,000 or 10,000.

ANSWERS Answers for the **Practice** part of the lesson appear in the separate **Practice and Reflect answer guide**.

Reflect

WAYS OF WORKING Pair work

IN FOCUS Children are required to use what they know about place value in a 5-digit number. They could use a place value grid or a part-whole model to help them identify the value of each digit.

ASSESSMENT CHECKPOINT Check that children recognise the difference in values between repeating digits in the number.

ANSWERS Answers for the **Reflect** part of the lesson appear in the separate **Practice and Reflect answer guide**.

After the lesson

- Can children confidently explain the value of each digit in numbers up to 100,000?
- Can they represent numbers with manipulables, in words and using part-whole models?

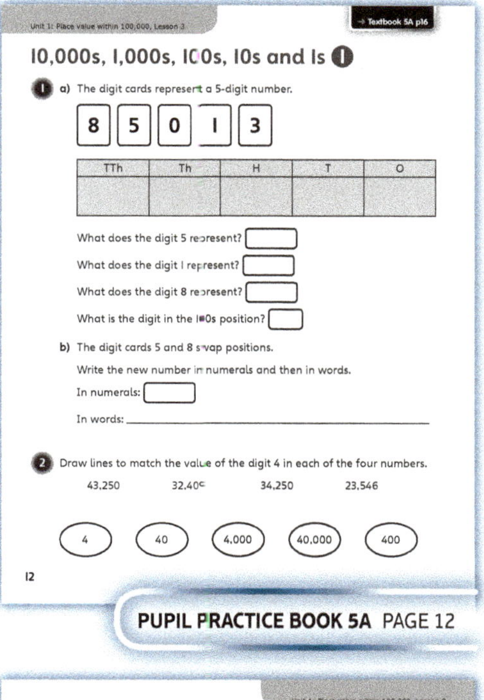

PUPIL PRACTICE BOOK 5A PAGE 12

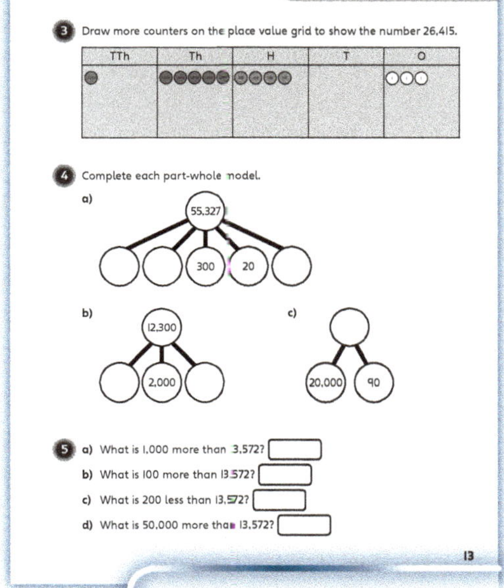

PUPIL PRACTICE BOOK 5A PAGE 13

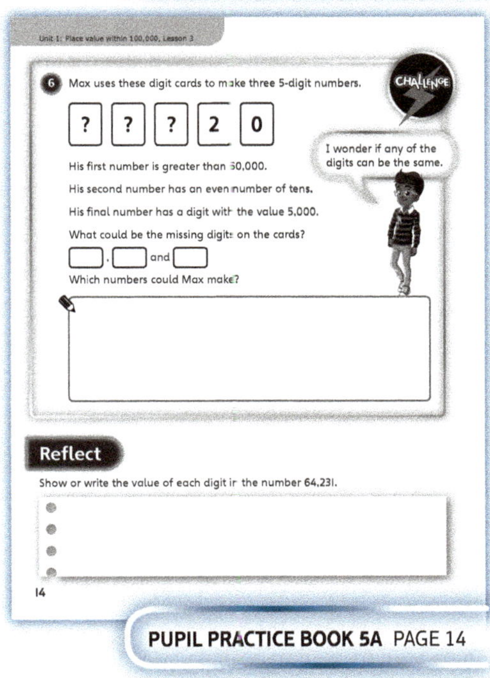

PUPIL PRACTICE BOOK 5A PAGE 14

10,000s, 1,000s, 100s, 10s and 1s ②

Learning focus

In this lesson, children will apply their understanding of the place value of numbers up to 100,000 to represent and partition them in different ways.

Small steps

- → Previous step: 10,000s, 1,000s, 100s, 10s and 1s (1)
- → **This step: 10,000s, 1,000s, 100s, 10s and 1s (2)**
- → Next step: The number line to 100,000

NATIONAL CURRICULUM LINKS

Year 5 Number – Number and Place Value

Solve number problems and practical problems that involve [reading, writing, comparing and ordering numbers up to 1,000,000].

ASSESSING MASTERY

Children can identify the whole from a given number of parts. Children can partition numbers flexibly using their understanding of equivalence and place value.

COMMON MISCONCEPTIONS

Children may assume that the numbers made by different partitionings should be different. Use place value counters to create different partitioning of the same number and ask:

- *What is the same and what is different each time?*

STRENGTHENING UNDERSTANDING

Use base 10 equipment to represent a 3- or 4-digit number, for example 542. Partition the number into 100s, 10s and 1s, recording this as 500 + 40 + 2. Move one block at a time, for example splitting the 100s into 400 and 100, and asking children to write the new partitioning sentence. Return to the initial arrangement and repeat with different blocks. Finally, join groups together to make, for example, 540 + 2 or 500 + 42 or 400 + 142. Check that children understand that the whole is the same each time, since no base 10 equipment has been added or taken away. Calculators are also a good way of proving the whole is the same each time.

GOING DEEPER

Ask children to explain if it is possible to partition any 4- or 5-digit number into four parts. Point out that some numbers have lots of 0s, for example 52,000, and ask children whether they can only partition it into two parts: 50,000 and 2,000. Children should explore the problem and discuss or show their thinking.

KEY LANGUAGE

In lesson: ten thousands (10,000s), thousands (1,000s), hundreds (100s), tens (10s), ones (1s), partition, **equivalent**, addition sentence

Other language to be used by the teacher: whole, part

STRUCTURES AND REPRESENTATIONS

place value grid, number line

RESOURCES

Optional: large place value grid, place value counters, base 10 equipment, calculators

 In the eTextbook of this lesson, you will find interactive links to a selection of teaching tools.

Before you teach 

- Can children confidently explain the value of each digit in a 5-digit number?
- Can they partition a 5-digit number into 10,000s, 1,000s, 100s, 10s and 1s?

Discover

WAYS OF WORKING Pair work

ASK

- Question ① a): *What are the masses of the five crates on the lorry? Which is heaviest? How do you know?*
- Question ① a): *How many digits will the total mass have? Will there be any 0s in the total mass?*
- Question ① b): *Is the mass on this lorry heavier or lighter than the mass on the lorry in the picture? How do you know?*
- Question ① b): *Are there any crates in the picture that you know are definitely not loaded onto this lorry?*

IN FOCUS The problems are set in the context of mass so that children have to apply their understanding of number. Question ① b) introduces children to representing and partitioning numbers in different ways, not always simply into the number of 10,000s, 1,000s, 100s, 10s and 1s. It provides the opportunity to explore equivalence – having the same value but not always looking the same.

PRACTICAL TIPS Give children small pieces of paper with the masses of the crates written on so that they can arrange them as required to help answer the questions.

ANSWERS

Question ① a): The total mass of the five crates on the lorry is 22,768 kg.

Question ① b): The crates on the lorry could be 10,000 kg, 9,000 kg, 100 kg and 40 kg. Alternatively, they could be 10,000 kg, 6,000 kg, 3,000 kg, 100 kg and 40 kg.

Share

WAYS OF WORKING Whole class teacher led

ASK

- Question ① a): *The two largest crates are 20,000 kg and 2,000 kg. Why is the first digit not 4?*
- Question ① a): *Can you represent the same number on a part-whole model? How many parts will you need to show the different values of the crates?*
- Question ① b): *Can you represent the number 19,140 with place value counters on a grid? Which counters will you need?*
- Question ① b): *Can you explain the different numbers on the number line? What does each one show?*

IN FOCUS Question ① a) includes two values that look very similar (20,000 kg and 2,000 kg) so the importance of the position of a digit in a number is revisited. Question ① b) has more than one possible solution, introducing more flexible partitioning. Children develop understanding that the total of the parts is still equivalent to the whole.

STRENGTHEN For question ① b), ask children to represent 19,140 with place value counters. Explore the second solution together by grouping the 9 thousands counters into two separate groups of 6,000 and 3,000 within the same column on the place value grid. Agree that this arrangement is still equivalent to 19,140 because the whole value is the same – the parts just look a little different. Confirm these as 10,000, 6,000, 3,000, 100 and 40.

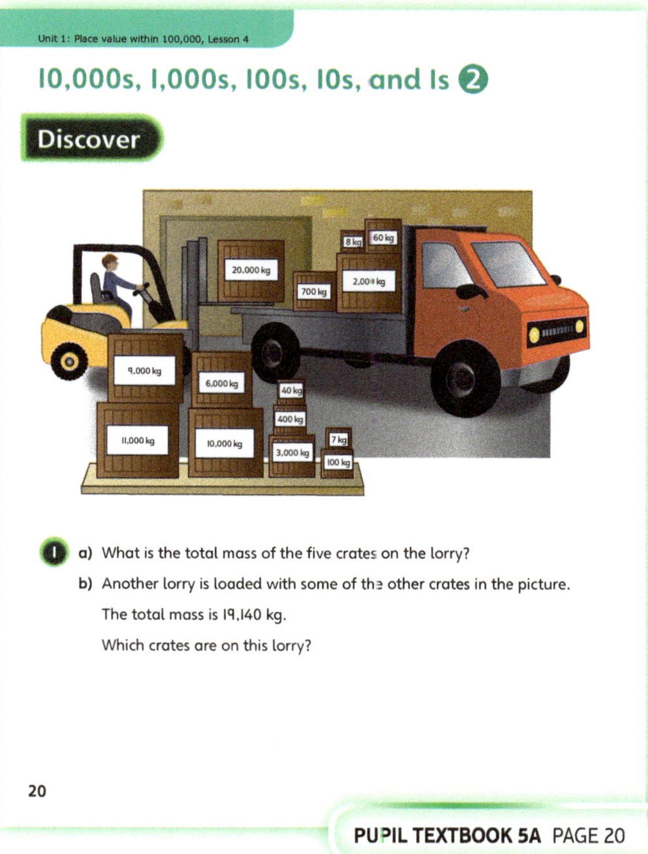

PUPIL TEXTBOOK 5A PAGE 20

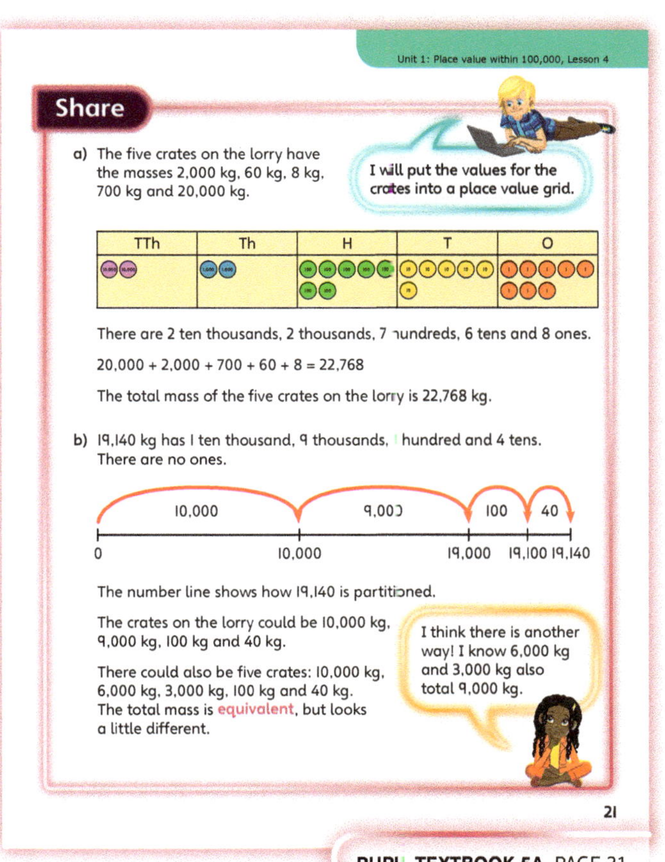

PUPIL TEXTBOOK 5A PAGE 21

Think together

WAYS OF WORKING Whole class teacher led (I do, We do, You do)

ASK

- Question ❶ : *Could you group some of the counters differently, but keeping them in the same columns? How would this help you fill in the missing numbers?*
- Question ❷ : *How can you work out the size of the second jump on the number line?*
- Question ❸ : *Is it possible for only two crates to have been loaded onto the lorry? How do you know?*
- Question ❸ : *How can you check that all of your solutions are correct?*

IN FOCUS In question ❶, children explore equivalence by partitioning the same number in different ways, supported by place value counters in a place value grid. The question provides opportunity for children to find more than one solution each time. Question ❸ continues the idea of multiple solutions, in the **Discover** context of loading crates onto a lorry. The inclusion of the 3,000 kg and 30 kg crates reinforces the importance of place value.

STRENGTHEN To strengthen understanding in question ❷, ask children to make 23,407 with place value counters. Then ask them to arrange the counters to show the value of each jump on the number line and explain why the final jump reaches 23,407.

DEEPEN When children have completed all of the questions, ask them to look again at question ❶ and to use a calculator to check that all of the partitioning is correct. Challenge them to explore other ways to partition 43,245 into two parts, three parts, four parts, five parts and so on, using a calculator to prove equivalence each time.

ASSESSMENT CHECKPOINT Use question ❶ to assess whether children can partition the same number in different ways. Check their understanding of equivalence and that all parts must be able to be recombined to equal the whole. Use question ❸ to check children's understanding of place value as they combine masses in different ways.

ANSWERS

Question ❶ : Possible solutions include:
 43,245 = 40,000 + 3,000 + 200 + 40 + 5
 43,245 = 30,000 + 10,000 + 3,000 + 200 + 40 + 5
 43,245 = 30,000 + 13,000 + 200 + 45
 43,245 = 43,000 + 200 + 45

Question ❷ :

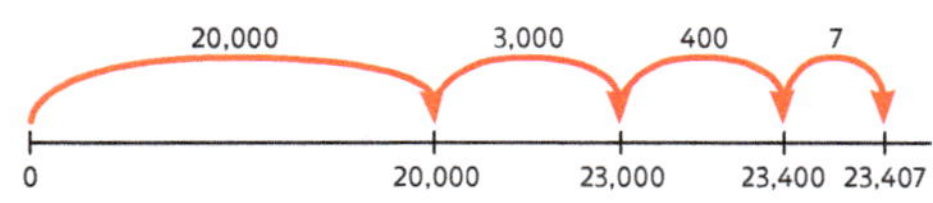

23,407 = 20,000 + 3,000 + 400 + 7

Question ❸ : Possible solutions include:
 20,000 kg, 6,000 kg and 30 kg
 20,000 kg, 6,000 kg, 20 kg and 10 kg
 20,000 kg, 3,000 kg, 3,000 kg, 30 kg
 20,000 kg, 3,000 kg, 3,000 kg, 20 kg and 10 kg
 10,000 kg, 10,000 kg, 6,000 kg and 30 kg
 10,000 kg, 10,000 kg, 3,000 kg, 3,000 kg, 20 kg and 10 kg
 11,000 kg, 9,000 kg, 6,000 kg and 30 kg

PUPIL TEXTBOOK 5A PAGE 22

PUPIL TEXTBOOK 5A PAGE 23

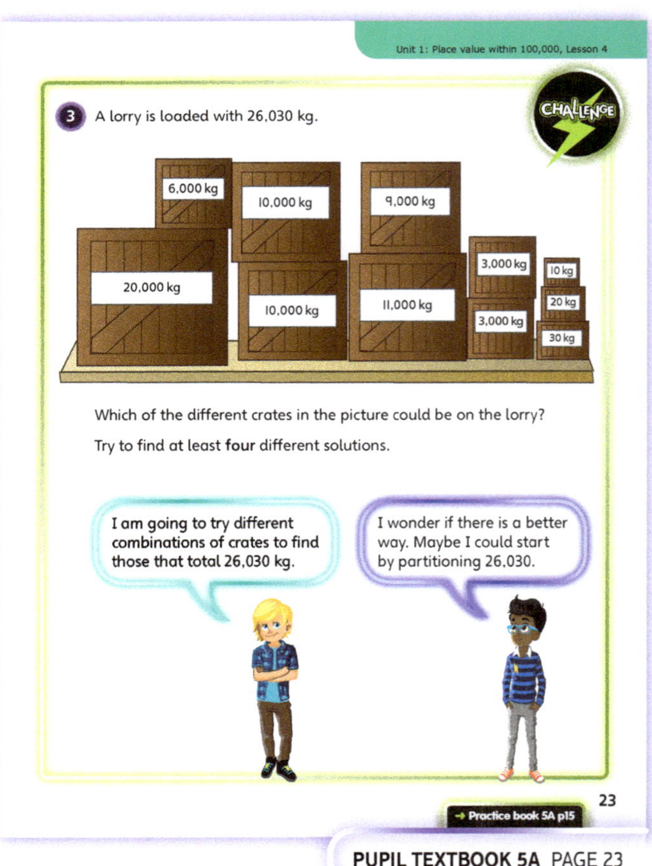

Practice

WAYS OF WORKING Independent thinking

IN FOCUS Question **2** is similar to question **1** in the teacher led part of the lesson, but includes a zero in the 10s position. Children will therefore need to think carefully about place value as they partition. In the more open-ended question **4**, children can share their partitioning and prove why their answer may be different from their partner's, but still equivalent. Question **6** requires children to partition in different ways but to also meet a given criterion each time. They must record their solutions in a table.

STRENGTHEN Encourage children to share solutions with their partner or to explore partitioning together. They should have access to place value counters so that they can arrange them in different ways to explore equivalence. Ask them to explain how they know that their partitioning is still equivalent each time (no counters have been added or taken away).

DEEPEN When children have completed question **6**, ask them to explain how the possible solutions would change if each ship needed a minimum of 2,000 passengers. Ask them whether they would have to start their calculations again or whether they would be able to use their previous solutions.

THINK DIFFERENTLY Question **5** requires children to apply their understanding of equivalence and partitioning with a limited number of options for the parts. In this situation, children's likely starting point of selecting 10,000 ml and 7,000 ml will lead to a wrong path as they go on to select 1,700 ml before realising that they still need 50 ml.

ASSESSMENT CHECKPOINT Use question **2** to assess whether children can partition the same number in different ways. Use question **3** to assess whether they can relate partitioning to the image of a number line, identifying the missing values. Use question **4** to check whether children can use what they already know to help re-partition, rather than starting from scratch each time, for example 68,000 + 359 and then 68,000 + 300 + 59.

ANSWERS Answers for the **Practice** part of the lesson appear in the separate **Practice and Reflect answer guide**.

Reflect

WAYS OF WORKING Independent thinking

IN FOCUS This task requires children to apply their understanding from the lesson to explain why two different examples of partitioning of the same number are correct.

ASSESSMENT CHECKPOINT Check that children recognise that both partitionings are correct because all of the parts combine to make the same whole. They are equivalent because they have the same value, although they may not look the same.

ANSWERS Answers for the **Reflect** part of the lesson appear in the separate **Practice and Reflect answer guide**.

After the lesson ⏸

- Can children explain why 34,257 = 20,000 + 2,000 + 2,200 + 7 is not true?
- Can they explain what the term 'equivalent' means?
- Can they use place value counters to show partitioning in different ways?

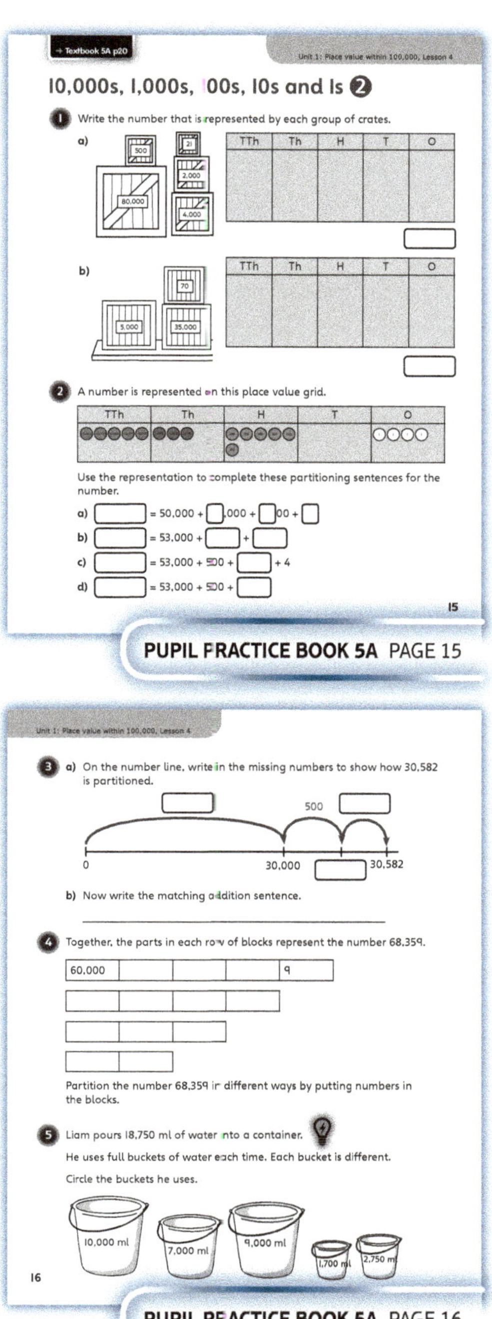

PUPIL PRACTICE BOOK 5A PAGE 15

PUPIL PRACTICE BOOK 5A PAGE 16

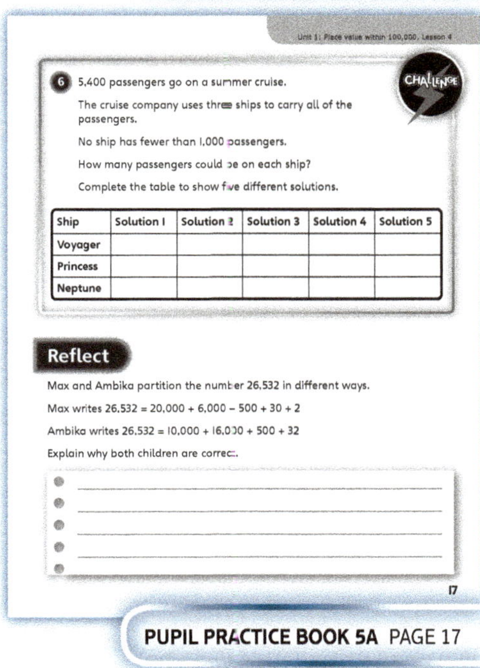

PUPIL PRACTICE BOOK 5A PAGE 17

The number line to 100,000

Learning focus

In this lesson, children will work with the number line to 100,000 and identify numbers that are between two points (for example multiples of 10,000), using mathematical language to describe the position. They will estimate both where a number should be placed on a number line and the number a point or label represents.

Small steps

→ Previous step: 10,000s, 1,000s, 100s, 10s and 1s (2)
→ **This step: The number line to 100,000**
→ Next step: Comparing and ordering numbers to 100,000

NATIONAL CURRICULUM LINKS

Year 5 Number – Number and Place Value

Read, write, order and compare numbers to at least 1,000,000 and determine the value of each digit.

ASSESSING MASTERY

Children can describe or estimate where a 5-digit number appears on a number line, using multiples of 10,000 to help make decisions. They can estimate the number that a position on a number line represents.

COMMON MISCONCEPTIONS

Children may make place value errors when identifying the half-way point between two multiples of 10, 100, 1,000 or 10,000, for example saying that 2,050 is half-way between 2,000 and 3,000. Use a number line or counting stick with ten divisions, working together to label the intervals and count up to the half-way point. Ask:
• *How much more than 2,000 is 2,500? How much less than 3,000 is 2,500? How do you know that 2,500 must be half-way?*

STRENGTHENING UNDERSTANDING

Use place value counters to build 5-digit numbers, for example 34,250. Discuss whether the number is more or less than 30,000. Partition the counters into 30,000 and 4,250 to show that the number is 4,250 more than 30,000. Ask which multiple of 10,000 they will reach next when counting up. Then discuss whether 34,250 is closer to 30,000 or 40,000, making a link to children's understanding of rounding. Finally, use the discussion to estimate the position of 34,250 on a sketched number line.

GOING DEEPER

Ask children to reason about the position of numbers on a range of vertical and horizontal number lines. Vary the interval size so that children need to think carefully about the given labels and how the scale should be used to help make decisions.

KEY LANGUAGE

In lesson: between, less than, more than, half-way, closer to, interval, estimate

Other language to be used by the teacher: next, previous

STRUCTURES AND REPRESENTATIONS

number line

RESOURCES

Optional: place value counters

 In the eTextbook of this lesson, you will find interactive links to a selection of teaching tools.

Before you teach 

• Can children confidently count on and back from any 4- or 5-digit number?
• Can they identify the two multiples of 10,000 that a number lies between?

Discover

WAYS OF WORKING Pair work

ASK

- Question ① a): *Is 55,000 more or less than half-way along the number line? How do you know? Is it a little more than half-way or a lot more than half-way?*
- Question ① a): *Which multiples of 10,000 is the number 55,000 between? Point to where you think it will be.*
- Question ① b): *If you start on 50,000 and count in 1,000s, how many steps to reach 53,000?*

IN FOCUS The questions require children to think carefully about the positions of numbers on the number line, using their proximity to multiples of 10,000 to help make decisions.

PRACTICAL TIPS Use a washing line to make a 0 to 100,000 number line. Children can start by pegging the multiples of 10,000 onto the line. Then give each child a number and ask them to go to their position on the number line.

ANSWERS

Question ① a): 55,000 is more than half-way along the number line.
55,000 lies between two multiples of 10,000
55,000 is 5,000 more than 50,000
55,000 is half-way between 50,000 and 60,000.

Question ① b): Both 53,000 and 56,200 are between 50,000 and 60,000.

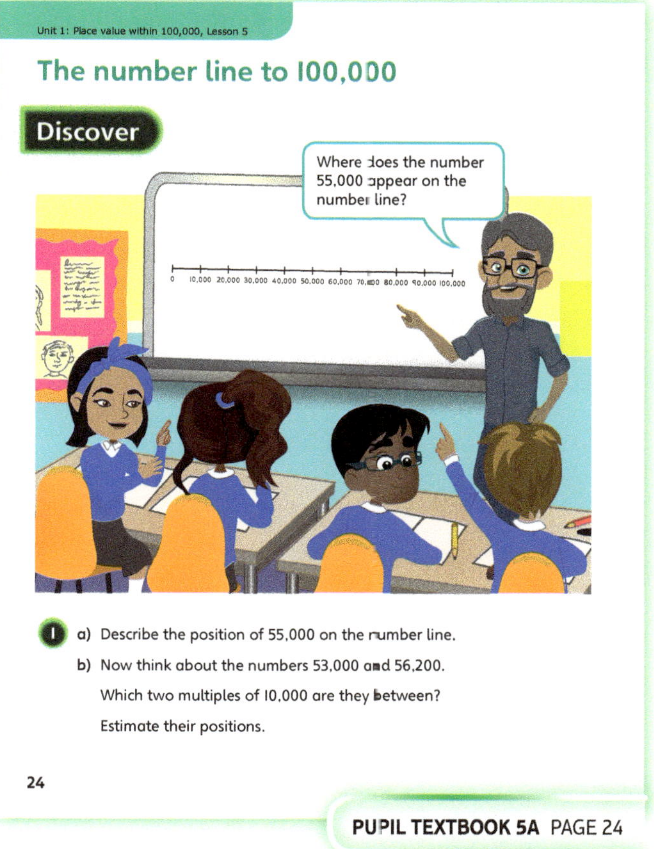

PUPIL TEXTBOOK 5A PAGE 24

Share

WAYS OF WORKING Whole class teacher led

ASK

- Question ① a): *How do you know that 55,000 is 5,000 more than 50,000?*
- Question ① a): *How many steps of 1,000 must you take from 50,000 to reach 60,000? What number would you land on if you only took five steps?*
- Question ① b): *How does Flo's suggestion help to position the numbers?*

IN FOCUS Question ① a) requires children to use the number line to help describe the position of a number on it. It encourages them to check the intervals and the labels to help make decisions. Children see how the position of a number can also be calculated as they use half of 10,000 to find the half-way point. The same skills are also applied when children work with statistical representations and with measurement. Through Flo, question ① b) introduces the idea of dividing the given intervals on the number line into 10 parts to help position numbers more accurately.

STRENGTHEN Ask children to suggest other numbers that are between 50,000 and 60,000. Then ask them to suggest numbers that are not between these multiples and explain why.

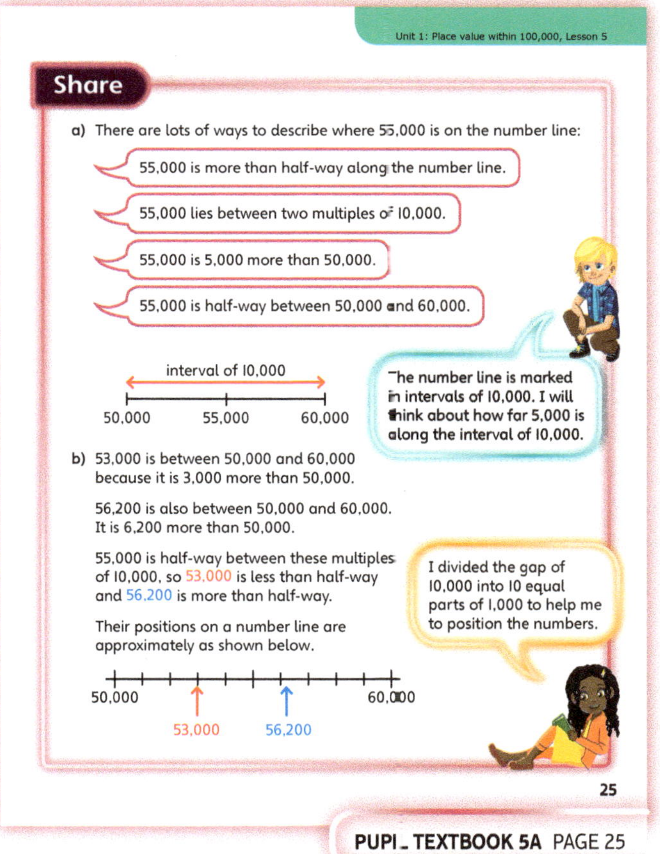

PUPIL TEXTBOOK 5A PAGE 25

Think together

WAYS OF WORKING Whole class teacher led (I do, We do, You do)

ASK

- Question ❶ : *How do you know that none of the numbers is a multiple of 10,000?*
- Question ❶ : *How could you estimate the number at C?*
- Question ❸ : *Which letter can you ignore straight away?*
- Question ❸ : *Is 14,324 more or less than half-way between 10,000 and 20,000? How do you know?*
- Question ❹ : *Astrid has made a mistake. Which number is half-way between 60,000 and 70,000?*

IN FOCUS For point C in question ❶, children need to find the half-way point between 80,000 and 85,000. They should notice that there are four intervals of 250 in each 1,000.

In question ❷, children should use appropriate mathematical language to explain their decisions.

Question ❹ provides children with four criteria to find a number, requiring them to narrow down the possibilities step by step.

STRENGTHEN To strengthen understanding in question ❹, encourage children to sketch a number line. Ask them to explain which multiples of 10,000 go at each end and to mark the line in intervals of 1,000 (as in question ❷).

DEEPEN Ask children to estimate the numbers represented by points A, B and D in question ❸. Ask them to explain their decisions and perhaps suggest other numbers that could appear between B and C.

ASSESSMENT CHECKPOINT Use question ❶ to assess whether children can estimate the values of numbers shown on a number line. Use question ❷ to assess whether they can estimate and position numbers, including those that are not half-way between two multiples of 10,000. Use question ❸ to check their confidence in reasoning about positions on a number line.

ANSWERS

Question ❶ : Point A is about 25,000 because it is half-way between 20,000 and 30,000.
Point B is about 75,000 because it is half-way between 70,000 and 80,000.
Point C is about 82,500 because it is half-way between 80,000 and 85,000.

Question ❷ a): The numbers are 31,000, 32,000, 33,000 and so on. (Each interval is 1,000.)

Question ❷ b): It should be half-way between 37,000 and 38,000.

Question ❷ c): 30,500 should be positioned half-way between 30,000 and 31,000.
34,500 should be positioned half-way between 34,000 and 35,000.

Question ❸ : The best estimate for 14,432 is C.

Question ❹ a): Any numbers from 68,000 to 68,999 inclusive.

Question ❹ b): Any number in the range 65,001 to 67,999 or 69,000 to 69,999.

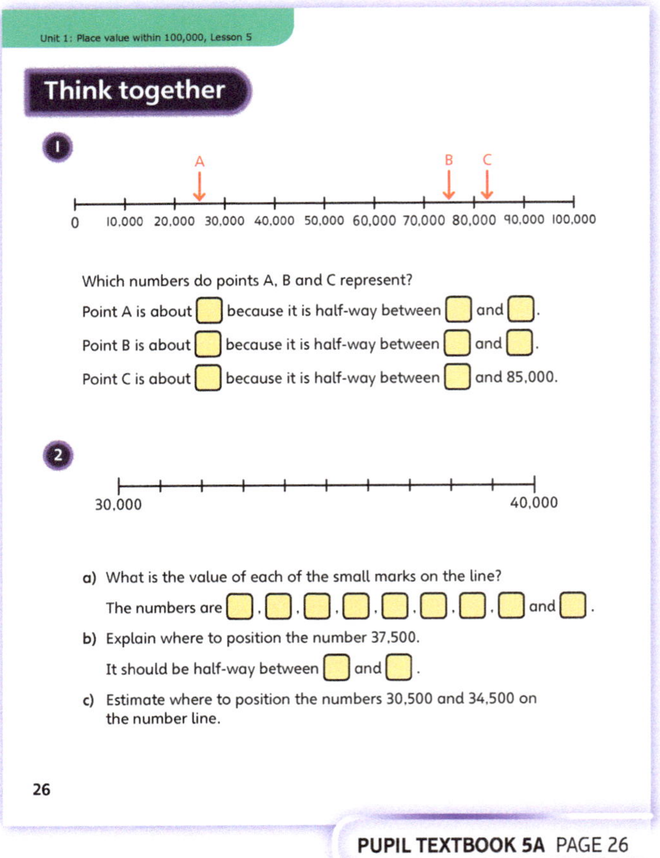

PUPIL TEXTBOOK 5A PAGE 26

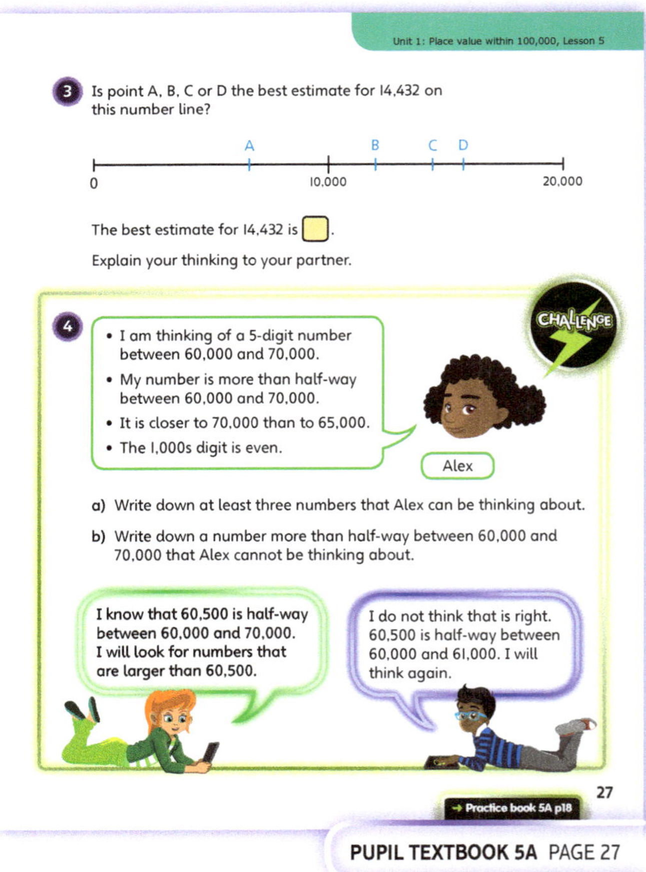

PUPIL TEXTBOOK 5A PAGE 27

Practice

WAYS OF WORKING Independent thinking

IN FOCUS In question **2**, the interval size has been chosen to be different from those used previously, using 2,000 instead of 1,000, and the scale is inverted. Question **6** requires children to reason about how to arrange five digit cards to make numbers to meet given criteria. They should use knowledge of place value to help make decisions about the choice and order of digits each time.

STRENGTHEN In question **5**, encourage children to annotate the scale on the vertical axis of the bar chart in the same way that they have on the number lines. Rotate the page so that children recognise that the orientation of the number line is irrelevant and that strategies to identify, position and estimate numbers can be applied in all examples.

DEEPEN Give children five digit cards and ask them to make up clues similar to those in question **6** for a partner to solve. They should provide the answer for each clue, so it can be compared with their partner's solution.

THINK DIFFERENTLY Question **5** requires children to apply their understanding of the number line to a statistical representation, linking back to their work on bar charts in Year 3 and Year 4. Remind children that they read the vertical axis in the same way they read a number line.

ASSESSMENT CHECKPOINT Use question **1** to assess whether children can estimate the values of numbers shown on a number line. Use question **2** to check that children use the labels on the number line to help make decisions rather than making assumptions about the scale based on what they have seen before. Use question **3** to assess whether children can work out the subdivisions of a non-standard number line in order to locate a number correctly.

ANSWERS Answers for the **Practice** part of the lesson appear in the separate **Practice and Reflect answer guide**.

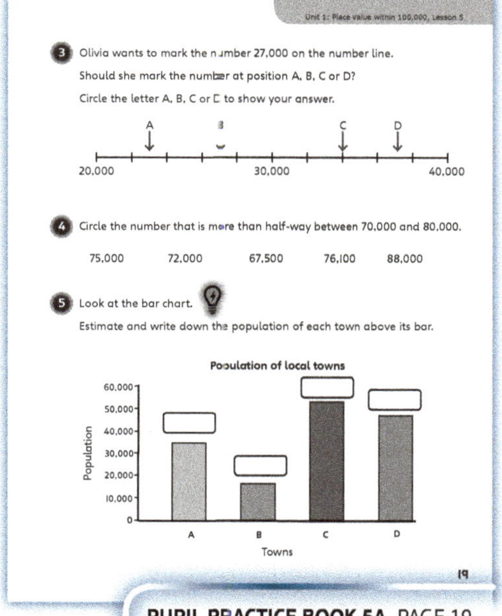

PUPIL PRACTICE BOOK 5A PAGE 18

PUPIL PRACTICE BOOK 5A PAGE 19

Reflect

WAYS OF WORKING Independent thinking

IN FOCUS Children apply their understanding from the lesson to reason about the similarities and differences between groups of numbers. The numbers have been chosen so that they are all between 40,000 and 50,000.

ASSESSMENT CHECKPOINT Check children recognise that all of the numbers are between 40,000 and 50,000. Assess their use of mathematical language to explain why 43,500 is closer to 40,000 while the others are closer to 50,000.

ANSWERS Answers for the **Reflect** part of the lesson appear in the separate **Practice and Reflect answer guide**.

After the lesson ▐▐

- Can children say the number that is half-way between two multiples of 10,000?
- Can they use what they know about an interval size to help estimate or identify a number that appears within it?

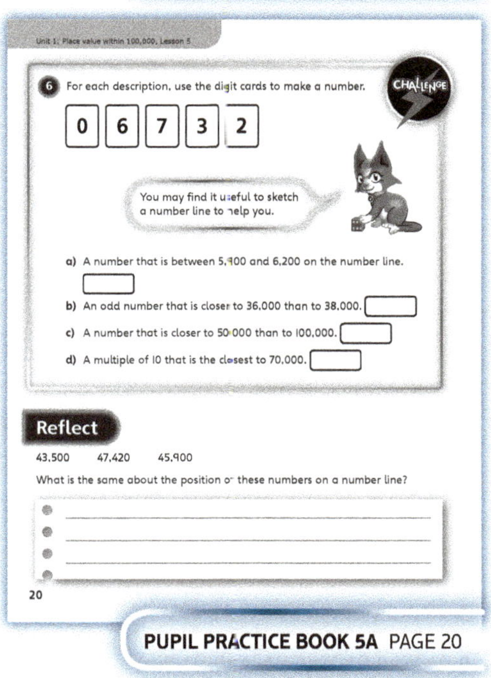

PUPIL PRACTICE BOOK 5A PAGE 20

Comparing and ordering numbers to 100,000

Learning focus

In this lesson, children will compare and order numbers to 100,000 using what they know about place value and will identify which digits they need to compare first each time, explaining what to do when the digits are the same. They will use the signs < and > to show comparisons and order.

Small steps

→ Previous step: The number line to 100,000
→ **This step: Comparing and ordering numbers to 100,000**
→ Next step: Rounding numbers within 100,000

NATIONAL CURRICULUM LINKS

Year 5 Number – Number and place value

Read, write, order and compare numbers to at least 1,000,000 and determine the value of each digit.

ASSESSING MASTERY

Children can compare pairs of numbers up to 100,000, explaining which is larger and why. They can write or represent a group of 4- and 5-digit numbers in ascending or descending order.

COMMON MISCONCEPTIONS

When comparing whole numbers, children may not check the number of digits first and look only at the size of the first digit. For example, when comparing the numbers 23,200 and 5,345, children may think that 5,345 is larger because 5 is larger than 2. Ask:

• *How would you represent these numbers on a place value grid? Which number has the digit in the position with the largest value? How does that help you decide which number is larger?*

STRENGTHENING UNDERSTANDING

Ask children to roll five dice (or the same dice five times) to create a 5-digit number. Ask them to arrange the digits to make the largest and then the smallest possible number, explaining their choices. Then ask them to make another number that is larger than the smallest number and another smaller than the largest, and finally a number that will go in the middle of the four numbers. Ask them to explain why the numbers are now in order.

GOING DEEPER

Give children measurements such as length or distance, mass or capacity to compare and order, asking them to explain how they can use what they know about ordering numbers to help them. Include some examples with different units, for example 34 km and 23,000 m so that children reason about why 34 km is further even though 23,000 is larger than 34.

KEY LANGUAGE

In lesson: compare, order, ascending, descending, less than (<), more than (>), smallest, largest, smaller, larger, highest, lowest

STRUCTURES AND REPRESENTATIONS

number line, place value grid

RESOURCES

Optional: place value counters

 In the eTextbook of this lesson, you will find interactive links to a selection of teaching tools.

Before you teach 

• Can children compare 3- and 4-digit numbers, explaining which is larger and why?
• Can they order a set of 3- or 4-digit numbers by comparing numbers within the set?

Discover

 Pair work

- Question ❶ a): *What do you notice about the digits in all of the scores? Can you find some digits that have the same value in different scores?*
- Question ❶ a): *How do you know immediately that Amal was not the winner?*
- Question ❶ b): *How do you know that Holly was not last? Which digit in the number tells you this?*

 Question ❶ a) and b) use the familiar context of a game show for children to compare and order 5-digit numbers, each containing similar digits, but in different positions within the numbers. The scores have been chosen so that only two scores need to be compared more closely in question ❶ a), as 56,725 and 55,276 have a smaller number of 10,000s.

 Provide pairs or small groups of children with a large place value grid and sufficient place value counters to represent all of the scores in the picture. When they have made all of the scores, they can use the place value counters to prove how they know that Amal is not the winner.

Question ❶ a): Toshi came first with 65,575.

Question ❶ b): In ascending order, the scores are 55,276, 56,725, 65,272, 65,575.

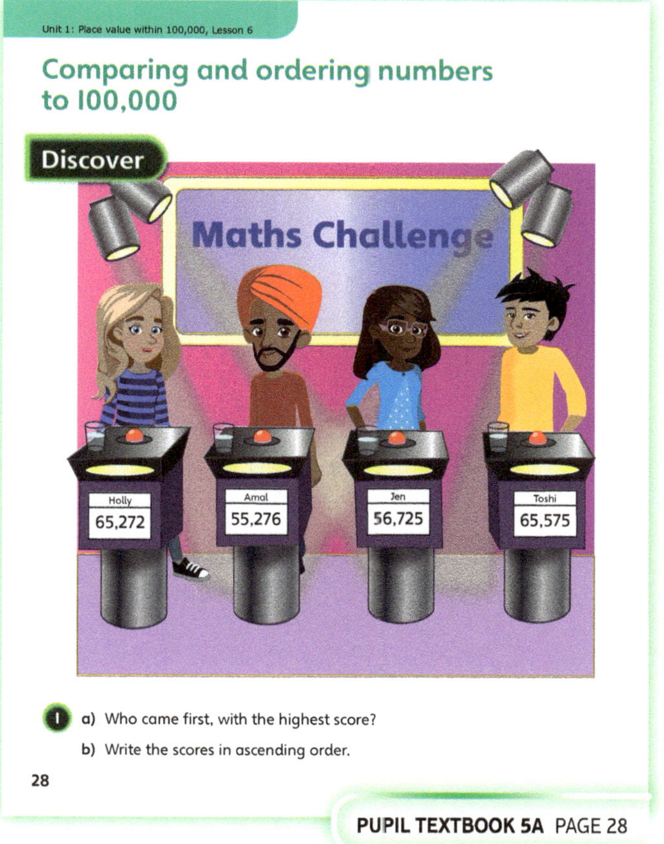

PUPIL TEXTBOOK 5A PAGE 28

Share

 Whole class teacher led

- Question ❶ a): *Why is it important to check how many digits each number has before comparing digits?*
- Question ❶ a): *Can you suggest a pair of 5-digit numbers where you need to check the 1,000s but not the 100s?*
- Question ❶ a): *What would you do if the number of 100s in the two scores had also been the same?*
- Question ❶ b): *Why do you only need to compare the numbers 55,276 and 56,725 now?*
- Question ❶ b): *How does representing the two numbers with place value counters help you compare them?*
- Question ❶ b): *How do the signs between the numbers tell you that they are in ascending order?*
- Question ❶ b): *What does descending mean? How could you write the numbers in descending order?*

 Question ❶ b) introduces the term 'ascending' to specify how the numbers should be ordered. The two lower scores have been chosen to have the same number of 10,000s, so children are required to compare the 1,000s.

 Ensure that children are confident to make sense of and use of the word 'ascending', recognising that 'descending' means in order from largest to smallest.

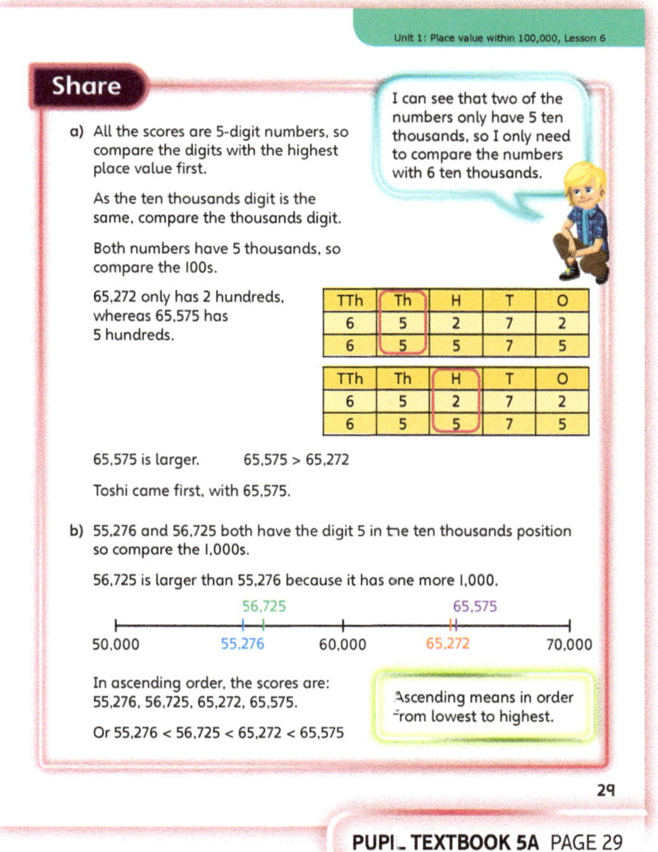

PUPIL TEXTBOOK 5A PAGE 29

Think together

WAYS OF WORKING Whole class teacher led (I do, We do, You do)

ASK

- Question **1** : *Bag B has more counters, but why might it not have the larger value?*
- Question **1** : *Which counters have the largest value? Why do you need to compare the numbers of 1,000s next?*
- Question **2** a) and b): *How do you know that you need to write the larger number in the first box each time?*
- Question **2** b): *One of the numbers starts with a larger digit. Does this mean that it must be the larger number? Why?*
- Question **3** : *How do you immediately know that the second number on the grid is the smallest?*

IN FOCUS Questions **1** and **2** reinforce the use of the signs < and >. Question **1** has been designed so there are more counters in Bag B but the actual value of the bag is less. In question **2** a), the numbers use the same digits but in a different order. In question **2** b), the numbers have been chosen to include a 4-digit and 5-digit number with the 4-digit number starting with a larger digit, requiring children to check the number of digits first.

STRENGTHEN To strengthen decision making in question **2**, ask children to represent each pair of numbers on a place value grid, using place value counters. Can they explain which number is larger and why?

DEEPEN In question **4**, ask children to explore further examples with a partner. Ask: *Is it always true that a number starting with a 6 is larger than a number starting with a 4?*

ASSESSMENT CHECKPOINT Use questions **1** and **2** to assess whether children can compare 5-digit numbers. Check that they understand the use of the signs < and >. Use question **3** to assess whether children can order 4- and 5-digit numbers. Use question **4** to check that they recognise when a number with a higher first digit will be larger and when it will not, they should be able to give examples and counter examples.

ANSWERS

Question **1** : Bag B has a smaller value because 22,512 < 23,110.

Question **2** a): 43,970 > 34,790

Question **2** b): 21,033 > 8,968

Question **3** : 20,932 > 20,923 > 8,560

Question **4** : Sometimes true – it will depend on the number of digits and therefore the place value of the digits 9 and 5. Children should show this using a range of examples for which it is true (for example, 93,245 and 56,278) or not true (for example, 9,375 and 54,267).

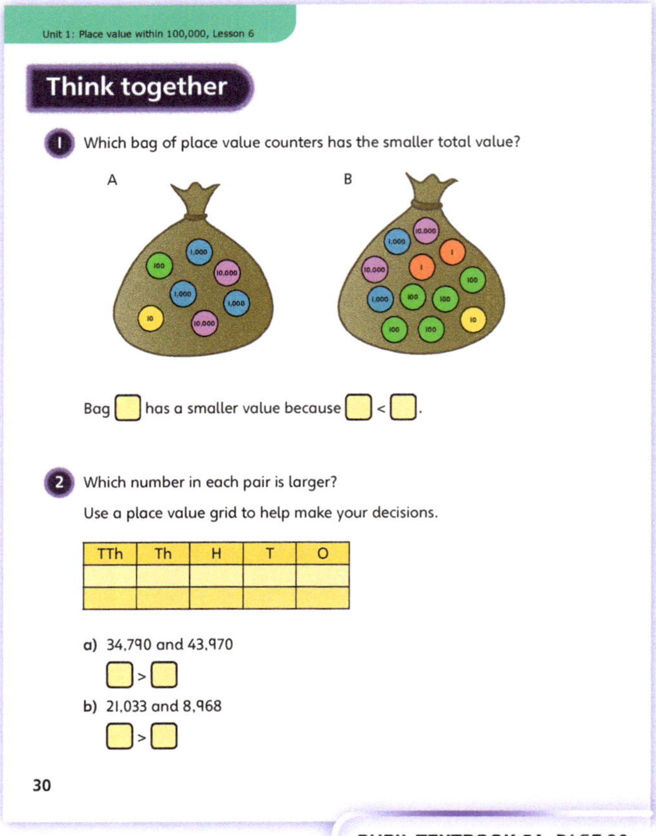

TTh	Th	H	T	O

PUPIL TEXTBOOK 5A PAGE 30

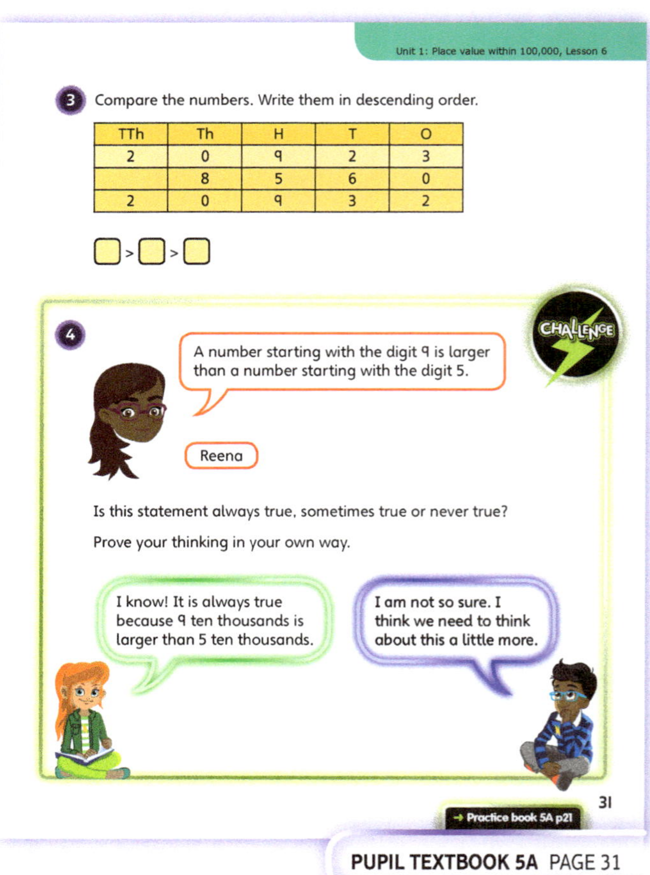

TTh	Th	H	T	O
2	0	9	2	3
	8	5	6	0
2	0	9	3	2

PUPIL TEXTBOOK 5A PAGE 31

Practice

WAYS OF WORKING Independent thinking

IN FOCUS In question **1**, the numbers have been chosen so that the value of the 10,000s, 1,000s and 100s are the same so children must check the 10s.

Question **4** involves comparing a 4-digit number with a 5-digit number, requiring children to check the number of digits first.

Question **6** requires children to reason about the positions of the missing digits. The problem is made more challenging as there are no 'lower' digit cards to make the 1,000s position in the first number less than 6,000.

STRENGTHEN Encourage children to use place value counters to check any comparisons or ordering of numbers they are not sure about. Look together at the use of the < and > signs to ensure that children are confident to use these appropriately and recognise why they can also be used for showing order.

DEEPEN Ask children to make up some true or false statements (similar to question **4**) for a partner to solve.

ASSESSMENT CHECKPOINT Use questions **1** and **2** to assess whether children can compare and order 5-digit numbers. Use question **4** to assess whether children can compare 4- and 5-digit numbers, checking the number of digits first. Check that children understand the terminology 'ascending' and 'descending' and the use of the signs < and >. Use questions **5** and **7** to check that children can apply their understanding of order to measurement and money.

ANSWERS Answers for the **Practice** part of the lesson appear in the separate **Practice and Reflect answer guide**.

Reflect

WAYS OF WORKING Independent learning

IN FOCUS Children could use place value counters or a grid with digits to help show the process they would use to order a set of 4- and 5-digit numbers. In this question, three numbers have the same 10,000s value and one number has a larger first digit but is only a 4-digit number, so children need to closely check the number of digits and their value.

ASSESSMENT CHECKPOINT Check that children know what to do next when two digits in the same position have the same value. Also, look for children who mistakenly think that 8,976 is larger because it starts with the digit 8.

ANSWERS Answers for the **Reflect** part of the lesson appear in the separate **Practice and Reflect answer guide**.

After the lesson ⏸

- Can children explain which number is larger in pairs of 5-digit and 4-digit numbers?
- Can they order numbers in ascending or descending order?
- Can they explain why 9,450 is smaller than 34,212 even though it starts with the digit 9?

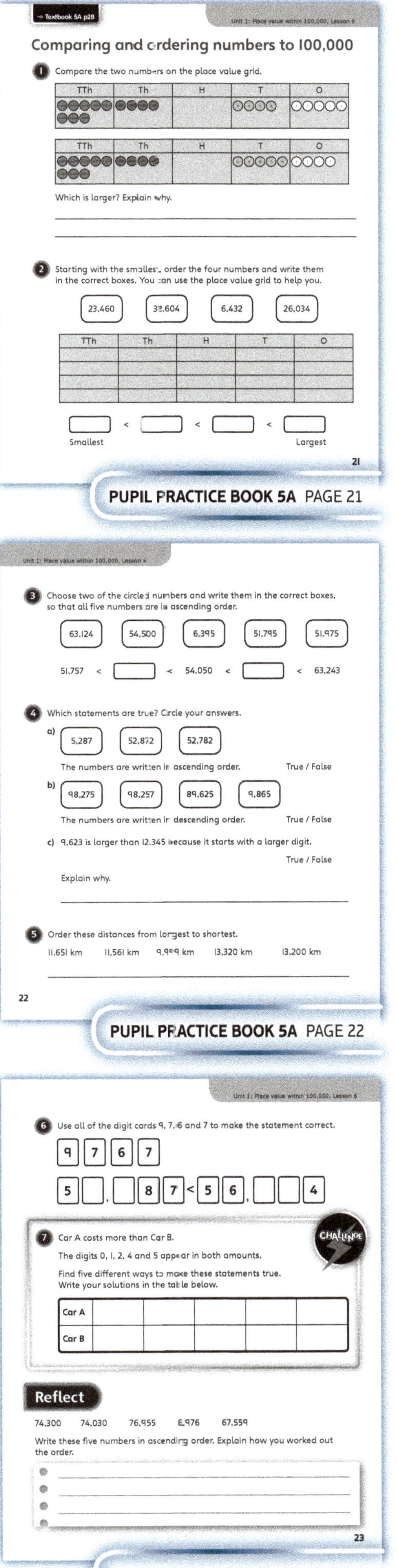

PUPIL PRACTICE BOOK 5A PAGE 21

PUPIL PRACTICE BOOK 5A PAGE 22

PUPIL PRACTICE BOOK 5A PAGE 23

Rounding numbers within 100,000

Learning focus

In this lesson, children will apply the rules for rounding to the nearest 10, 100 and 1,000 to round 5-digit numbers. They will also learn to round to the nearest 10,000 and identify the next and previous multiple, reasoning about which digit to check in a number to help make decisions on rounding.

Small steps

→ Previous step: Comparing and ordering numbers to 100,000
→ **This step: Rounding numbers within 100,000**
→ Next step: Roman numerals to 10,000

NATIONAL CURRICULUM LINKS

Year 5 Number – Number and Place Value

Round any number up to 1,000,000 to the nearest 10, 100, 1,000, 10,000 and 100,000.

ASSESSING MASTERY

Children can round 5-digit numbers to the nearest 10,000, 1,000, 100 or 10, explaining their decision each time. They can identify the digit they need to check in the number, depending on the degree of rounding.

COMMON MISCONCEPTIONS

When checking digits to help rounding, children often assume that the digit to check must also be the value that the number is to be rounded to, for example the 1,000s digit to round to the nearest 1,000. Ask:
• *Which multiples of 1,000 is 23,605 between? Which multiple is it closer to? Which digit shows that you need to round up?*

STRENGTHENING UNDERSTANDING

Encourage children to explain which digit in a number determines how the number should be rounded each time. Use place value counters to practise rounding a 5-digit number to the nearest 10,000, 1,000, 100 or 10, looking at how this affects the digit to check. Discuss how they need to round the digit 5, reminding them that the convention is to round up.

GOING DEEPER

Ask children to explain why, for example, 30,004 rounds to the same 10,000, 1,000, 100 and 10. Encourage them to find other numbers that round in a similar way. Challenge them to make generalisations about numbers that round in this way and ask them to explore numbers like 29,995.

KEY LANGUAGE

In lesson: round, multiple, digit, ten thousands (10,000s), thousands (1,000s), hundreds (100s), tens (10s), ones (1s), nearest, next, previous, closer to

STRUCTURES AND REPRESENTATIONS

place value grid, number line

RESOURCES

Optional: place value counters

 In the eTextbook of this lesson, you will find interactive links to a selection of teaching tools.

Before you teach 

• Can children say the two multiples of 10,000 that a number lies between?
• Can children confidently explain the rules for rounding to the nearest 10, 100 and 1,000? Can they explain which digit will help them?

Discover

 Pair work

- Question **1** a): *What can you remember about the rules for rounding? Which digit do you need to check to round to the nearest 10,000?*
- Question **1** a): *Why do you not look at the digit 9 in the 10,000s position to round to the nearest 10,000?*
- Question **1** b): *Which two multiples of 100 is the number 77,735 between? How do you know?*

 In question **1** a) the number is very close to 100,000 to provide an opportunity to recognise that both 90,000 and 100,000 are multiples of 10,000.

 Ask children to read all of the numbers aloud and explain the value of each digit. Practise counting in steps of 10,000 up to 100,000 and back again. Discuss some other numbers that will or will not round to the same numbers and use place value counters to help visualise the values of numbers.

Question **1** a): 98,275 litres rounds up to 100,000 litres to the nearest 10,000 litres.

Question **1** b): The oval pool holds 77,700 litres of water to the nearest 100 litres.

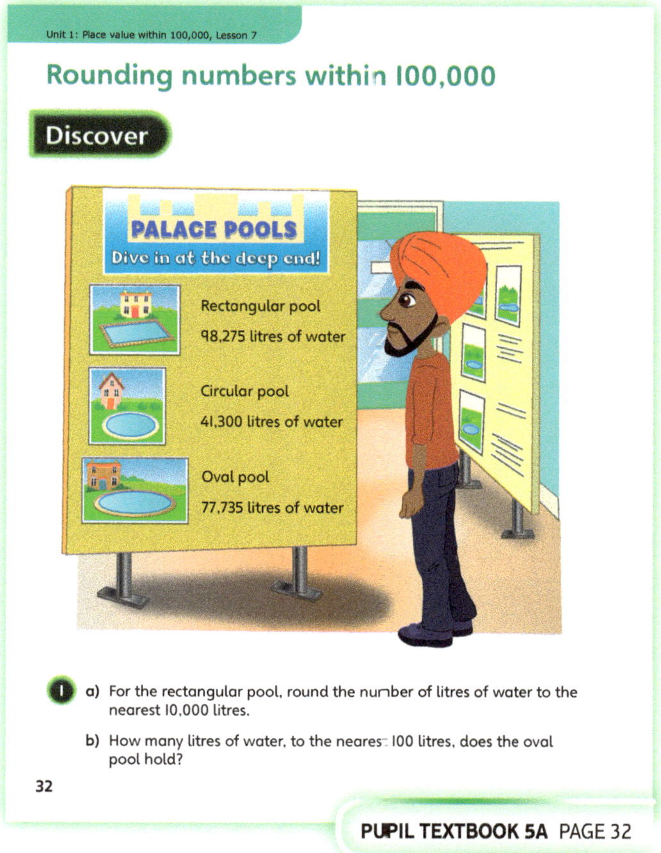

PUPIL TEXTBOOK 5A PAGE 32

Share

 Whole class teacher led

- Question **1** a): *Why do you think Astrid made this mistake?*
- Question **1** a): *What numbers can you think of that will round to 90,000? How about 95,000?*
- Question **1** b): *Why do you need to check the 10s digit when rounding to the nearest 100?*
- Question **1** b): *What other numbers will also round to the same 100?*

 In question **1** a) children are reminded about the use of the number line and the position of the number between two multiples to help make decisions about rounding up or down. Question **1** b) focuses more on the digits that should be checked each time to help round, developing the ability to round without a number line. The number line is provided for support, but the reasoning is based on the digit position and value.

 Use place value counters to reinforce that the 1,000s digit shows whether the number is closer to the next or previous multiple of 10,000. Ask children to make 18,000 and then ask them to use their place value counters to show the previous and next multiples of 10,000, explaining what they need to do. They should realise that they need to take away the eight 1,000 counters to go down to 10,000 or add two 10,000 counters to go up to 20,000, demonstrating that the digit to check is the 1,000s digit.

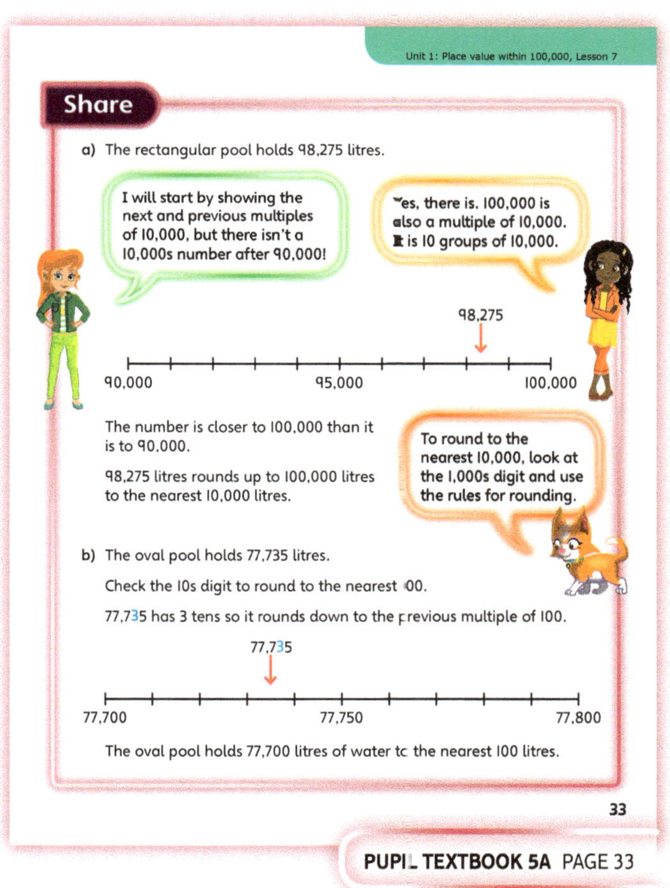

PUPIL TEXTBOOK 5A PAGE 33

Think together

Unit 1: Place value within 100,000, Lesson 7

Think together

WAYS OF WORKING Whole class teacher led (I do, We do, You do)

ASK

- Question **1** a): *How do you know that 41,500 is half-way between 41,000 and 42,000?*
- Question **2** : *How can you use a number line to show whether Aki is correct or not?*
- Question **3** : *Which digit do you need to check to find any mistakes that Jamilla has made?*

IN FOCUS In question **1** a), children round 41,500 to the nearest 1,000 and then to the nearest 10,000. The number has been chosen because it is half-way between the two multiples of 1,000. Question **1** d) asks children to look at a rounded number, identify which digit has been used, and hence work out the accuracy of the rounding.

Question **3** provides an opportunity to address any misconceptions, as children identify the digit that should be checked to help round to the nearest 1,000.

STRENGTHEN Encourage children to sketch a number line and write the appropriate multiples at each end. Ask them to use their number line to explain whether the number to be rounded is closer to the next or previous multiple.

DEEPEN In question **3**, ask children to work out where some other numbers would go in Jamilla's rounding table, justifying their decision to their partner each time.

ASSESSMENT CHECKPOINT Use question **1** a) to assess whether children can round a 5-digit number to the nearest 1,000, stating the next and previous multiple of 10 for a given number. Use question **1** c) to assess whether children can round to the nearest 10,000.

ANSWERS

Question **1** a): 41,300 is between 41,000 and 42,000 on a number line.
41,300 is closer to 41,000.
41,300 litres rounds to 41,000 litres to the nearest 1,000 litres.

Question **1** b): Max needs to look at the 1,000s digit.

Question **1** c): 41,300 litres rounds to 40,000 litres to the nearest 10,000 litres.

Question **1** d): 77,735 litres rounded to the nearest 10 litres is 77,740 litres.

Question **2** : Aki is not correct because 98,275 rounds to 98,280 to the nearest 10.

Question **3** : 34,615 rounds up to the nearest 1,000 so is in the wrong box.
26,322, 85,488, 5,259 all round down to the nearest 1,000.
It appears that Jamilla may have looked at the 1,000s digit instead of the 100s digit each time.

Think together

1 a) For the circular pool, round the number of litres of water to the nearest 1,000 litres.

41,300

```
├──┼──┼──┼──┼──┼──┼──┼──┼──┤
41,000            41,500            42,000
```

The circular pool holds 41,300 litres of water.

41,300 is between ☐,000 and ☐,000 on a number line.

41,300 is closer to ☐ .

41,300 litres rounds to ☐ litres to the nearest 1,000 litres.

b) Max rounds 41,300 to the nearest 10,000 without using a number line.

Which digit does Max need to look at?

Max needs to look at the ☐ digit.

c) Now round 41,300 to the nearest 10,000.

41,300 litres rounds to ☐ litres to the nearest 10,000 litres.

d) Complete the sentence.

77,735 litres rounded to the nearest ☐ litres is 77,740 litres.

34

PUPIL TEXTBOOK 5A PAGE 34

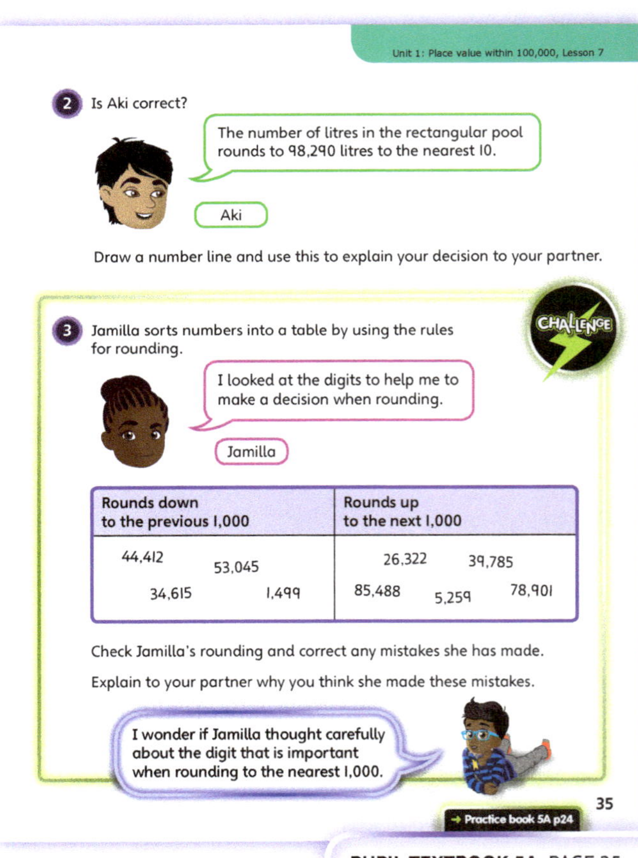

Rounds down to the previous 1,000	Rounds up to the next 1,000
44,412 53,045	26,322 39,785
34,615 1,499	85,488 5,259 78,901

PUPIL TEXTBOOK 5A PAGE 35

Practice

WAYS OF WORKING Independent thinking

IN FOCUS For question **2**, children can model the question using place value counters to help them round the number in different ways and complete the table. The number has been chosen so that the previous multiple of 10 is also a multiple of 100, so the number rounded to the nearest 100 is the same as the number rounded to the nearest 10.

Question **4** offers an opportunity for abstract fluency as no representations are provided.

In question **5**, children reason about missing digits using rules for rounding to help them. The final part involves two missing digits and requires the number to be rounded to suit two criteria.

STRENGTHEN For question **7**, suggest that children look at the first clue initially and write down a number that matches this clue. Then ask them to think about what needs to be changed so the second clue is also met. They can continue in this way, changing one digit at a time as they look at each clue.

DEEPEN When children have completed question **7**, ask them to discuss the strategy they used to solve it with their partner. Ask them whether they made decisions about each digit to help them: the 1,000s digit must be 5 or more, the 100s digit 4 or less, the 10s digit 4 or less and the 1s digit 5 or more. Then ask them to make up a similar challenge for their partner.

ASSESSMENT CHECKPOINT Use questions **2** and **4** to assess whether children can round to the nearest 10, 100, 1,000 and 10,000. Use question **6** to check the accuracy of children's rounding when applied to a context.

ANSWERS Answers for the **Practice** part of the lesson appear in the separate **Practice and Reflect answer guide**.

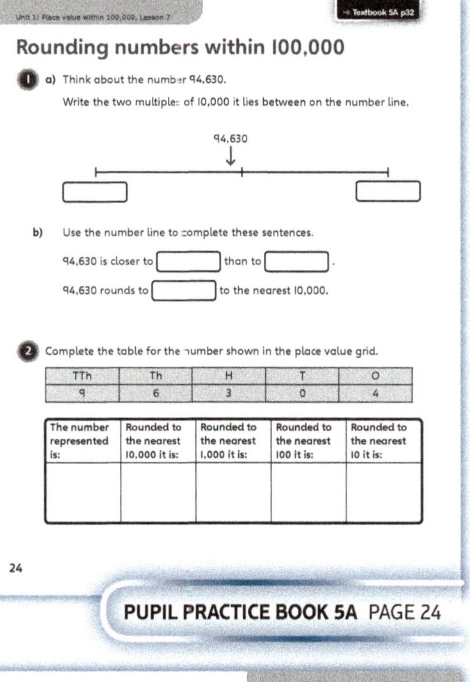

PUPIL PRACTICE BOOK 5A PAGE 24

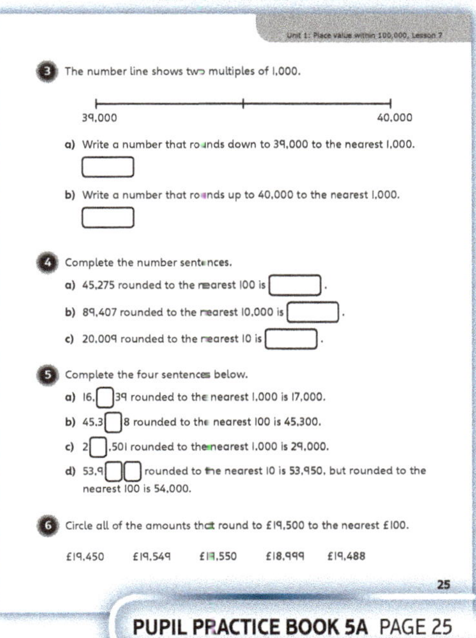

PUPIL PRACTICE BOOK 5A PAGE 25

Reflect

WAYS OF WORKING Independent thinking

IN FOCUS The task provides an opportunity to re-focus on the digit that is important to check when rounding to a given multiple.

ASSESSMENT CHECKPOINT Check that children are not confusing the multiple a number is to be rounded to with the digit to be checked, for example not checking the 1,000s digit to round to the nearest 1,000.

ANSWERS Answers for the **Reflect** part of the lesson appear in the separate **Practice and Reflect answer guide**.

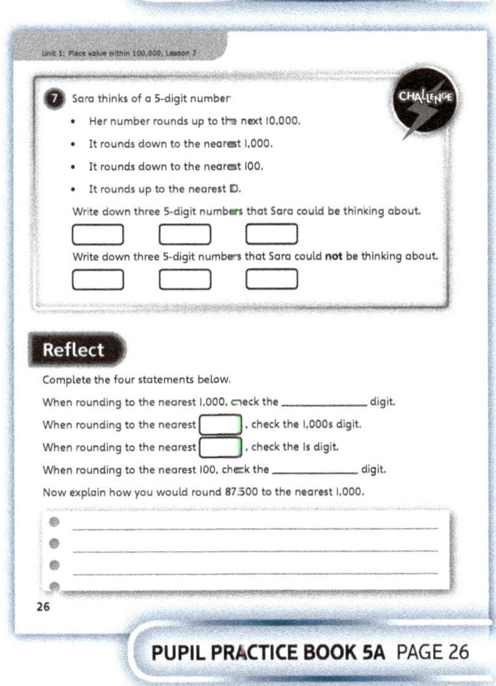

PUPIL PRACTICE BOOK 5A PAGE 26

After the lesson 

- Can children explain which digit should be checked each time in a number to be rounded to 10, 100, 1,000 or 10,000?
- Can children reason about numbers that do or do not round to a given multiple?
- Do children know which multiple of 10,000 comes after 90,000?

Roman numerals to 10,000

Learning focus

In this lesson, children will revisit Roman numerals to 100 and learn the numerals M (1,000) and D (500). They will explore reading and writing numbers using Roman numerals, and will use the numerals M and D to recognise and represent years.

Small steps

→ Previous step: Rounding numbers within 100,000
→ **This step: Roman numerals to 10,000**
→ Next step: 100,000s, 10,000s, 1,000s, 100s, 10s and 1s (1)

NATIONAL CURRICULUM LINKS

Year 5 Number – Number and Place Value

Read Roman numerals to 1,000 (M) and recognise years written in Roman numerals.

ASSESSING MASTERY

Children can read and write Roman numerals using M, D, C, L, X, V and I. They can recognise and represent a year written in Roman numerals, and talk about other ways they are still used in everyday life.

COMMON MISCONCEPTIONS

Children may use numerals that are too small before a larger Roman numeral, for example for the number 99, they may see this as 100 subtract 1 so show this as IC rather than representing this as 90 + 9 using XCIX. The only pairs of numbers that are used for this subtraction rule are IV, IX, XL, XC, CD and CM. So only I, X and C can be used in this way and they can only come before the two numbers above them in value. Ask:

• *Which Roman numerals can you use to subtract from a larger number? Can you make up a rule to remind you?*

STRENGTHENING UNDERSTANDING

Encourage children to use part-whole models to break up larger numbers into recognisable parts. Introduce children to a range of numbers written using Roman numerals for children to identify patterns. Ensure that children understand that the maximum number of repeats is three before the subtraction rule is applied, for example X, XX, XXX, XL. Present them with 400 written as CCCC and ask them to identify the mistake.

GOING DEEPER

Encourage children to think about the ways that repeating numerals can be read as a group rather than individually, for example MCCC as 1,000 + 300 = 1,300. Ask children to represent MCCCL using a part-whole model with only three parts.

KEY LANGUAGE

In lesson: Roman numerals, ascending, descending

STRUCTURES AND REPRESENTATIONS

part-whole model

RESOURCES

Optional: sets of cards showing I, I, I, V, X, X, X, L, C, C, C, D, M, set of matching cards for 4, 9, 40, 90, 400, 900 in numbers and Roman numerals

 In the eTextbook of this lesson, you will find interactive links to a selection of teaching tools.

Before you teach 

• Can children match Roman numerals to the numbers on a clock face?
• Can they explain what is different about the numerals IX, X and XI?
• Can children recognise and write Roman numerals for numbers within 100?

Discover

WAYS OF WORKING Pair work

ASK

- Question ① a): *Where have you seen Roman numerals being used in everyday life?*
- Question ① a): *How is the Roman numeral I used with X to represent 9, 11 and 19?*
- Question ① a): *Which of Ebo's Roman numerals do you recognise? What are their values?*
- Question ① b): *Which Roman numerals does Jamie have so far? What do they represent?*

IN FOCUS Questions ① a) and b) are important as they remind children of the Roman numerals they encountered in Year 4. Use the poster to revisit the use of a smaller numeral before a larger, for example 9 as 10 subtract 1 and 40 as 50 subtract 10.

PRACTICAL TIPS Ask children to count in steps of 10 to 100 and use the poster on the wall in the picture to discuss any patterns they notice within the Roman numerals, particularly how 40 and 90 are represented.

Give pairs of children a set of nine cards with the Roman numerals: I, I, I, V, X, X, X, L, C. Ask them to make some different numbers to 100, for example 37, 95 and 44. Later add C, C, D and M so children can make higher numbers.

ANSWERS

Question ① a): Ebo's Roman numerals represent the number 1,690.

Question ① b): Jamie needs to cut out the Roman numeral I. 74 in Roman numerals is LXXIV.

Share

WAYS OF WORKING Whole class teacher led

ASK

- Question ① a): *Why would you not repeat the Roman numeral D to make DD (1,000)? What would you use instead?*
- Question ① a): *What would CX mean?*
- Question ① b): *Why should you not represent 74 in Roman numerals as LXXIIII?*
- Question ① b): *How would Jamie use the I to show 76?*

IN FOCUS Question ① a) introduces the numerals M for 1,000 and D for 500. Children are led through the process of working out a number from its Roman numerals, breaking them down into parts to find the value of each, and revisiting how placing a smaller numeral before a larger one reduces its value (in this case XC for 100 − 10 = 90).

STRENGTHEN Ask children to use the cards on their table to show Ebo's number. Use a part-whole model to show the value of each of the parts: M, D, C and XC.

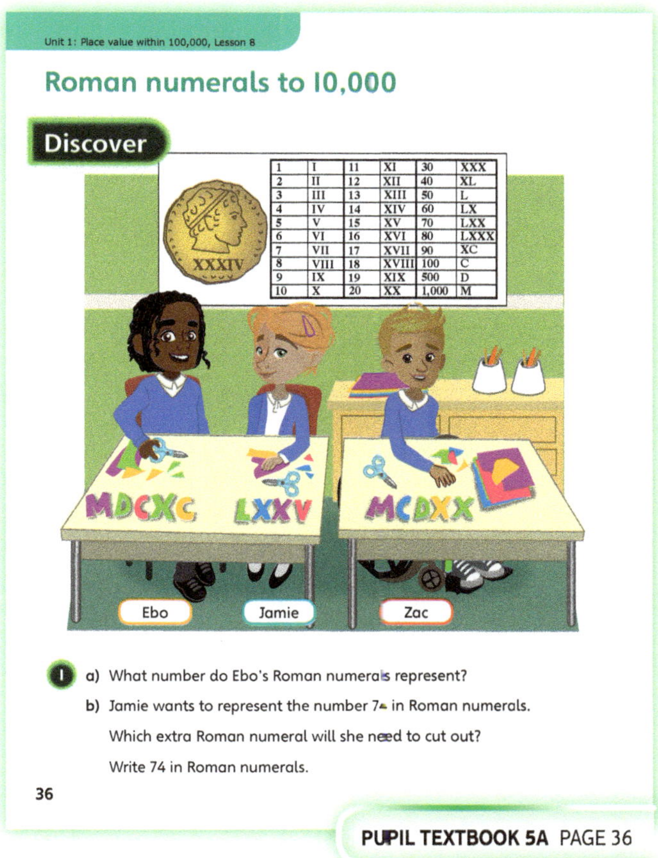

PUPIL TEXTBOOK 5A PAGE 36

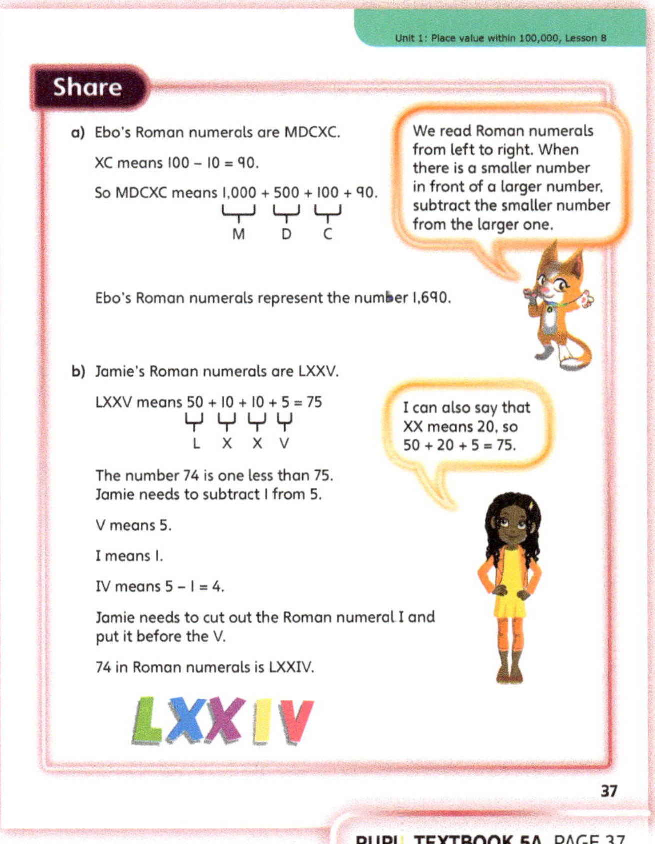

PUPIL TEXTBOOK 5A PAGE 37

Think together

WAYS OF WORKING Whole class teacher led (I do, We do, You do)

ASK

- Question **1** : *What is the value of each part? How do you know?*
- Question **1** : *What does it mean when C is used before D? What if C is used after D?*
- Question **2** a): *I think the baseball poster shows the number 117. What mistake have I made?*
- Question **3** : *Can you use a part-whole model to help find the missing digits? What parts do you already know?*
- Question **4** : *How do you know that the answer to calculation b) must be wrong?*

IN FOCUS In question **1**, a part-whole model is shown to help children recognise the value of parts that are used to build up numbers in Roman numerals. Question **2** demonstrates the use of Roman numerals in different real-life contexts.

Question **4** includes incorrect calculations using Roman numerals. It emphasises the need to be meticulous with Roman numerals. For example, part b) repeats the numeral V for X. In part e), children may recognise that XC add X is simply C as 100 − 10 + 10 = 100.

STRENGTHEN Ensure that children are secure with the six pairs of Roman numerals used to show the numbers 4, 9, 40, 90, 400, 900 (IV, IX, XL, XC, CD and CM). Play a matching game based on cards showing numbers and Roman numerals to build up children's familiarity with these forms.

DEEPEN Ask children to explain why some numerals are repeated (such as X) and some are not (such as L). For example, L is not repeated because LL would be 100.

ASSESSMENT CHECKPOINT Use question **2** to assess whether children can move fluently between numbers and Roman numerals. Use question **4** to look for children who can break Roman numerals down into the right groups.

ANSWERS

Question **1** : M means 1,000
CD means 500 − 100 = 400
XX means 10 + 10 = 20
1,000 + 400 + 20 = 1,420
Zac's number is 1,420.

Question **2** a): XCVII means 97

Question **2** b): 450 means CDL

Question **2** c): MMIX means 2009

Question **2** d): 1791 means MDCCXCI

Question **3** a): MC**M**XX means 1,920.

Question **3** b): DCC**L**IV means 754.

Question **3** c): **XCI**X means 99.

Question **4** : Corrected answers: b) LXXXII, c) CC, e) DCVI

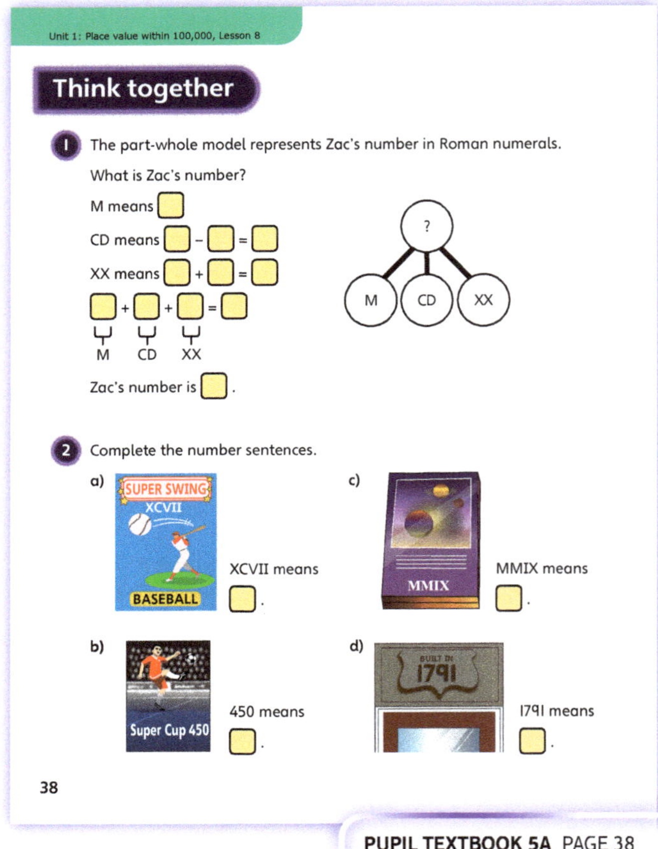

PUPIL TEXTBOOK 5A PAGE 38

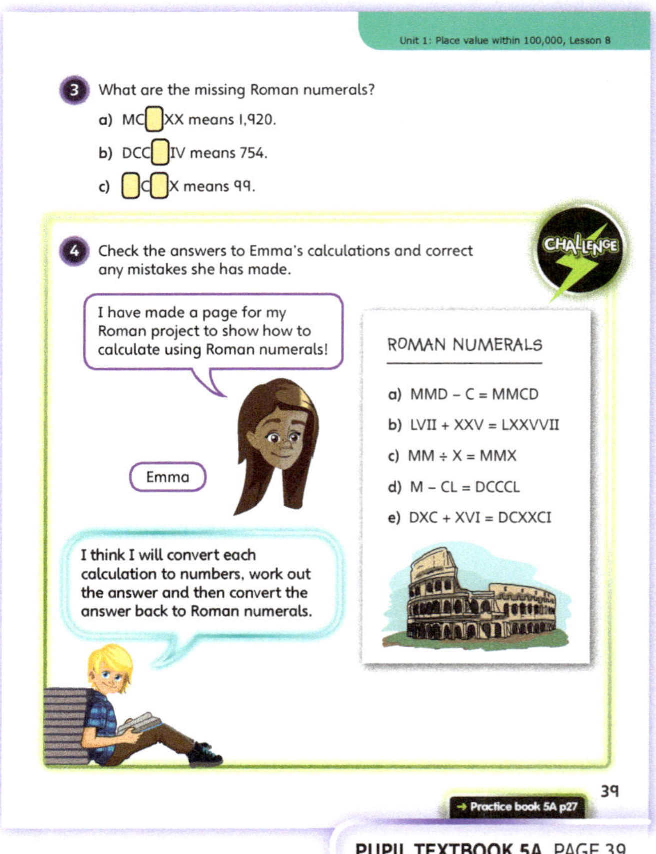

PUPIL TEXTBOOK 5A PAGE 39

Practice

WAYS OF WORKING Independent thinking

IN FOCUS In question **2**, the numbers begin with those that can be simply partitioned using the value of the Roman numerals. Part c) requires children to recognise that two numerals (XC) are grouped together to represent 90. Question **3** also requires Roman numerals to be grouped together to represent either larger parts (such as LXX) or because a smaller Roman numeral is needed before a larger one to show that the smaller value is subtracted (such as XL).

Questions **6** and **7** both require children to reason about Roman numerals that can and cannot be before or after other numerals.

STRENGTHEN Encourage children to use the digit cards and part-whole models to help make decisions. They may find it useful to use this strategy to support thinking in question **6**, perhaps by first randomly trying a few digits and then trying to partition the numerals, checking if they make sense or not.

DEEPEN When children have completed all of the questions, ask them to make up a table to show the multiples of 250 from 0 to 2,000 in Roman numerals. Encourage them to look for patterns in their table.

ASSESSMENT CHECKPOINT Use question **1** to assess whether children can write the multiples of 100 to 1,000 in Roman numerals, so they are able to build larger numbers confidently. Use questions **3**, **4** and **5** to check that children can convert between numbers and Roman numerals. Use questions **6** and **7** to check that children know the possible order that Roman numerals can be written in and why, for example, DMCXCV is not possible because D cannot come before M.

ANSWERS Answers for the **Practice** part of the lesson appear in the separate **Practice and Reflect answer guide**.

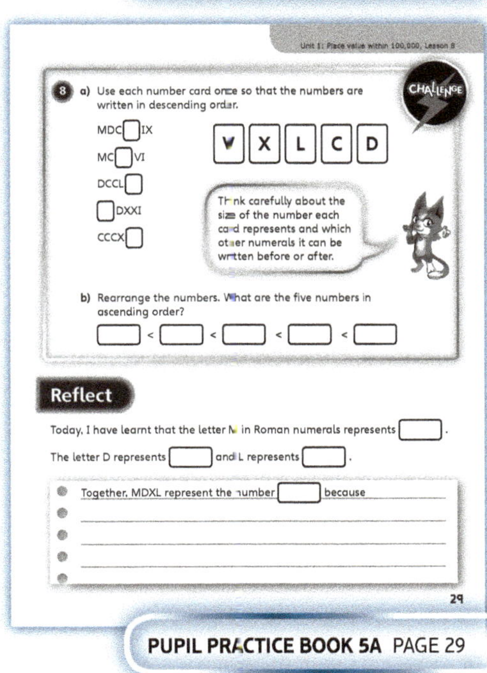

PUPIL PRACTICE BOOK 5A PAGE 27

PUPIL PRACTICE BOOK 5A PAGE 28

Reflect

WAYS OF WORKING Independent thinking

IN FOCUS The task revisits the numerals M and D that were introduced in the main lesson, together with L. Children are required to give the value of each and explain how to find the number represented by MDXL.

ASSESSMENT CHECKPOINT Check that children know the value of each numeral and can explain that to find the value of MDXL they need to subtract 10 from 50 because the X comes before the L.

ANSWERS Answers for the **Reflect** part of the lesson appear in the separate **Practice and Reflect answer guide**.

After the lesson 

- Can children explain the use of the numeral C in the numbers MC, MCC and MCM?
- Can children write the current year in Roman numerals?
- Can they explain why we do not use the Roman numerals MDCCCC to represent 1900?

PUPIL PRACTICE BOOK 5A PAGE 29

End of unit check

> **Don't forget the *Power Maths* unit assessment grid on p26.**

 Group work adult led

 In question **3**, children check the value of each digit in abstract numbers to find the one that matches the representation, being careful not to make place value errors where numbers possess similar digits.

Question **5** is designed to make children think about a good estimate for a position on a number line, using the labels and proximity of numbers between multiples of 10,000.

Question **8** is a SATs-style question on Roman numerals. The number 2019 requires children to demonstrate that they know when to put a smaller value letter before a larger value.

Children who have mastered the concepts in this unit will know the value of each digit in numbers up to 100,000. They will identify the two multiples of 10, 100 or 1,000 that a number lies between and will apply this understanding to make decisions about rounding. They will flexibly partition numbers, appreciating that the combined parts must still be equivalent to the whole. Children will apply their knowledge of place value and the number line to help compare and order numbers. Children will convert between numbers and Roman numerals up to M.

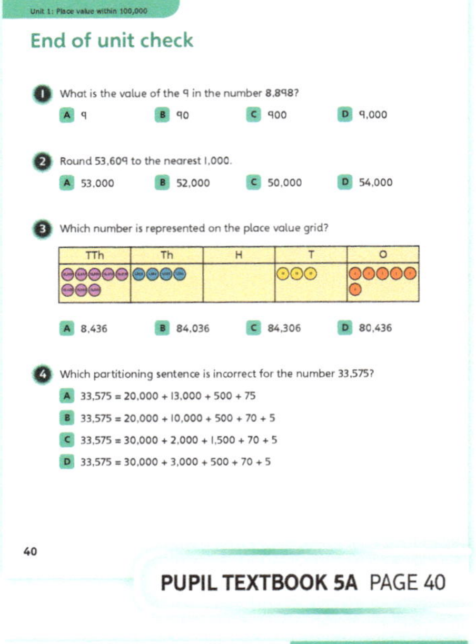

PUPIL TEXTBOOK 5A PAGE 40

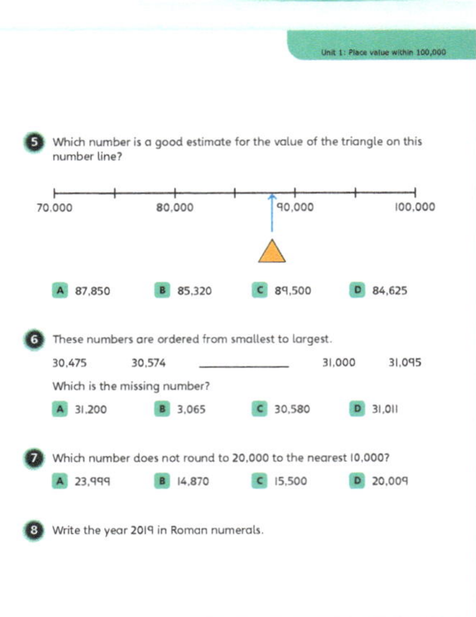

PUPIL TEXTBOOK 5A PAGE 41

Q	A	WRONG ANSWERS AND MISCONCEPTIONS	STRENGTHENING UNDERSTANDING
1	B	A, C and D suggest children do not know the value of digits within a number.	Explore equivalence by building numbers with place value counters, identifying and writing the value of each digit as you go along. Put the counters into part-whole models, then move them to make a different number of parts by reducing or increasing the size of different parts. Build pairs of numbers with a varying number of digits using base 10 equipment or place value counters.
2	D	A suggests that children have looked at the 1,000s digit to round. C suggests they have looked at the 1,000s digit to round to the previous digit.	
3	B	A indicates that children have not understood that 0 is used as a place holder. C and D suggest that they are unsure of the value of each place.	
4	B	A suggests that children think there must be five parts for a 5-digit number. C indicates they think each part must be a multiple of a place value. D indicates that they do not understand equivalence.	
5	A	B and C suggest that children are unable to subdivide the interval. D indicates that they do not recognise that the number must be closer to 90,000 than 80,000.	
6	C	A and D suggest that children may not be comparing the value of each digit, making decisions based on only the first two digits.	
7	B	A and D show that children are less secure with rounding down. C suggests they are uncertain which way to round when the digit is 5.	
8	MXIX	Some children may not remember that the 9 is written as 10 − 1, so I must come before the last X.	

My journal

WAYS OF WORKING Independent thinking

ANSWERS AND COMMENTARY

Children may describe the number 12,546 in many ways. For example:
- 12,546 is a 5-digit number because it has a digit in the 10,000s place
- 12,546 is 546 more than 12,000
- 12,546 is between the multiples 12,000 and 13,000
- 12,546 is a little more than half-way between 12,000 and 13,000
- 12,546 rounds to 13,000 to the nearest 1,000. 12,546 rounds to 10,000 to the nearest 10,000.

If children need support using the key language, ask:
- *Is 12,546 closer to 12,000 or 13,000?*
- *How many 10,000s, 1,000s, 100s, 10s and 1s can you see in 12,546?*

Representations could include place value grids and partitioning in part-whole models, on number lines or as abstract number sentences.

If children need support representing the number in different ways, ask:
- *How have you represented numbers in this unit? What models have you used?*
- *How many columns will you need in a place value grid to represent 12,546?*

Power check

WAYS OF WORKING Independent thinking

ASK

- *What do you know about the different values of digits in a 5-digit number?*
- *Can you say the multiples of 1,000 or 10,000 that a number lies between?*
- *Do you know how to round a number to the nearest 1,000?*
- *How confident do you feel ordering a group of 4- and 5-digit numbers?*

Power play

WAYS OF WORKING Pair work

IN FOCUS This game involves children applying their learning from the unit. They need to justify their answer each time, both players agreeing on the answer before the counter is placed. This is also a strategic game as children try to cover four squares in a row, column or diagonal, while also blocking their partner from doing the same and winning.

ANSWERS AND COMMENTARY Supply each pair with a 1–6 dice for the game and two sets of coloured counters, one colour for each player. Children could play in teams to support decision making. Listen out for children providing reasoning as they give answers to the rounding questions. Do children require number lines or place value grids to help or are they able to visualise the answers?

After the unit 

- How well did the prompts and questions promote learning, and what were children's responses to them?
- Are children ready to move on with their learning? Following on from the end of unit assessment, how confident are they working with numbers within 100,000?

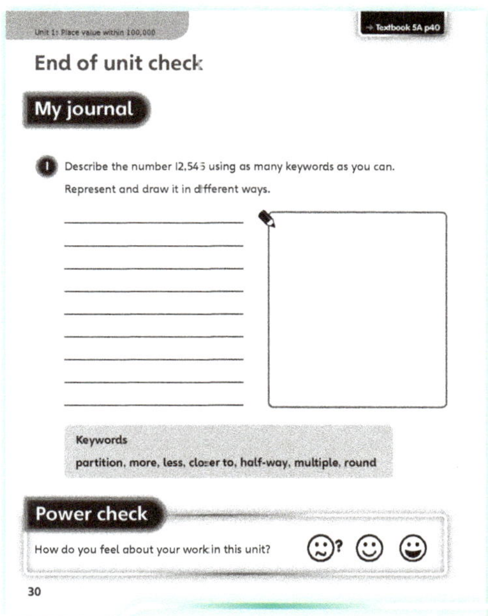

→ Textbook 5A p40

End of unit check

My journal

1. Describe the number 12,546 using as many keywords as you can. Represent and draw it in different ways.

Keywords

partition, more, less, closer to, half-way, multiple, round

Power check

How do you feel about your work in this unit?

30

PUPIL PRACTICE BOOK 5A PAGE 30

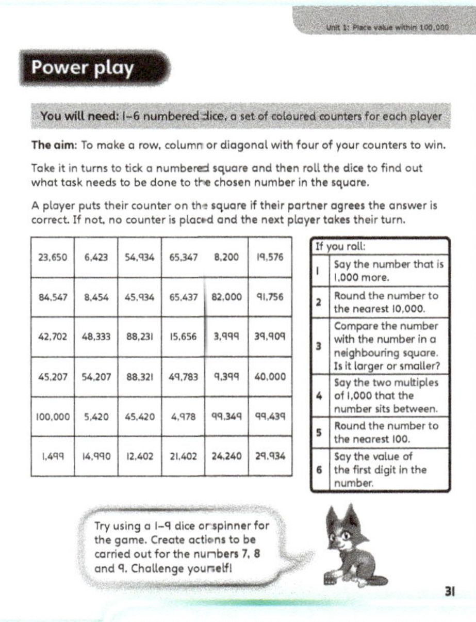

Power play

You will need: 1–6 numbered dice, a set of coloured counters for each player

The aim: To make a row, column or diagonal with four of your counters to win.

Take it in turns to tick a numbered square and then roll the dice to find out what task needs to be done to the chosen number in the square.

A player puts their counter on the square if their partner agrees the answer is correct. If not, no counter is placed and the next player takes their turn.

23,650	6,423	54,934	65,347	8,200	19,576
84,547	8,454	45,934	65,437	82,000	91,756
42,702	48,333	88,231	15,656	3,999	39,909
45,207	54,207	88,321	49,783	9,399	40,000
100,000	5,420	45,420	4,978	99,349	99,439
1,499	14,990	12,402	21,402	24,240	29,934

If you roll:	
1	Say the number that is 1,000 more.
2	Round the number to the nearest 10,000.
3	Compare the number with the number in a neighbouring square. Is it larger or smaller?
4	Say the two multiples of 1,000 that the number sits between.
5	Round the number to the nearest 100.
6	Say the value of the first digit in the number.

31

PUPIL PRACTICE BOOK 5A PAGE 31

Strengthen and **Deepen** activities for this unit can be found in the *Power Maths* online subscription.

Unit 2
Place value within 1,000,000

Don't forget to watch the Unit 2 video!

WHY THIS UNIT IS IMPORTANT

In this unit, children will recap their understanding of numbers up to 10,000 and begin using numbers in the 100,000s. They will develop their fluency with these numbers by partitioning and recombining them in different ways. Children will develop their understanding of number by using a number line to find and identify numbers up to 1,000,000, going on to compare and order them. They will then develop their understanding of rounding numbers, with particular focus on numbers in the 100,000s. Children will develop their understanding of negative numbers and how they relate and compare to positive numbers. They will use this understanding to calculate with both negative and positive numbers. Finally, they will use their ability to count in regular steps to help them identify rules in number sequences and accurately follow and complete them.

WHERE THIS UNIT FITS

→ Unit 1: Place value within 100,000
→ **Unit 2: Place value within 1,000,000**
→ Unit 3: Addition and subtraction

This unit builds on children's work in Unit 1, where they explored numbers to 100,000 and their representation on place value grids and number lines. It also builds on previous experience of comparing, ordering and rounding numbers from Unit 1. This unit provides preparation for using place value grids to add and subtract 5-digit numbers.

Before they start this unit, it is expected that children:
- can recognise 1s, 10s, 100s, 1,000s and 10,000s
- can use a place value grid to the 10,000s and can recognise and use a number line
- can partition and recombine numbers.

ASSESSING MASTERY

Children who have mastered this unit will be able to demonstrate fluent understanding of place value in numbers up to 1,000,000. They will be able to name, partition and write the names of numbers accurately and use this, and their understanding of place value, to compare and order numbers up to 1,000,000. Children will use number lines confidently and will be able to complete partial number lines. They will be able to use their understanding of place value to round up and down to the nearest multiple of ten and will be able to recognise and use negative numbers down to ⁻1,000,000. Finally, they will be able to identify the rules that govern number sequences and use this knowledge to complete them.

COMMON MISCONCEPTIONS	STRENGTHENING UNDERSTANDING	GOING DEEPER
Children may need help to bridge into the next place value column when adding 10, 100, 1,000 or 10,000.	Show children different representations of the numbers they are counting, for example using base 10 equipment.	Ask children to create inverse puzzles for each other, working out a starting number from a final number and a series of steps.
Children may miscalculate the difference between positive and negative numbers.	Show children the two numbers on a number line. Ask: *Why isn't the difference 1?*	Encourage children to write word problems that use negative numbers as their context.

Unit 2: Place value within 1,000,000

PUPIL TEXTBOOK 5A PAGE 42

WAYS OF WORKING

Use these pages to introduce the unit, with teacher-led discussion. Encourage children to think back to the previous unit and work out how the place value grid has changed. Challenge children to explain as many of the keywords as they can.

STRUCTURES AND REPRESENTATIONS

Place value grid: Place value grids are used in this unit to help children read numbers and recognise the value of each digit in numbers up to 1,000,000.

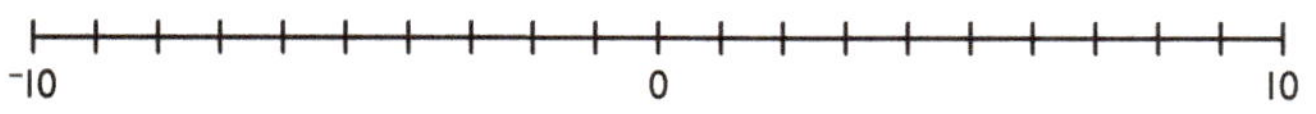

HTh	TTh	Th	H	T	O

Number line: Number lines are used to help children plot numbers from 0 to 1,000,000 and to show negative numbers.

−10   0   10

Place value counters: Place value counters are used to help children create and partition numbers.

KEY LANGUAGE

Here is some key language that children will need to know as part of the learning in this unit:

→ place value
→ ones (1s), tens (10s), hundreds (100s), thousands (1,000s), ten thousands (10,000s), hundred thousands (100,000s), million (1,000,000)
→ partition, partitioning
→ number line, count
→ negative number, positive number
→ minus
→ rounding, round up, round down
→ estimate
→ compare, order
→ sequence, rule
→ ascending, descending
→ less than (<), greater than (>), nearest.

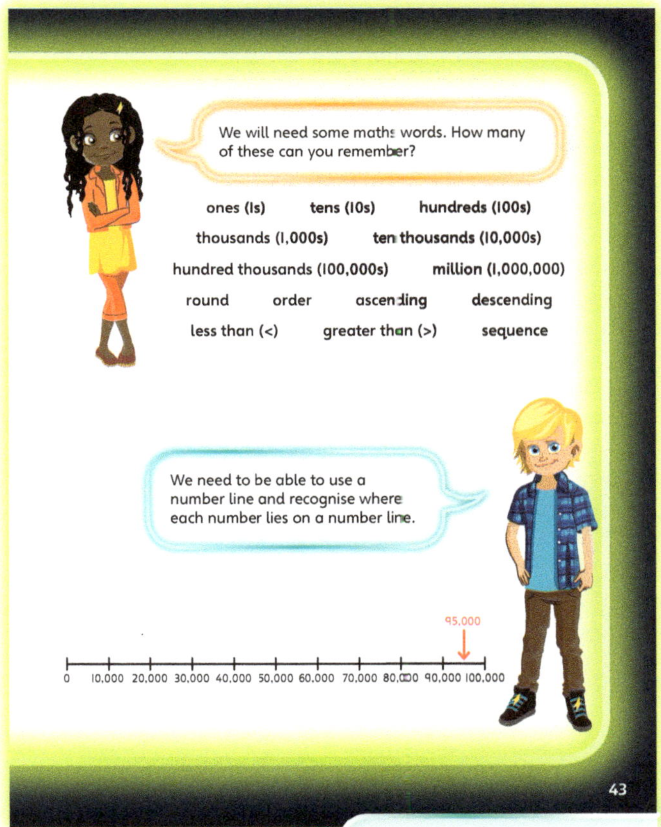

PUPIL TEXTBOOK 5A PAGE 43

100,000s, 10,000s, 1,000s, 100s, 10s and 1s ①

Learning focus

In this lesson, children will develop their understanding of place value up to the 100,000s, learning to read and write numbers accurately. They will count in steps of 100,000, 10,000, 1,000, 100, 10 and 1.

Small steps

→ Previous step: Roman numerals to 10,000

→ **This step: 100,000s, 10,000s, 1,000s, 100s, 10s and 1s (1)**

→ Next step: 100,000s, 10,000s, 1,000s, 100s, 10s and 1s (2)

NATIONAL CURRICULUM LINKS

Year 5 Number – Number and Place Value

Read, write, order and compare numbers to at least 1,000,000 and determine the value of each digit.

ASSESSING MASTERY

Children can use their knowledge of place value to recognise and name numbers. They are able to count fluently in regular steps of 100,000, 10,000, 1,000, 100, 10 and 1.

COMMON MISCONCEPTIONS

Children may misspell the mathematical vocabulary when trying to write the names of numbers. It may be beneficial to have the vocabulary displayed prominently on the classroom wall for children to refer to. Ask:
• *How do you write 1,000 in words? What about 12,000? Now try writing 400,000 in words.*

STRENGTHENING UNDERSTANDING

Provide opportunities to investigate the place value of numbers up to hundreds of thousands before the lesson. This could be achieved by encouraging children to use base 10 equipment to make numbers, or place value cards to make and partition numbers.

GOING DEEPER

Children could be encouraged to investigate what happens when they add another hundred thousand to a number they have created or been given. Ask: *What happens if you add 100,000 to 15,678? Use resources or a picture to show what happens when you add 10,000 to 345,987.*

KEY LANGUAGE

In lesson: digit, ones (1s), tens (10s), hundreds (100s), thousands (1,000s), ten thousands (10,000s), hundred thousands (100,000s), place value

STRUCTURES AND REPRESENTATIONS

place value grid

RESOURCES

Mandatory: place value counters

Optional: base 10 equipment, place value cards

 In the eTextbook of this lesson, you will find interactive links to a selection of teaching tools.

Before you teach 

• How confident are children in their use of a place value grid up to 1,000s?
• What concrete representations will you use to ensure children's deep conceptual understanding of the place value of 6-digit numbers?

Discover

WAYS OF WORKING Pair work

ASK

• Question ❶ a): *How many sweets does one container hold? How could you represent your thinking? How many sweets would seven containers hold? What about eight? Can you spot any patterns?*
• Question ❶ b): *What could you use to clearly represent the place value of each digit in the number?*

IN FOCUS Question ❶ a) gives children an opportunity to begin counting in regular steps of 100,000. Encourage children to predict how the numbers will change with different numbers of containers. Question ❶ b) gives children their first opportunity to investigate the place value of 6-digit numbers.

PRACTICAL TIPS Place value counters could be used to represent the containers of sweets and to help children count up and down in 100,000s, 10,000s and 1,000s.

ANSWERS

Question ❶ a): Six containers can hold 600,000 sweets.

Question ❶ b): The board says that 461,905 sweets have
been made today.
There are 4 hundred thousands.
There are 6 ten thousands.
There is 1 thousand.
There are 9 hundreds.
There are 0 tens and 5 ones.

PUPIL TEXTBOOK 5A PAGE 44

Share

WAYS OF WORKING Whole class teacher led

ASK

• Question ❶ a): *What is each container worth? How does knowing that help you count? What resources did you find useful when representing 100,000? Could they help you to predict how many sweets seven containers would contain, or eight containers?*
• Question ❶ b): *How is this place value grid different from the ones you have used before?*

IN FOCUS When looking at question ❶ b), it will be essential to allow children the time to consider how the place value grid is different. Ask children to discuss the headings of the place value grid and, if necessary, remind them of the meanings of the headings they have seen before. Ask children what the new heading could mean and ensure they understand that 4 counters in the hundred thousands column means there are 4 hundred thousands.

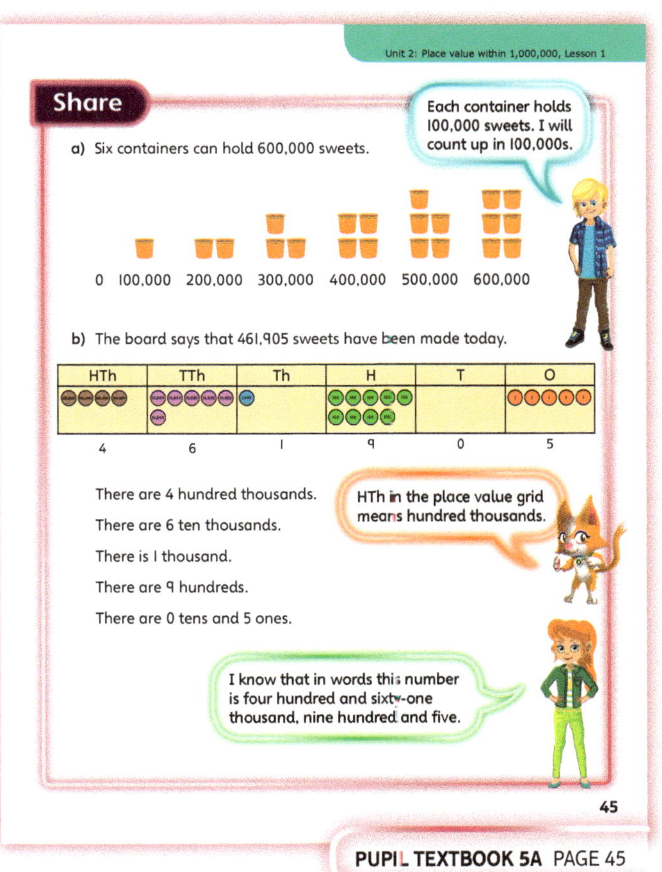

PUPIL TEXTBOOK 5A PAGE 45

Think together

WAYS OF WORKING Whole class teacher led (I do, We do, You do)

ASK

- Question ❶ : *How can you represent the containers of sweets? How does your representation help you to identify the number of sweets accurately?*
- Question ❷ : *What mathematical vocabulary will you need to use to write the number in words?*
- Question ❸ : *What mathematical structure will help you to identify the value of each 5?*

IN FOCUS In question ❶, children will develop their ability to count in steps of 100,000. They should be encouraged to use resources to support their counting. Question ❷ will develop children's fluency with the written names of numbers up to the 100,000s. Encouraging children to use place value grids to support them with question ❸ will develop their understanding of the place value of 6-digit numbers while helping to develop their fluency with the new place value column they have met in the lesson.

STRENGTHEN If children need help writing the names of the numbers in question ❷, have the key mathematical vocabulary available to them in written form. This could be as flash cards, a wall display or number lines that show both numerals and written number names.

DEEPEN Use Sparks's suggestion in question ❸ to deepen children's understanding of the number concepts they have covered in the lesson so far. Ask: *Is it easier to read or to write a 4-, 5- or 6-digit number? Explain why.* Challenge children to write a 6-digit number that is quicker to write in words than a 4-digit number and ask them how it is possible.

ASSESSMENT CHECKPOINT Can children count in steps of 100,000? Do children use the correct mathematical vocabulary when reading and writing the numbers in question ❷? In question ❸, look out for clear understanding that the 5 in each number is worth a different amount.

ANSWERS

Question ❶ : There are 900,000 sweets.

Question ❷ a): Seven hundred and twenty-eight thousand, six hundred and eleven

Question ❷ b): 370,938

Question ❸ a): The digit 5 represents: 50,000

Question ❸ b): The digit 5 represents: 5

Question ❸ c): The digit 5 represents: 500

Question ❸ d): The digit 5 represents: 5,000

Question ❸ e): The digit 5 represents: 50

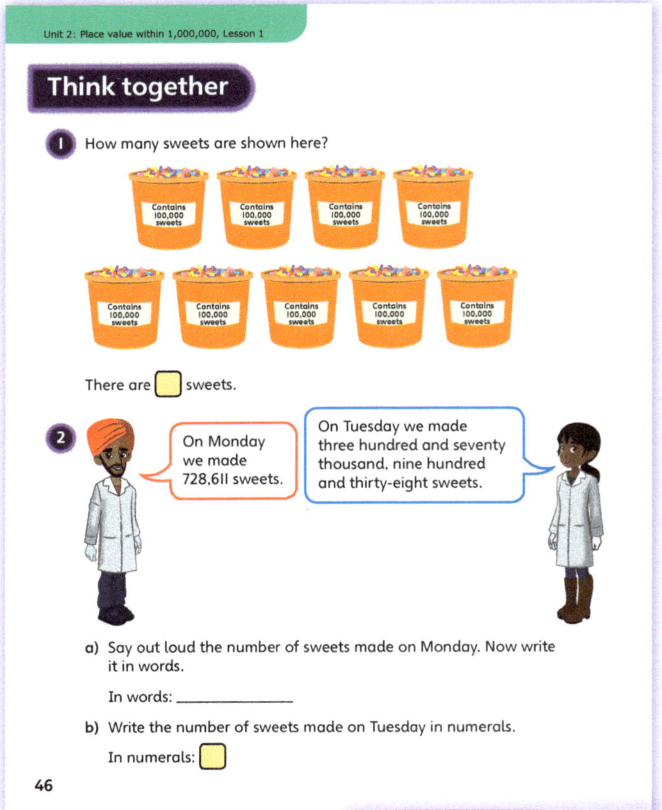

PUPIL TEXTBOOK 5A PAGE 46

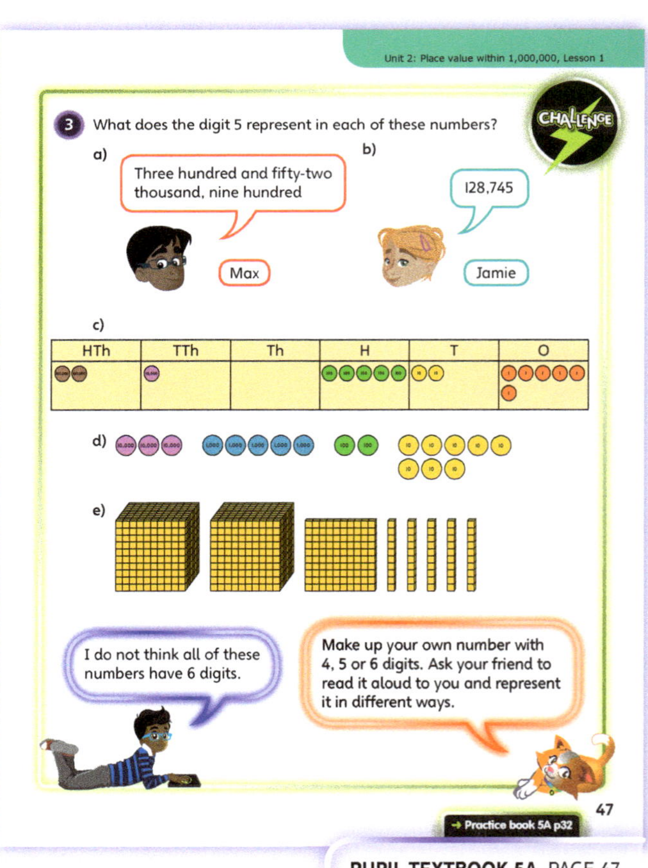

PUPIL TEXTBOOK 5A PAGE 47

Practice

WAYS OF WORKING Independent thinking

IN FOCUS Question ② allows children to demonstrate their ability to count in steps of 100,000 up to 1,000,000. Questions ③ and ④ provide an opportunity for children to practise reading and writing the key vocabulary from the lesson, using different pictorial and abstract representations.

STRENGTHEN To strengthen understanding of writing the numbers in question ③ in words, it may help to use place value cards to build the numbers first. Ask: *What does each digit represent? How would you write the name of the number that digit is worth?*

DEEPEN Question ⑥ deepens children's understanding of place value by requiring them to use their reasoning ability to demonstrate flexibility with place value. Ask: *Are there only five possible solutions? How can you prove you have found all of the possibilities?*

THINK DIFFERENTLY In question ⑤, children have an opportunity to further develop their understanding of place value. The question requires children to recognise that although the numbers might have some digits in common and in the same order, the value of the 4 in each number may still be different.

ASSESSMENT CHECKPOINT Can children accurately read and write 4-, 5- and 6-digit numbers in words? In question ⑤, can children correctly value the 4 in each number? Do they recognise that the 4 in 240 and 500,240 is worth the same, but the 4 in 314,912 and 77,314 is not?

ANSWERS Answers for the **Practice** part of the lesson appear in the separate **Practice and Reflect answer guide.**

Reflect

WAYS OF WORKING Pair work

IN FOCUS This **Reflect** question will offer children a final opportunity to practise reading and writing 6-digit numbers and using a place value grid up to 100,000s.

ASSESSMENT CHECKPOINT Look particularly for children's understanding of the place value grid and the hundred thousands column that has been introduced in this lesson.

ANSWERS Answers for the **Reflect** part of the lesson appear in the separate **Practice and Reflect answer guide.**

After the lesson ⏸

- Can children confidently value the digits in 4-, 5- and 6-digit numbers?
- Are there any written words that children need to become more fluent with?
- How can you support children's future use of the written mathematical vocabulary?

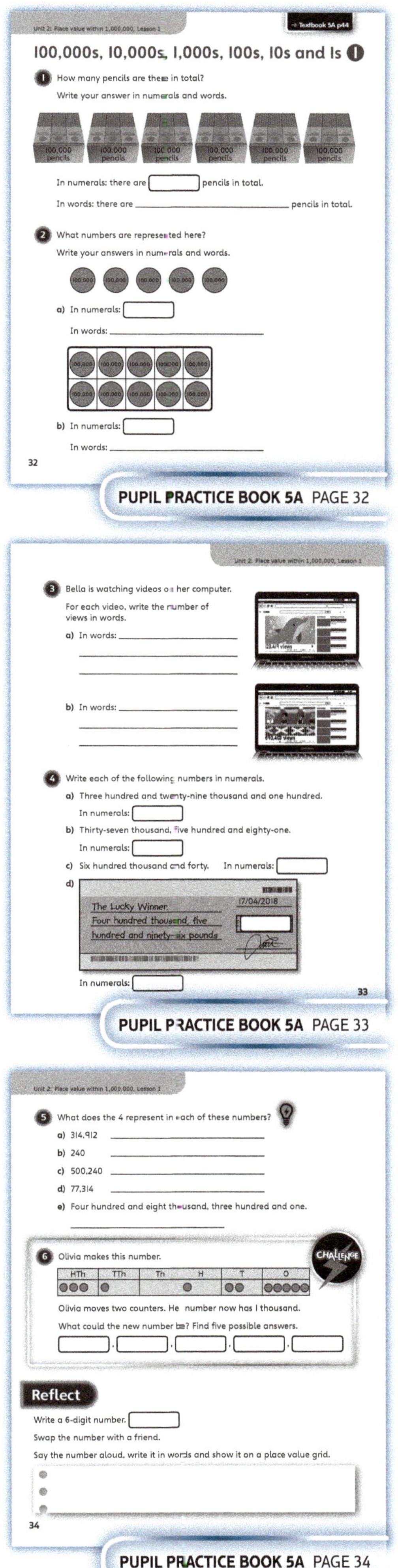

PUPIL PRACTICE BOOK 5A PAGE 32

PUPIL PRACTICE BOOK 5A PAGE 33

PUPIL PRACTICE BOOK 5A PAGE 34

100,000s, 10,000s, 1,000s, 100s, 10s and 1s ②

Learning focus

In this lesson, children will use their understanding of the place value of numbers up to 6 digits, to partition and recombine numbers to solve number problems.

Small steps

→ Previous step: 100,000s, 10,000s, 1,000s, 100s, 10s and 1s (1)

→ **This step: 100,000s, 10,000s, 1,000s, 100s, 10s and 1s (2)**

→ Next step: Number line to 1,000,000

NATIONAL CURRICULUM LINKS

Year 5 Number – Number and Place Value

Solve number problems and practical problems that involve [reading, writing, ordering and comparing numbers to at least 1,000,000 and determining the value of each digit].

ASSESSING MASTERY

Children can recognise, describe, read and write numbers of up to 6 digits. They can use what they know about place value to describe the value of each digit in a number and can solve problems using this understanding.

COMMON MISCONCEPTIONS

Children may assume that the only way to partition a number is into its place value headings (i.e. 136 can only be partitioned as 100, 30 and 6). Base 10 equipment or place value counters may help children to become more flexible in the way they partition. Ask:
• *Is that the only way you can break this number into partitions? Show me another way using a resource.*

STRENGTHENING UNDERSTANDING

Before starting the lesson, children could strengthen their fluency with 4-, 5- and 6-digit numbers and their understanding of place value, by counting up and down in different steps from a given number. Ask: *Can you count up in 10,000s from 453? Can you count down in 100,000s from 876,232?*

GOING DEEPER

Encourage children to create their own missing number equalities to challenge their partner with. For example, 300,000 + ? + 6,000 + 40 + 5 = 376,045. Ask:
• *Create a challenge like this for your partner. How can you make your question easier or trickier?*

KEY LANGUAGE

In lesson: ones (1s), tens (10s), hundreds (100s), thousands (1,000s), ten thousands (10,000s), hundred thousands (100,000s), partition, place value, digit

Other language to be used by the teacher: re-combine

STRUCTURES AND REPRESENTATIONS

place value grid

RESOURCES

Mandatory: place value counters

Optional: place value arrow cards, play money

 In the eTextbook of this lesson, you will find interactive links to a selection of teaching tools.

Before you teach 

• Are there any misconceptions that you need to plan for, based on children's work in the previous lesson?
• How will you develop children's ability to generalise in this lesson?

Discover

 Pair work

ASK

- Question ① a): *How can you tell how much money is in the bank vault?*
- Question ① b): *What is the simplest way to partition £360,400? Have you found the same solution as your partner?*

IN FOCUS Question ① a) will give children an opportunity to begin recognising numbers when partitioned and recombining them into their total amount. Question ① b) will demonstrate to children that a number can be partitioned in different ways, not just into its place value columns.

PRACTICAL TIPS Encourage children to investigate different ways of making a given amount, using a variety of denominations of play money.

ANSWERS

Question ① a): In digits: £236,253 is in the bank vault.
In words: two hundred and thirty-six thousand, two hundred and fifty-three pounds

Question ① b): Look for examples of partitioning that work within the context of money (using denominations of £100,000, £10,000, £1,000 and £100). For example:
3 bags of £100,000, 6 bags of £10,000 and 4 bags of £100
3 bags of £100,000, 5 bags of £10,000, 10 bags of £1,000 and 4 bags of £100

PUPIL TEXTBOOK 5A PAGE 48

Share

 Whole class teacher led

ASK

- Question ① a): *How can a place value grid help for this question? What headings would the place value grid need? How do you know?*
- Question ① b): *Can you show different ways of partitioning this number using place value counters? How will you write your partitioning?*

IN FOCUS Question ① a) provides a link between children's understanding of place value grids from the previous lesson to their learning in this lesson. Question ① b) will help to reinforce the understanding that a number can be partitioned in different ways. It will be important to display the different solutions children found and discuss what is the same and different about them to highlight how they all total the same number.

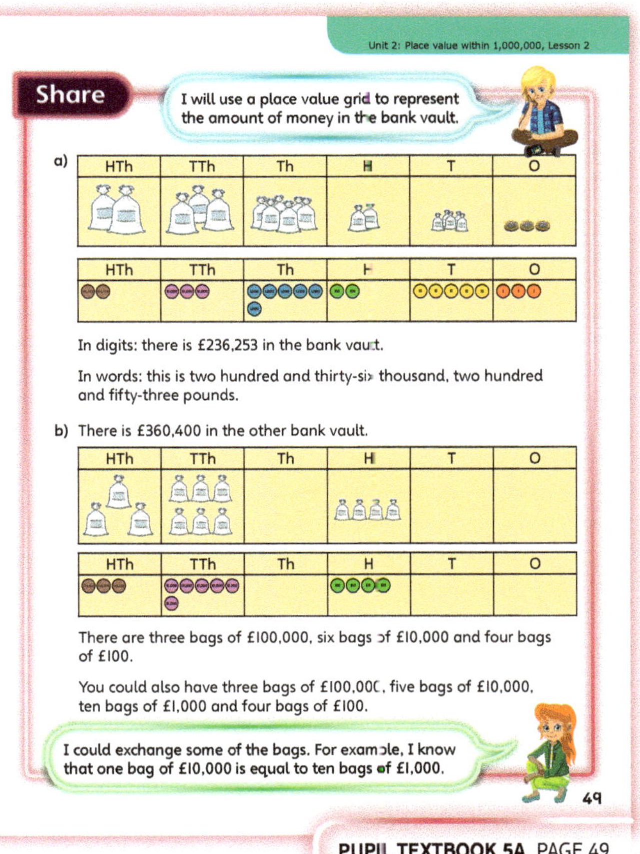

PUPIL TEXTBOOK 5A PAGE 49

Think together

WAYS OF WORKING Whole class teacher led (I do, We do, You do)

ASK

- Question **1** : *How are the amounts represented differently? Which is clearer and why?*
- Question **2** : *Use resources to prove your solution. How many solutions can you find?*
- Question **3** : *How could you use a place value grid to help you with this question?*

IN FOCUS Question **1** will help scaffold children's understanding of partitioning 6-digit numbers by withdrawing the support of the place value grid in part b), encouraging more independent thinking.

Question **2** offers children an opportunity to partition a 6-digit number and reinforces the point that a number can be partitioned in a variety of ways.

STRENGTHEN Help children partition the number in question **2** by suggesting they use place value counters or place value cards to make the number first and to easily see the parts it is made up of. Ask: *How can these representations help you partition the number?*

DEEPEN For question **3**, to deepen children's understanding of place value and develop their ability to partition, ask: *Is there only one solution for each question? How many solutions do you think you could find for part a)?*

ASSESSMENT CHECKPOINT Assess children's recognition of place value in a number – are they fluent with place value? Are children able to recombine a number that has been partitioned? For question **3**, assess whether children can partition a 6-digit number. Look for clear understanding that numbers can be partitioned in different ways.

ANSWERS

Question **1** a): In digits: there is £407,361 in the bank vault.

In words: this is four hundred and seven thousand, three hundred and sixty-one pounds.

Question **1** b): In digits: there is £245,730 in the bank vault.

In words: this is two hundred and forty-five thousand, seven hundred and thirty pounds.

Question **2** :   5 × £100,000, 7 × £10,000, 7 × £1,000, 1 × £100, 8 × £10, 2 × £1

Accept other alternatives that total £577,182.

Question **3** a): 726,140 is equal to 7 hundred thousands, 2 ten thousands, 6 thousands, 1 hundred and 4 tens.

Question **3** b): 58,415 is equal to 5 ten thousands, 8 thousands, 4 hundreds, 1 ten and 5 ones.

Question **3** c): 6 hundred thousands, 4 thousands and 2 hundreds = 604,200

Question **3** d): 951,618 = 900,000 + 50,000 + 1,000 + 600 + 10 + 8

Question **3** e): 300,000 + 500 + 60 + 2 = 300,562

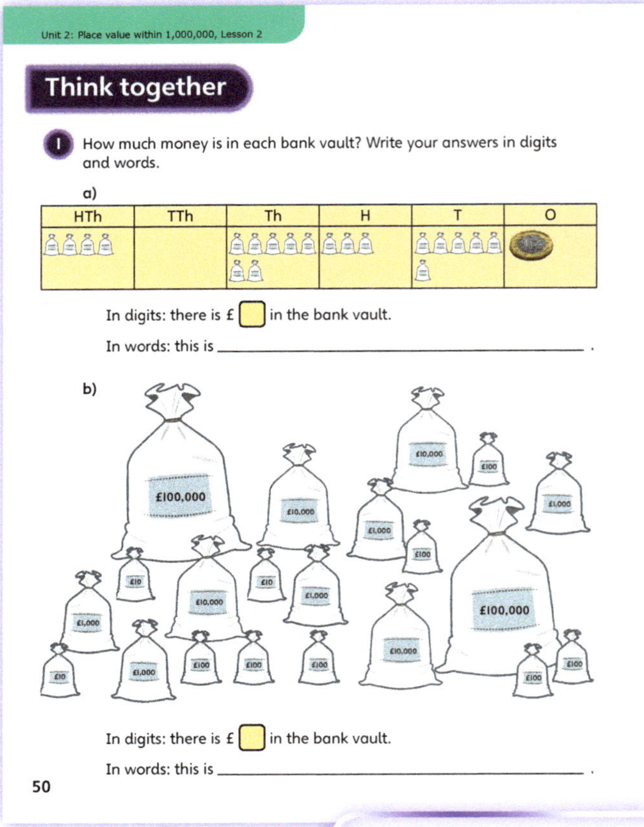

PUPIL TEXTBOOK 5A PAGE 50

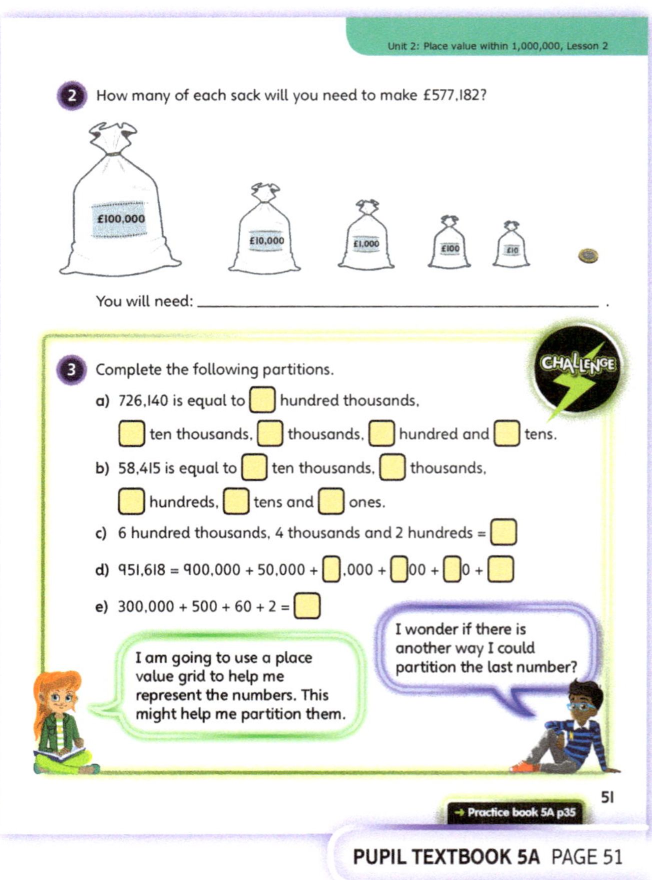

PUPIL TEXTBOOK 5A PAGE 51

Practice

WAYS OF WORKING Independent thinking

IN FOCUS Question ❶ provides an opportunity for children to recognise a partitioned number within a different context and independently recombine the number to find the total. Question ❹ has been designed to allow children to develop their fluency when partitioning and recombining 4-, 5- and 6-digit numbers recorded in an abstract way. It is scaffolded to enable children to work through parts a) to g) independently.

STRENGTHEN To strengthen understanding when recombining the numbers shown in question ❺, ask: *What structures or representations have you used previously to show a number? How could you use them here to help you?*

DEEPEN Question ❻ will test children's fluency with partitioned numbers. As some of the partitioned numbers have not been ordered logically, children will need to ensure they check carefully when finding the solution. Ask: *What makes recombining these numbers trickier? How could you make it easier for yourself?*

ASSESSMENT CHECKPOINT Can children fluently partition and recombine numbers with up to 6-digits? Are children able to demonstrate clear understanding of place value by recombining partitioned numbers listed out of order?

ANSWERS Answers for the **Practice** part of the lesson appear in the separate **Practice and Reflect answer guide.**

Reflect

WAYS OF WORKING Independent thinking

IN FOCUS This **Reflect** question will offer children the opportunity to develop further their fluency and flexibility when partitioning numbers. Allow time for children to compare their solutions with a partner and encourage them to discuss why their ideas are the same, or different.

ASSESSMENT CHECKPOINT Assess whether children recognise that the number can be partitioned in numerous ways. Are they able to demonstrate some of these ways?

ANSWERS Answers for the **Reflect** part of the lesson appear in the separate **Practice and Reflect answer guide.**

After the lesson 

- How fluent were children at recognising that numbers can be partitioned in multiple ways?
- Were children equally confident in partitioning and recombining numbers? If not, how will you support their understanding in the concept they find trickier?
- Can children use both words and digits to write out the value of each digit of a 6-digit number?

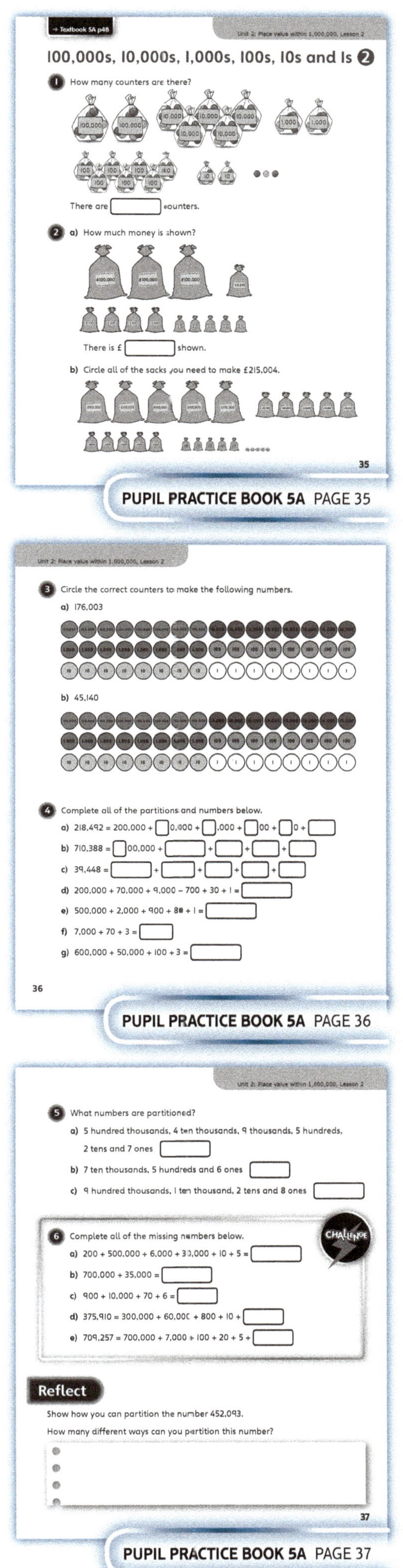

PUPIL PRACTICE BOOK 5A PAGE 35

PUPIL PRACTICE BOOK 5A PAGE 36

PUPIL PRACTICE BOOK 5A PAGE 37

Number line to 1,000,000

Learning focus

In this lesson, children will estimate and accurately identify where numbers to 1,000,000 would lie on a number line. They will use their understanding of place value to help them achieve this.

Small steps

→ Previous step: 100,000s, 10,000s, 1,000s, 100s, 10s and 1s (2)
→ **This step: Number line to 1,000,000**
→ Next step: Comparing and ordering numbers to 1,000,000

NATIONAL CURRICULUM LINKS

Year 5 Number – Number and Place Value

Read, write, order and compare numbers to at least 1,000,000 and determine the value of each digit.

ASSESSING MASTERY

Children can use their understanding of place value to help them accurately identify, or estimate, where a number would lie on a number line for numbers up to 1,000,000.

COMMON MISCONCEPTIONS

Children may miscalculate, and so misinterpret, the unlabelled divisions on a number line. Ask:
• *If you count in those steps, does your number line work? Show me.*

STRENGTHENING UNDERSTANDING

Provide children with number cards up to 1,000,000. Encourage them to peg the cards along a piece of string in order of size, smallest first. Ask: *How do you know which number comes first?*

Discuss with children whether the gap between each number should be the same. If not, why not? To show this practically, use four or five numbers (to 50) along a bead string. Label the appropriate beads and discuss why the gaps between them are different sizes.

GOING DEEPER

Children can draw their own number lines to 1,000,000 and place arrows along them. Suggest they challenge their partner to identify what number each arrow is pointing to.

KEY LANGUAGE

In lesson: intervals, **million**, hundred thousand, ten thousand, thousand, hundred, ten, one

Other language to be used by the teacher: estimate, pattern

STRUCTURES AND REPRESENTATIONS

number line

RESOURCES

Optional: bead strings, number cards

 In the eTextbook of this lesson, you will find interactive links to a selection of teaching tools.

Before you teach ⏸

• How confident are children in their use of number lines?
• Can children count fluently in 1,000s, 10,000s and 100,000s?
• What other real-life contexts can you apply to number lines in the lesson?

Discover

WAYS OF WORKING   Pair work

ASK

- Question ❶ a): *What is each interval worth? How do you know? What is different about the positions of the minimum price and maximum price? Which is easier to read and why?*
- Question ❶ b): *Why is it trickier to find £720,000 on this number line?*

IN FOCUS   Question ❶ a) encourages children to explain how a number line is organised and ensures children develop their ability to recognise and calculate the value of unlabelled intervals. Question ❶ b) encourages children to accurately estimate where numbers appear between unlabelled intervals along a number line.

PRACTICAL TIPS   When introducing this concept, provide children with number cards labelled with numbers up to 1,000,000. Ask children to order these numbers along a blank number line, placing them at regular intervals (for example, going up in steps of 100,000). This could be made trickier by including some numbers that don't fit the pattern.

ANSWERS

Question ❶ a): The minimum house price is £200,000.

The maximum house price is £850,000.

Question ❶ b):

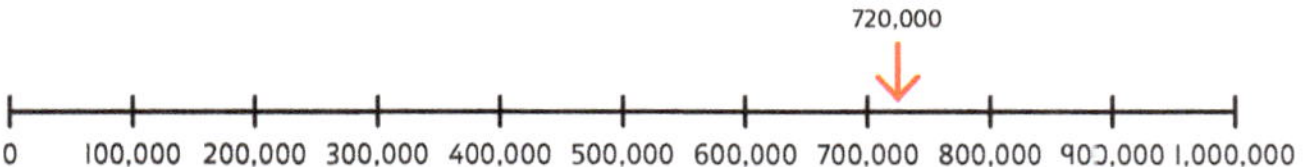

Share

WAYS OF WORKING   Whole class teacher led

ASK

- Question ❶ a): *How can you test that the number line goes up in intervals of 100,000? How can you be sure that the maximum price is £850,000?*
- Question ❶ b): *Can you use this method to find £480,000?*

IN FOCUS   At this point in the lesson it will be important to ensure that children can calculate unlabelled intervals along a number line, and question ❶ a) will support this learning. Question ❶ b) will then develop children's learning as they visualise and calculate unlabelled intervals.

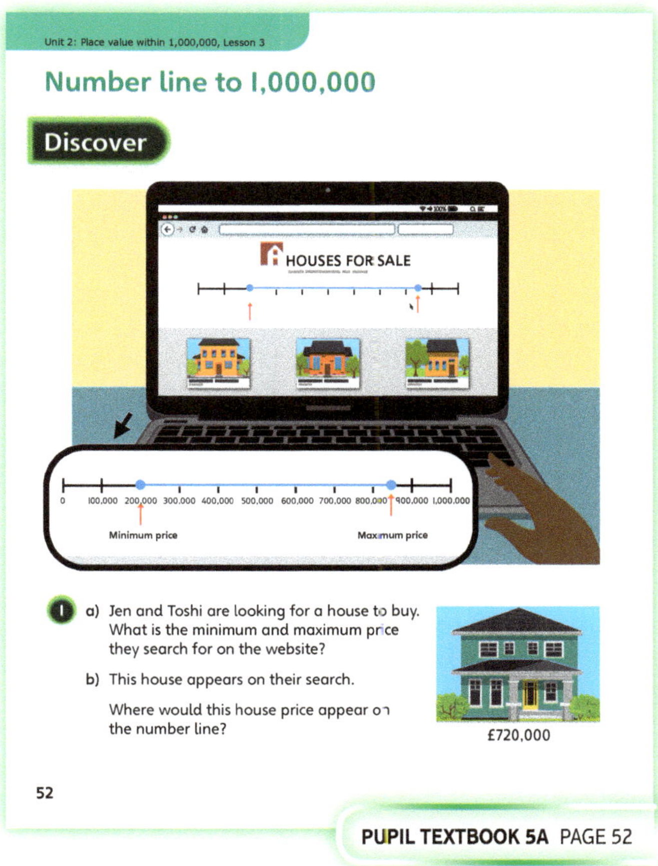

PUPIL TEXTBOOK 5A PAGE 52

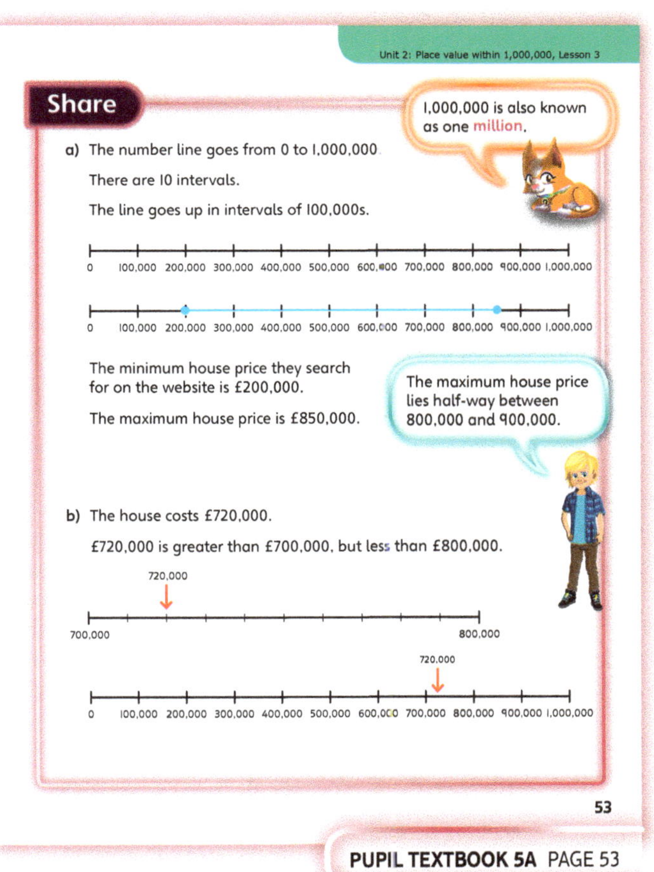

PUPIL TEXTBOOK 5A PAGE 53

Think together

WAYS OF WORKING Whole class teacher led (I do, We do, You do)

ASK

- Question **1** : *How can you work out what to count up in? How can you check how accurate you have been?*
- Question **2** : *How can you work out what numbers are being pointed to?*
- Question **3** : *How are the number lines the same? How are they different? How will the position of the number change in each number line?*

IN FOCUS Question **1** will help children to develop their ability to use number lines, recognising patterns and using them to calculate unlabelled intervals. Question **2** has been included to develop children's fluency with number lines. They will need to calculate what the unlabelled intervals represent and then use this to find A, B and C along the number line.

STRENGTHEN To strengthen understanding when recognising the missing numbers in question **1**, suggest that children read aloud the numbers that are already on the number line. Ask: *What patterns can you spot that will help you to find the missing numbers?*

DEEPEN Question **3** deepens children's understanding as it demonstrates how changing the start and end point of a number line can change where a number will be placed along it. Ask: *Is there any way of predicting where the number will be on each number line before you work it out?*

ASSESSMENT CHECKPOINT All questions in this section assess children's reasoning and fluency when finding numbers along a number line. Check children are counting up each number line accurately and using any patterns they spot to calculate the unlabelled intervals.

ANSWERS

Question **1** a): Missing numbers: 100,000, 300,000, 400,000, 700,000, 900,000

Question **1** b): Missing numbers: 230,000, 240,000, 250,000, 260,000, 270,000, 290,000

Question **1** c): Missing numbers: 418,100, 418,200, 418,300, 418,400, 418,500, 418,600, 418,700, 418,800, 418,900

Question **2** : A = 353,000
B = 355,000
C = 359,000

Question **3** a):

Question **3** b):

Question **3** c):

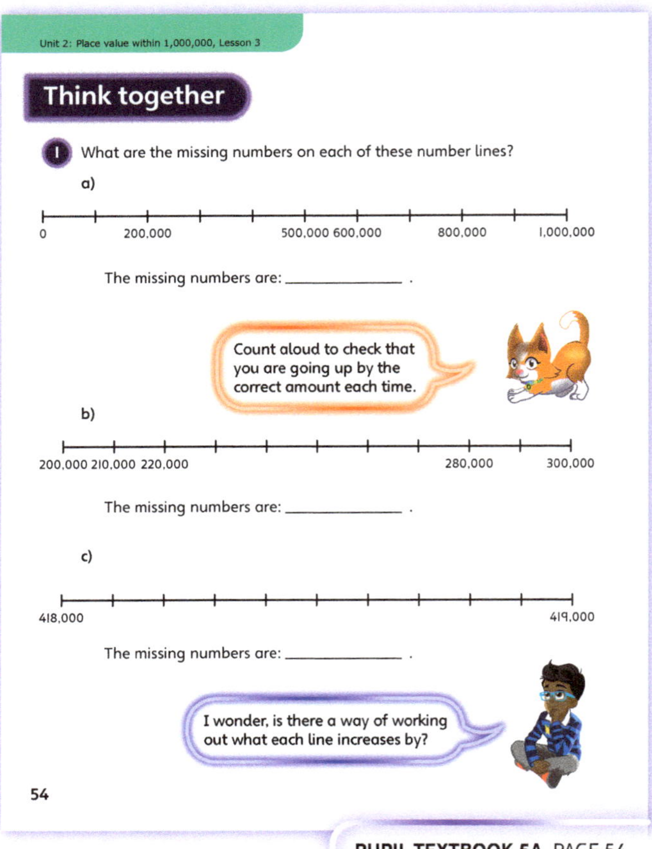

PUPIL TEXTBOOK 5A PAGE 54

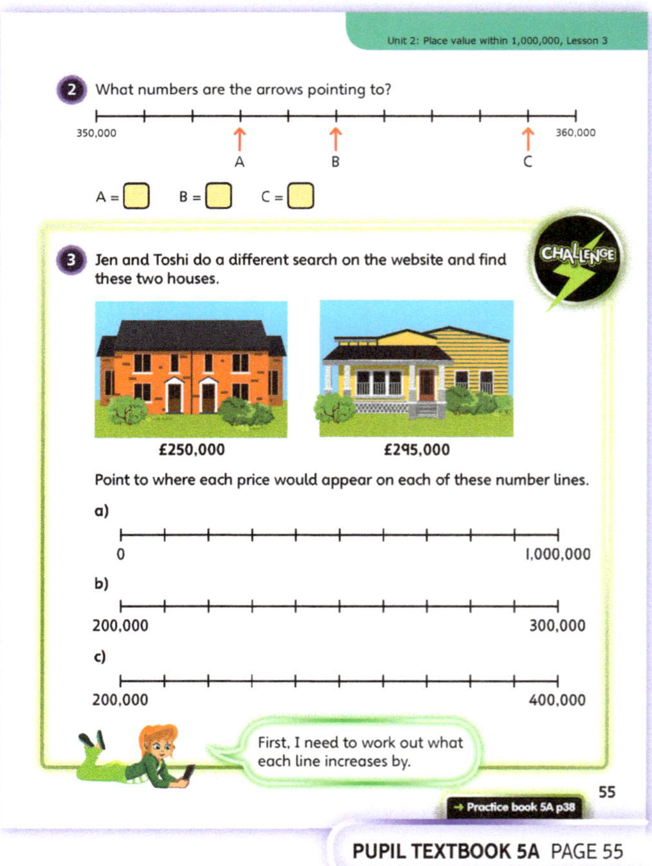

PUPIL TEXTBOOK 5A PAGE 55

Practice

WAYS OF WORKING Independent thinking

IN FOCUS In questions ❶ and ❷, children will practise calculating unlabelled intervals and finding missing numbers on a number line. Question ❸ offers children an opportunity to estimate where numbers lie along a number line. This will be trickier in this question as there are no marked intervals.

STRENGTHEN If children are finding it difficult to estimate the location of the numbers in question ❸, ask: *Is it useful to find any particular points first? Explain which, and why. How does knowing half-way help you estimate the locations of the numbers?*

DEEPEN Question ❺ will enhance children's understanding of the similarities and differences between number lines. Extend this by asking: *Which number line is it easiest to plot 330,000 on? Why? Describe a number line where this number would not be found. Explain what makes the number line you described different from those in the question.*

THINK DIFFERENTLY Question ❹ develops children's problem solving with number lines. It is a two-step problem as they will need to find out the value of point A and point B and then use this knowledge to identify which numbers from the list are found between the two points. Ask: *What is stopping you from identifying the numbers straight away?*

ASSESSMENT CHECKPOINT Check that children are accurately calculating the unlabelled intervals on a number line before they begin to work out any missing numbers. In question ❸, assess children's ability to recognise where numbers are positioned along a number line, looking in particular for an understanding of their proportional size based on where they are along the line. In question ❺, check that children recognise that the start and end point of a number line will determine where a number is positioned along the line.

ANSWERS Answers for the **Practice** part of the lesson appear in the separate **Practice and Reflect answer guide.**

Reflect

WAYS OF WORKING Independent thinking

IN FOCUS This **Reflect** question allows children to demonstrate the depth of their understanding by asking them to explain how they divided a given number line into intervals.

ASSESSMENT CHECKPOINT Assess whether children have divided the number line into an appropriate number of intervals and that the intervals are of equal size. Also, check that they have positioned both numbers in approximately the right place.

ANSWERS Answers for the **Reflect** part of the lesson appear in the separate **Practice and Reflect answer guide.**

After the lesson ⏸

- Were children able to plot numbers fluently on a number line with no marked intervals? Were they able to clearly explain ways of making this possible?
- How was children's understanding of place value developed in this lesson? Was the link between place value and number lines made explicit?

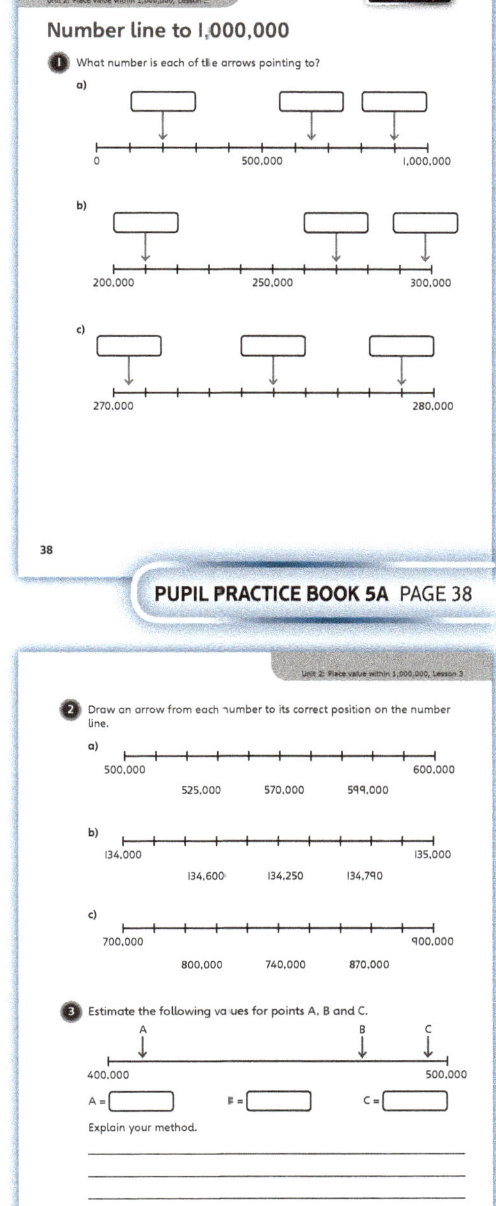

PUPIL PRACTICE BOOK 5A PAGE 38

PUPIL PRACTICE BOOK 5A PAGE 39

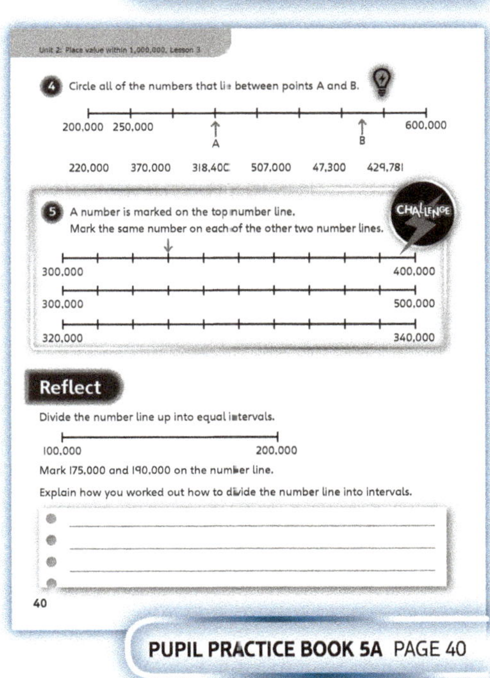

PUPIL PRACTICE BOOK 5A PAGE 40

Comparing and ordering numbers to 1,000,000

Learning focus

In this lesson, children will use their understanding of place value and numbers up to 1,000,000 to compare and order numbers.

Small steps

→ Previous step: Number line to 1,000,000
→ **This step: Comparing and ordering numbers to 1,000,000**
→ Next step: Rounding numbers to 1,000,000

NATIONAL CURRICULUM LINKS

Year 5 Number – Number and Place Value

Read, write, order and compare numbers to at least 1,000,000 and determine the value of each digit.

ASSESSING MASTERY

Children can use their understanding of place value and numbers to 1,000,000 to explain how they know a number is greater or less than another, make accurate comparisons between numbers and order them correctly.

COMMON MISCONCEPTIONS

Children may order numbers incorrectly, for example ordering in descending order instead of ascending order as the question may dictate. Ask:
• *How will you prove that you have put the numbers in that order?*

STRENGTHENING UNDERSTANDING

Children should be given opportunities to compare and order numbers up to 100,000 (last met in Year 4). This will give children the chance to remind themselves of the process of comparing and ordering numbers before they move on to bigger numbers that include more place value headings.

GOING DEEPER

Children can play a game in groups of three or four. In turn, each child picks a number of digit cards and places them anywhere they like in a place value grid (up to 100,000s), with the aim of making the largest number (or smallest, depending on the rules) compared to the rest of their group. (Provide enough digit cards for each child to make a number in the 100,000s.) Ask: *Who has the biggest number? Who has the smallest? Use a number line to order the numbers.*

KEY LANGUAGE

In lesson: ascending, descending, compare, greater than (>), less than (<), largest, order, place value, smallest

Other language to be used by the teacher: equal to, most, number line, ones (1s), tens (10s), hundreds (100s), thousands (1,000s), ten thousands (10,000s), hundred thousands (100,000s), partition, partitioning

STRUCTURES AND REPRESENTATIONS

place value grid

RESOURCES

Mandatory: place value counters

Optional: digit cards, place value arrow cards

 In the eTextbook of this lesson, you will find interactive links to a selection of teaching tools.

Before you teach 

• Are all children comfortable using the new column, hundred thousands (HTh, 100,000s), in the place value grid?

Discover

 Pair work

ASK

- Question **1** a): *What have you learnt in previous lessons that can help you to compare these numbers? What is the same about the two numbers? What is different?*
- Question **1** b): *What does 'ascending' mean? How could a place value grid help you to order these numbers?*

IN FOCUS In question **1** a), children begin to investigate comparative sizes of numbers, and it will be important to link this to their prior learning about place value. Question **1** b) requires children to use their understanding of numbers to arrange them in ascending order for the first time in the lesson.

PRACTICAL TIPS Suggest children use digit cards or place value arrow cards to create the number shown. Encourage them to discuss which numbers are bigger, which are smaller and how they can tell.

ANSWERS

Question **1** a): Oxford has more 10,000s and so has a greater population.

150,200 > 123,900

Question **1** b): The populations in ascending order are Durham, Cambridge, Oxford, Sunderland and Bristol.

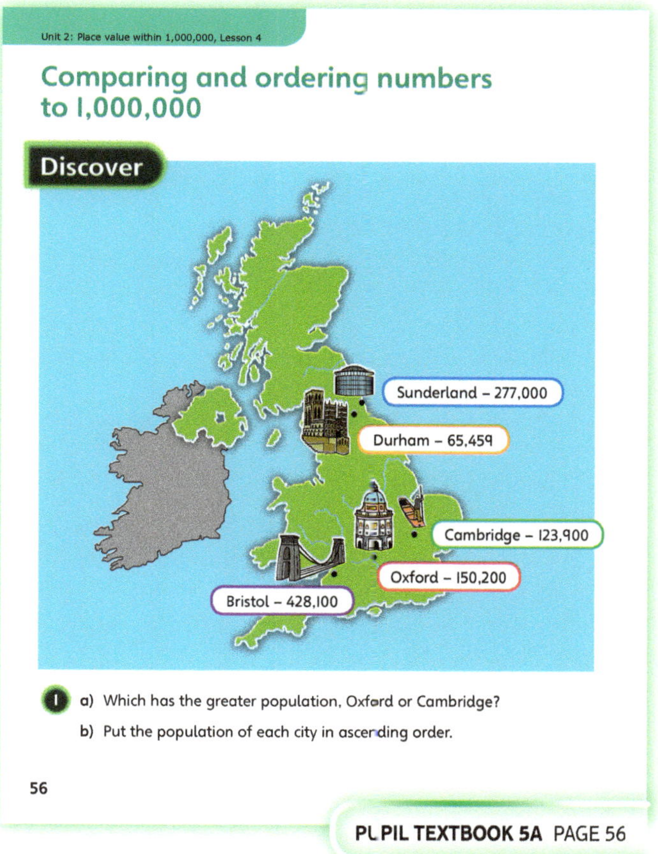

PL PIL TEXTBOOK 5A PAGE 56

Share

 Whole class teacher led

ASK

- Question **1** a): *Show the number using a place value grid or place value counters. What other ways could you use to show the value of each number?*
- Question **1** b): *How does the place value grid make the comparison clearer? What order did you put the populations into? Why? What end of the place value grid did you compare first? Why?*

IN FOCUS For question **1** b) it is important to make explicit how the place value grid can help with comparing the value of numbers and how it can be used to order numbers easily. Make sure children recognise how to work through the place value grid, comparing the larger columns first.

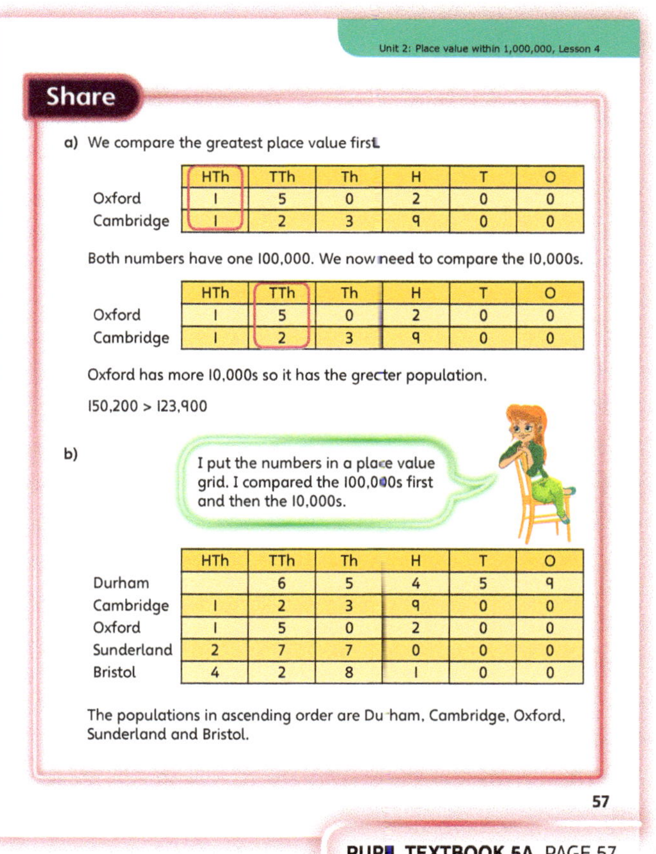

	HTh	TTh	Th	H	T	O
Oxford	1	5	0	2	0	0
Cambridge	1	2	3	9	0	0

	HTh	TTh	Th	H	T	O
Oxford	1	5	0	2	0	0
Cambridge	1	2	3	9	0	0

	HTh	TTh	Th	H	T	O
Durham		6	5	4	5	9
Cambridge	1	2	3	9	0	0
Oxford	1	5	0	2	0	0
Sunderland	2	7	7	0	0	0
Bristol	4	2	8	1	0	0

PUPIL TEXTBOOK 5A PAGE 57

Think together

WAYS OF WORKING Whole class teacher led (I do, We do, You do)

ASK

- Question **1** : *What will help you to compare these two numbers? What place value column should you begin your comparison at?*
- Question **2** : *How is this question different from those you have seen before? What will you need to find before you can find the third largest population?*
- Question **3** : *How will you compare the numbers that have been written in different ways?*
- Question **4** : *What will you do to begin investigating this question?*

IN FOCUS Question **1** will develop children's fluency when comparing 6-digit numbers. Question **2** will then take this a step further by requiring children to order the numbers before finding the answer.

STRENGTHEN In question **3** , where children are asked to compare the numbers represented in different ways, encourage children to think about another way that they could represent the numbers to make them easier to compare. Ask: *What did you learn previously about the written names of numbers? Can you convert them to numerals?*

DEEPEN Deepen children's reasoning around comparing and ordering numbers in question **4** by offering them a problem with multiple solutions. Ask: *How many solutions are there to this problem? How can you prove that you have found them all?*

ASSESSMENT CHECKPOINT Check that children are able to compare two or more 6-digit numbers and use the information to order them correctly. Are children confident with the key language from the lesson, such as 'compare' and 'ascending'?

ANSWERS

Question **1** : 291,080 < 291,804

Stockport has the smaller population.

Question **2** : Aberdeen has the third largest population.

Question **3** : 195,311, 99,999, 308,000, seventy-nine thousand, two hundred

Question **4** : Look for any combination that works. For example:

72,500, 126,091, 126,470, 133,904, 133,912

72,500, 126,191, 127,470, 133,904, 133,952

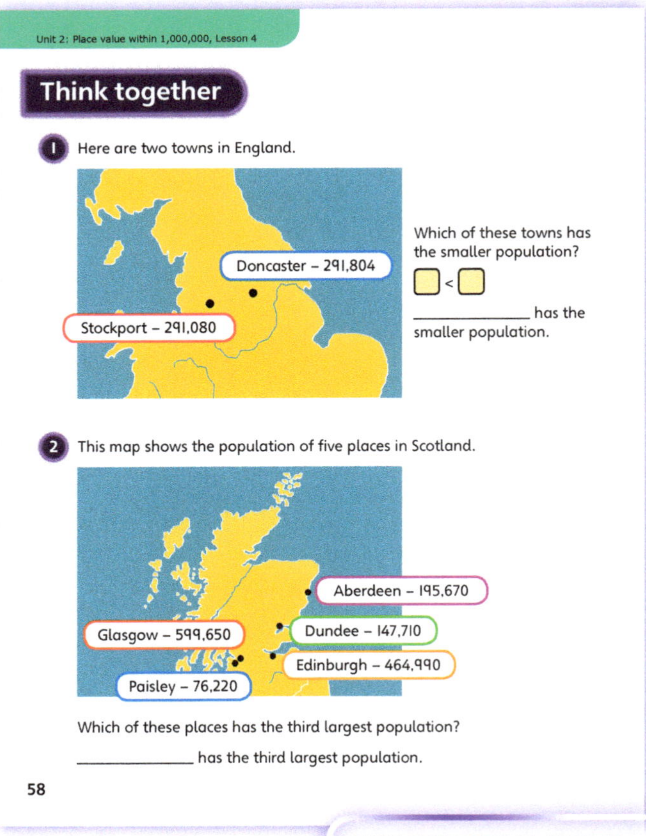

PUPIL TEXTBOOK 5A PAGE 58

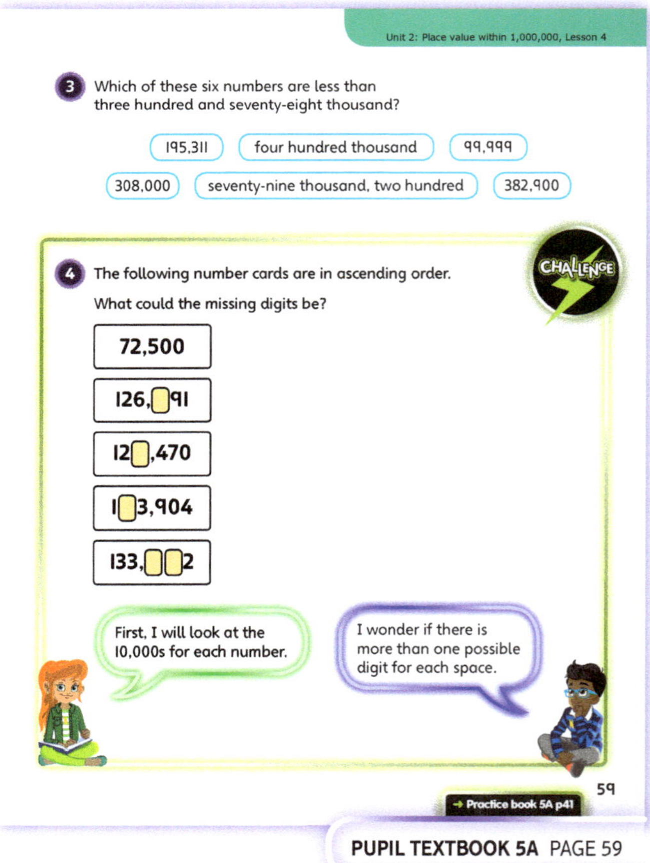

PUPIL TEXTBOOK 5A PAGE 59

Practice

WAYS OF WORKING Independent thinking

IN FOCUS Question ① offers children an opportunity to begin comparing numbers and identifying the largest, linking concrete representations of number to children's understanding of comparing and ordering numbers. Question ③ links children's use of the mathematical notation for comparison to their learning in this lesson. Question ④ requires children to order numbers in descending order for the first time in the lesson.

STRENGTHEN Provide place value grids if extra help is needed to compare pairs of numbers in question ①. Ask: *Show me how you would use a place value grid to compare the two numbers.*

DEEPEN Question ⑥ allows children to develop their understanding of the concepts of the lesson by requiring them to use reasoning to work out the missing digit in each case, looking at the largest place value first.

THINK DIFFERENTLY Question ⑤ offers children the opportunity to investigate different solutions for the same problem. Ask: *Is there only one solution to each statement? How can you prove that you have found all of the possible solutions?*

ASSESSMENT CHECKPOINT Can children compare numbers and order them correctly? Are they using the place value grid correctly to support them? In question ③, check children are correctly using the mathematical notation for number comparison. Question ④ assesses children's ability to compare and order numbers in descending order. Check that children have correctly started with the largest number.

ANSWERS Answers for the **Practice** part of the lesson appear in the separate **Practice and Reflect answer guide**.

Reflect

WAYS OF WORKING Independent thinking

IN FOCUS This **Reflect** question offers a good opportunity to assess children's understanding of how to compare numbers accurately. It may be interesting to ask children to compare their methods. What is the same and different about their approaches?

ASSESSMENT CHECKPOINT Children may refer to their use of the structures and representations covered in this unit, for example the place value grid. Look for recognition of why it is important to begin comparing at the larger end of the place value grid.

ANSWERS Answers for the **Reflect** part of the lesson appear in the separate **Practice and Reflect answer guide**.

After the lesson ❚❚

- Were children able to fluently and accurately order numbers in both ascending and descending order?
- What opportunities can you offer children to use these concepts in other areas of the curriculum?
- How many children made good progress in this lesson? What support will you offer those children who did not?

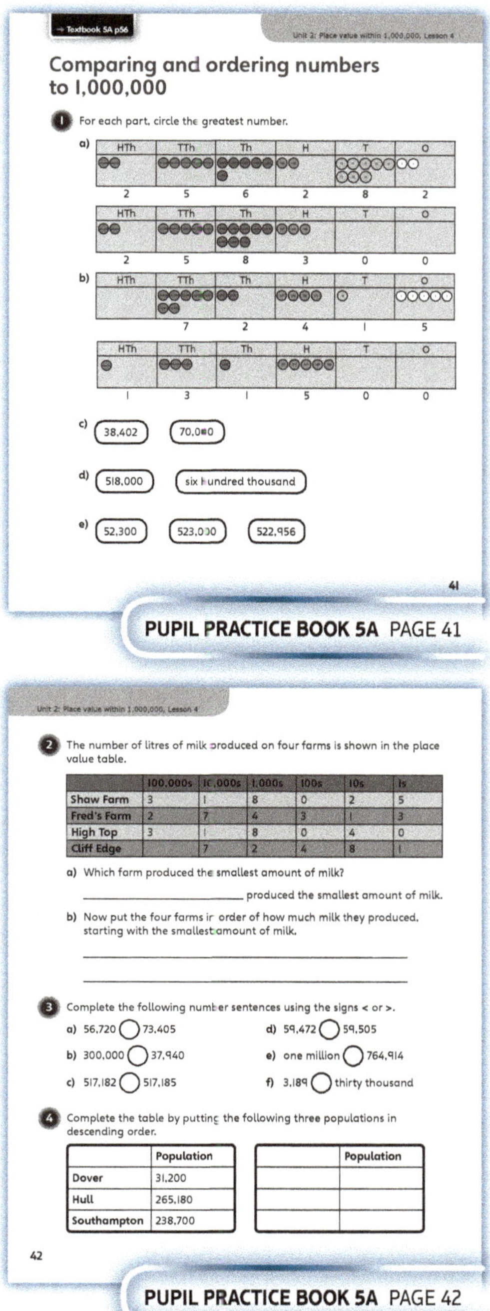

PUPIL PRACTICE BOOK 5A PAGE 41

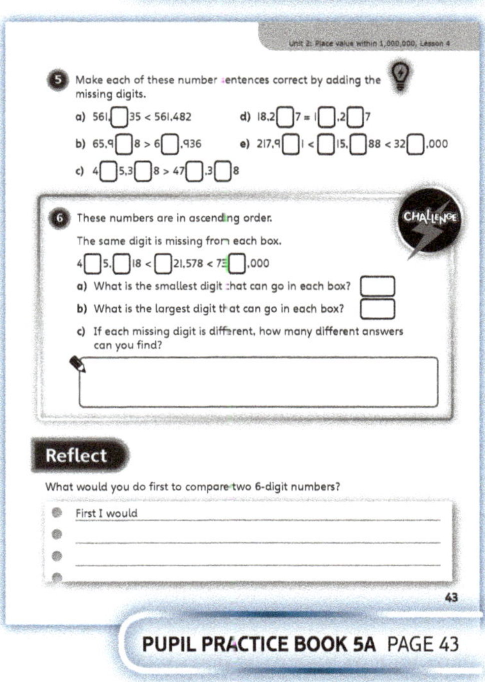

PUPIL PRACTICE BOOK 5A PAGE 42

PUPIL PRACTICE BOOK 5A PAGE 43

Rounding numbers to 1,000,000

Learning focus

In this lesson, children will use their understanding of place value to help them round numbers up to 1,000,000. They will discuss when rounding is appropriate and which multiple of 10 to round to in a given context.

Small steps

→ Previous step: Comparing and ordering numbers to 1,000,000
→ **This step: Rounding numbers to 1,000,000**
→ Next step: Negative numbers

NATIONAL CURRICULUM LINKS

Year 5 Number – Number and Place Value

Round any number up to 1,000,000 to the nearest 10, 100, 1,000, 10,000 and 100,000.

ASSESSING MASTERY

Children can recognise when to round numbers and explain how to do so fluently. They can apply their understanding of rounding to larger numbers and can reliably round up and down to the nearest 10, 100, 1,000, 10,000, 100,000 and 1,000,000.

COMMON MISCONCEPTIONS

Children may assume that rounding only affects the place value column it references (for example, that rounding to the nearest 100 will only affect the digits in the hundreds, tens and ones columns). Ask:

• *If you round 1972 to the nearest 100, what happens? Show me how the number will change using a picture or resources. What happens when you count on one more hundred from 900?*

STRENGTHENING UNDERSTANDING

Before the lesson, children who may need more help should be encouraged to recap their understanding of rounding. Encourage children to round numbers to the nearest 10,000. Ask: *What does it mean to round a number? What numbers can you round to? Try showing me how to round X to the nearest X?*

GOING DEEPER

Provide children with six digit cards. Ask: *What numbers can you make with these digit cards? What numbers in the 100,000s can you make that will round down to the nearest 1,000? What numbers in the 100,000s can you make that will round up to the nearest 1,000? Is there more of one type than the other? Why?*

KEY LANGUAGE

In lesson: compare, greater, greatest, place value, round, trial and error

Other language to be used by the teacher: round down, round up

STRUCTURES AND REPRESENTATIONS

number line

RESOURCES

Mandatory: digit cards

Optional: place value counters

 In the eTextbook of this lesson, you will find interactive links to a selection of teaching tools.

Before you teach 

• Can children recall the rules for rounding?
• Are children confident when using a number line?
• How will you help children understand the importance of place value when rounding?

Discover

WAYS OF WORKING Pair work

ASK

- Question ❶ a): *How will you know whether to round up or down? Which digit will you need to look at when rounding to the nearest 100?*
- Question ❶ b): *What number could Jamie not have made? Explain how you know.*

IN FOCUS Question ❶ a) gives children their first opportunity in this lesson to round up and down. Question ❶ b) allows children to begin generalising about the rules for rounding, which will help them later in the lesson.

PRACTICAL TIPS Provide children with digit cards, as in the picture. Suggest they investigate the different numbers that can be made with the same five or six digits. Challenge them to find out what numbers they can make that can be rounded to the same hundred, thousand, ten thousand and so on.

ANSWERS

Question ❶ a): 712,458 rounded to the nearest 100,000 is 700,000.
712,458 rounded to the nearest 100 is 712,500.

Question ❶ b): Jamie could have made many numbers, for example 45,178, 47,812, 48,271, 51,478, 54,782.

Share

WAYS OF WORKING Whole class teacher led

ASK

- Question ❶ a): *What digit did you look at to know how to round the number? Why? How does the number line show this?*
- Question ❶ b): *How did you prove your numbers rounded to 50,000? What rules for rounding helped you identify what numbers would and would not work?*

IN FOCUS For questions ❶ a) and b) it is important to make sure children are given the opportunity to identify the rules that apply to rounding to any multiple of ten. Remind children to look at the place value column before the one they are rounding to (for example the thousands column if they are rounding to the nearest 10,000). If the number is less than 5 they round down. If the number is 5 or more they round up.

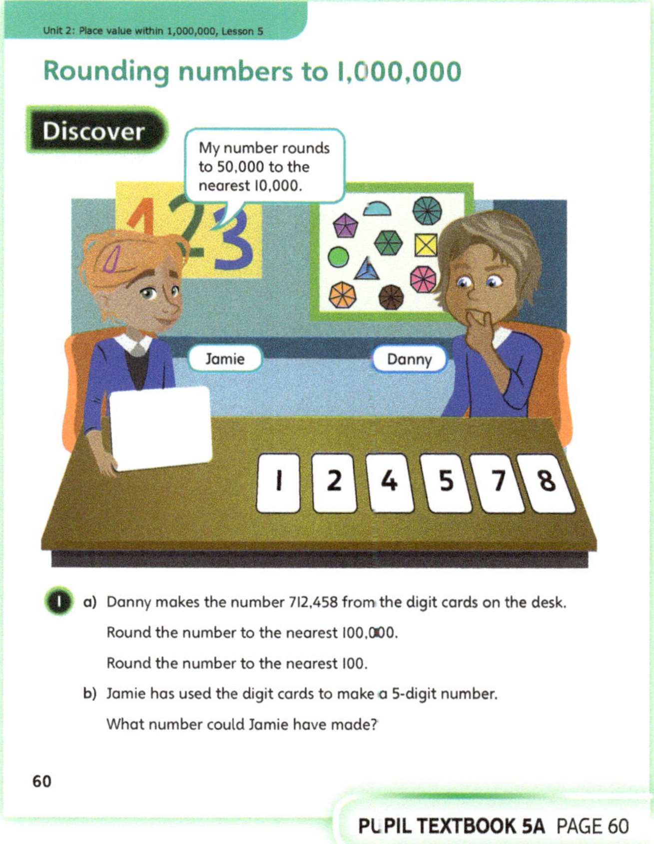

PUPIL TEXTBOOK 5A PAGE 60

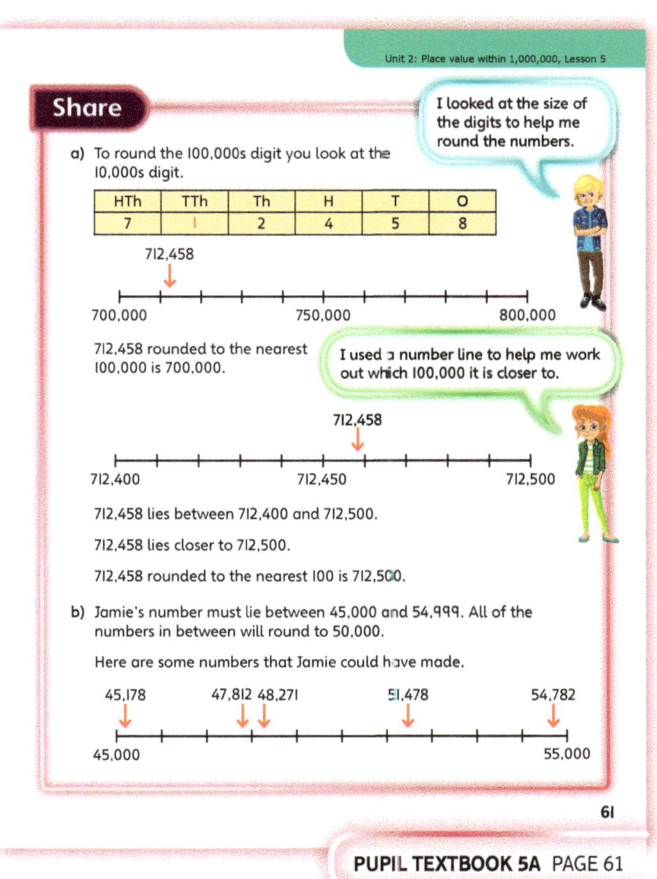

PUPIL TEXTBOOK 5A PAGE 61

Think together

WAYS OF WORKING Whole class teacher led (I do, We do, You do)

ASK

- Question ❶ : *What digit will you need to look at to round to the nearest 100,000 or 1,000?*
- Question ❷ : *What is different about rounding to the nearest 1,000 and the nearest 100,000?*
- Question ❸ b): *What needs to be the same about the two numbers to make sure they both round to the same 1,000?*

IN FOCUS Question ❶ offers children the opportunity to round numbers to different multiples of ten and links the concept of rounding to the more visual representation of a number line, which shows which multiple of ten the number they are rounding is closer to. Question ❷ provides an opportunity for children to investigate how rounding to a different multiple of ten can affect a number in different ways, helping children to strengthen their understanding that a number can round up or down depending on the value of each digit.

STRENGTHEN For question ❶, when children are working with the number lines, ask: *Which thousand is your number closer to? How can the 100s digit help you?*

DEEPEN Question ❸ deepens children's understanding of rounding by requiring them to demonstrate their fluency and problem solving in relation to the concept. Ask: *How can you prove how many possible solutions there are? What is the same about all of the solutions? What is different?*

ASSESSMENT CHECKPOINT Are children able to use a number line to support them when rounding up or down? Are they confident when deciding whether to round up or down? Check that children understand the rule for rounding. Use question ❷ to assess whether children can correctly round a number up or down to the nearest 10, 100, 1,000, 10,000 and 100,000.

ANSWERS

Question ❶ a): 500,000

Question ❶ b): 458,000

Question ❷ :

Danny's number, rounded to the nearest ...				
100,000	10,000	1,000	100	10
100,000	130,000	128,000	127,900	127,850

Question ❸ a): The first missing digit must be 5. Any remaining digit can be used in the tens column: 915,702, 915,732, 915,742, 915,762, 915,782

Question ❸ b): Children should make two numbers that fall between 6,950 and 7,049 inclusive.

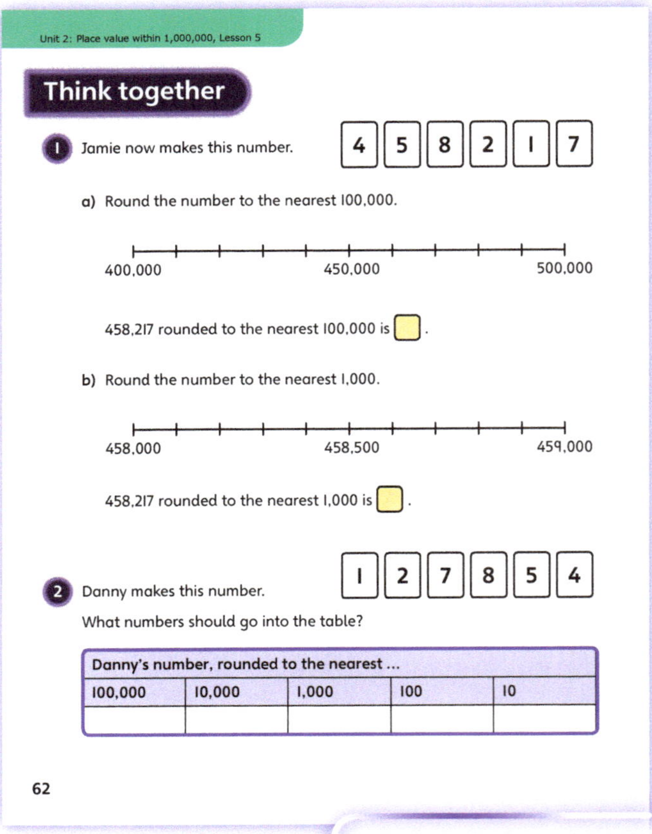

PUPIL TEXTBOOK 5A PAGE 62

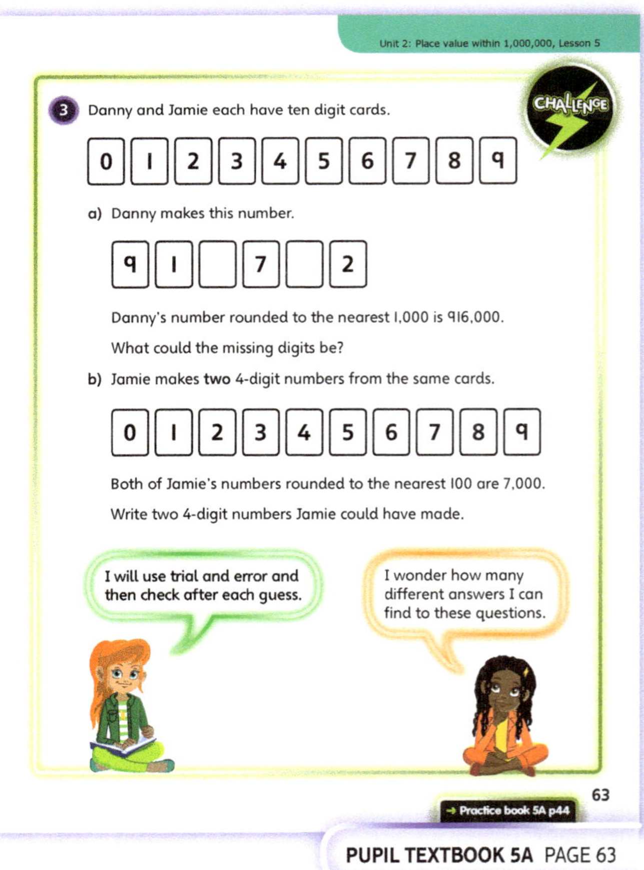

PUPIL TEXTBOOK 5A PAGE 63

Practice

WAYS OF WORKING Independent thinking

IN FOCUS Question **1** helps children to independently develop their ability to round numbers to the nearest 100,000, using the visual representation of a number line to help support their understanding. Question **2** offers children an opportunity to round numbers to the nearest 10,000. The early parts of the question offer a number line as a scaffold for their thinking, but this is removed from question **2** c) onwards to help develop children's fluency.

STRENGTHEN To strengthen understanding in question **3** c), ask: *What happens if you put all eight counters into one column? Could that method work? Prove it. Could you solve this using trial and error? Explain why or why not. Is there anything you know for definite about Danny's number?*

DEEPEN Question **6** encourages children to generalise about the properties of numbers and how these properties can influence how a number changes when rounded. Ask: *What do you notice about numbers where all the digits change? Does this apply when rounding to other multiples of ten? Explain your answer.*

THINK DIFFERENTLY Question **5** develops children's fluency when rounding numbers by asking them to work backwards from the number being rounded to. This often produces more than one solution. Ask: *Is there only one solution for each question? Are there more solutions for one of the questions than the others? Why might this be?*

ASSESSMENT CHECKPOINT Check children are rounding to the nearest 10, 1,000, 10,000 and 100,000 accurately. For question **5**, are children able to work backwards from rounded numbers to find the original starting number? Question **6** assesses children's recognition of how the properties of a number can affect how it is rounded, in particular how multiple 9s in a number can produce a chain bridging the multiples of 10.

ANSWERS Answers for the **Practice** part of the lesson appear in the separate **Practice and Reflect answer guide**.

Reflect

WAYS OF WORKING Independent thinking

IN FOCUS This **Reflect** question offers a final opportunity to assess children's understanding of the rules for rounding.

ASSESSMENT CHECKPOINT Look for clear understanding from children that they need to consider a specific digit when rounding, for example, the 10,000s digit when rounding to the nearest 100,000.

ANSWERS Answers for the **Reflect** part of the lesson appear in the separate **Practice and Reflect answer guide**.

After the lesson

- Are children able to clearly explain the rules of rounding?
- What percentage of the class mastered rounding numbers to 1,000,000?

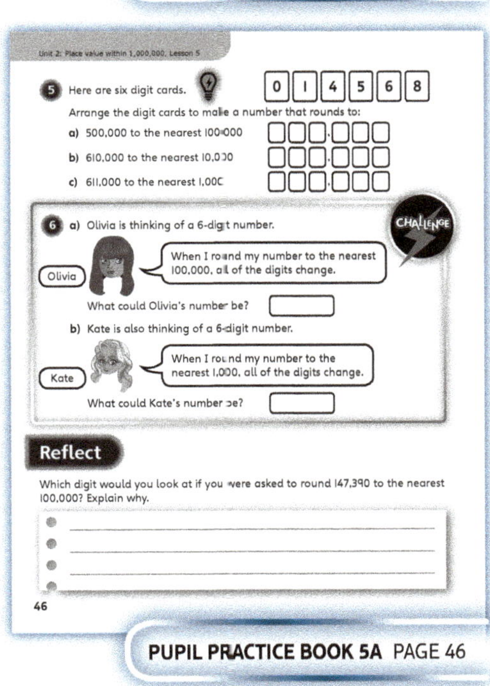

PUPIL PRACTICE BOOK 5A PAGE 44

Number	Rounded to the nearest 10,000	Rounded to the nearest 1,000	Rounded to the nearest 10
239,145			
128,783			
758,007			
		632,000	632,180
			825,430
6☐7,14☐	630,000		627,150
☐35,☐72	640,000		

PUPIL PRACTICE BOOK 5A PAGE 45

PUPIL PRACTICE BOOK 5A PAGE 46

Negative numbers

Learning focus

In this lesson, children will learn about negative numbers and their relationship with positive numbers. They will use a number line to identify negative numbers and begin calculating with them.

Small steps

→ Previous step: Rounding numbers to 1,000,000
→ **This step: Negative numbers**
→ Next step: Counting in 10s, 100s, 1,000s and 10,000s

NATIONAL CURRICULUM LINKS

Year 5 Number – Number and Place Value

Interpret negative numbers in context, count forwards and backwards with positive and negative whole numbers, including through zero.

ASSESSING MASTERY

Children can identify and explain what a negative number is and how negative numbers are similar to and different from positive numbers. They can reliably find negative numbers on a number line and begin calculating with them.

COMMON MISCONCEPTIONS

Children may assume that negative numbers work in a similar way to positive numbers, for example that ⁻5 must be greater than ⁻1 because 5 is greater than 1. Show children a number line and ask:
• *Plot the numbers from ⁻10 to 10 on this number line. Which end shows the greatest number? Which end shows the smallest? What can you say about the difference between ⁻5 and ⁻1? Explain how you know.*

Children may miscalculate the difference between positive and negative numbers (for example, they may calculate the difference between 7 and ⁻6 as 1). Ask:
• *Show me these two numbers on a number line. Count the difference between them.*

STRENGTHENING UNDERSTANDING

Children's fluency with this concept could be strengthened by giving them plenty of opportunities to count forwards and backwards through positive and negative numbers. Ask: *Can you count back from 21 in twos? What happens when you get to 1?*

GOING DEEPER

Encourage children to create their own word problems requiring the use of negative numbers. Discuss with children the ways in which negative numbers are used in real-life contexts (linked to the **Discover** practical ideas) and ask how they might use these ideas to create a word problem.

KEY LANGUAGE

In lesson: °C, minus (⁻), negative

Other language to be used by the teacher: positive (⁺)

STRUCTURES AND REPRESENTATIONS

number line

RESOURCES

Optional: laminated paper thermometers, number cards

 In the eTextbook of this lesson, you will find interactive links to a selection of teaching tools.

Before you teach

• In what contexts might children have met negative numbers before?

Discover

WAYS OF WORKING Pair work

ASK

- Question **1** a): *How is a thermometer similar to a number line? How is it different? How can you use the thermometer to find the difference between the two temperatures?*
- Question **1** b): *How do you know which two temperatures have the greatest difference?*

IN FOCUS This section offers children an opportunity to begin calculating with negative numbers in context, in this instance finding the difference between two temperatures. The thermometers can be used like a number line and provide scaffolding. This will help children to recall what they learnt about negative numbers in Year 4.

PRACTICAL TIPS Provide children with laminated thermometers to colour in with dry-wipe pens. Give them a temperature to mark on their thermometer and ask them to show what the thermometer would look like if it got X degrees colder. Discuss what happens when the temperature drops below 0 °C.

ANSWERS

Question **1** a): In Tomsk, May is 13 °C warmer than March.

Question **1** b): The two months that have the greatest temperature difference are January and July.

The difference in temperature between January and July is 27 °C.

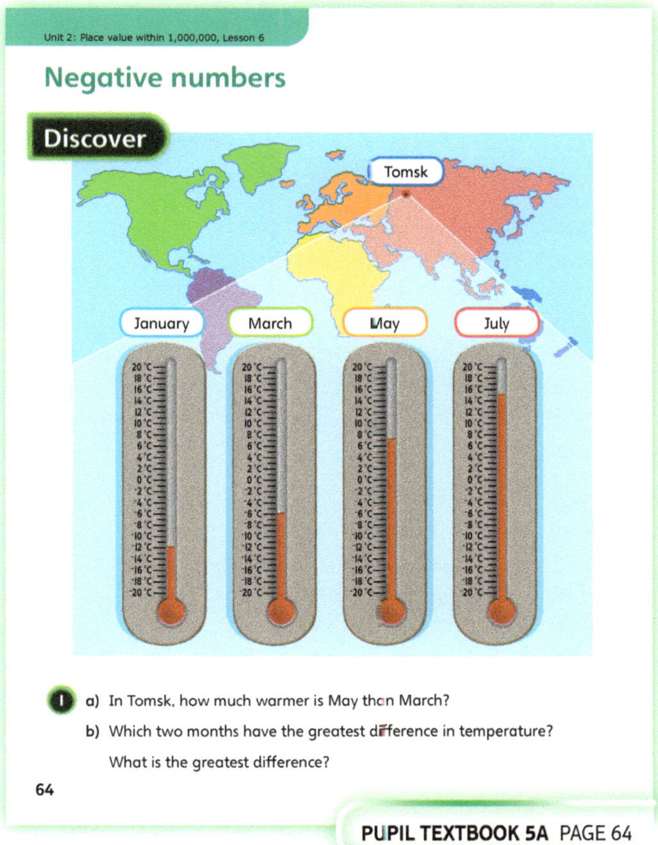

PUPIL TEXTBOOK 5A PAGE 64

Share

WAYS OF WORKING Whole class teacher led

ASK

- Question **1** a): *What difference did you find? Why was the difference not 1 °C?*
- Question **1** b): *How did you know which temperature was greatest and which was least? How did you find the difference between the two temperatures? Why does Dexter's method work?*

IN FOCUS Questions **1** a) and **1** b) link the thermometer seen in **Discover** to number lines which children are very familiar with. At this point in the lesson it will be important to discuss the difference between positive and negative numbers. Children need to recognise that the numbers are mirrored, therefore numbers they recognise as larger will be smaller as negative numbers.

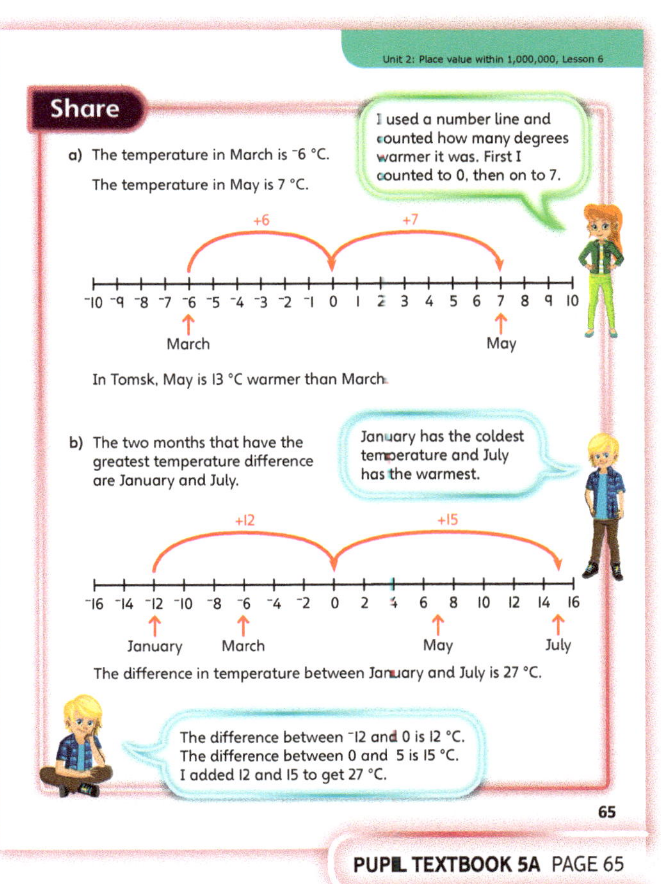

PUPIL TEXTBOOK 5A PAGE 65

Think together

WAYS OF WORKING Whole class teacher led (I do, We do, You do)

ASK

- Question ❶ : *How could you find the difference between these temperatures using a number line? Which temperature is warmer? How do you know?*
- Question ❷ : *How many floors do you predict she will have travelled down? Explain your prediction.*
- Question ❸ : *How could you represent the temperatures in a different way? Can you find the difference between two of the temperatures?*

IN FOCUS Question ❸ deepens children's understanding as it gives them an opportunity to draw their own conclusions from a set of data that includes negative numbers. Children should be encouraged to justify their ideas with evidence or proof.

STRENGTHEN Encourage children to recognise that there is a number line hidden within the floor list in question ❷, ask: *Could you show (or find) these numbers on a number line? What is the same about the number line and the lift buttons? What is different? How could you use a number line to find the difference between the two floors?*

DEEPEN For those children who are mastering negative numbers quickly, provide several digit cards of positive and negative numbers. Ask them to draw a vertical number line and mark the numbers from their cards on it. Encourage them to explain how they decided on a start and finish number for their number line.

ASSESSMENT CHECKPOINT Check that children can find the difference between a negative number and a positive number. Use children's conclusions in question ❸ to assess whether they can find differences between numbers, including two negative numbers.

ANSWERS

Question ❶ : Cairo is 21 °C warmer than New York.

Question ❷ : Mrs Dean travels down 19 floors.

Question ❸ : Children may note that the temperature rises during the morning and then descends in the afternoon. They may also note the greatest difference as 32 °C or that the individual changes in temperature are $^+$11 °C, $^+$21 °C, $^-$15 °C (in that order).

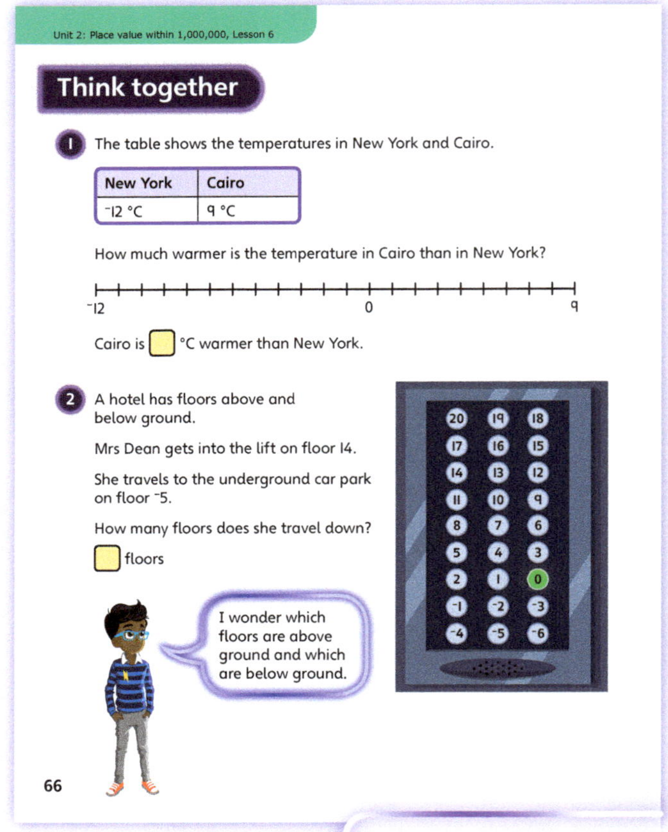

PUPIL TEXTBOOK 5A PAGE 66

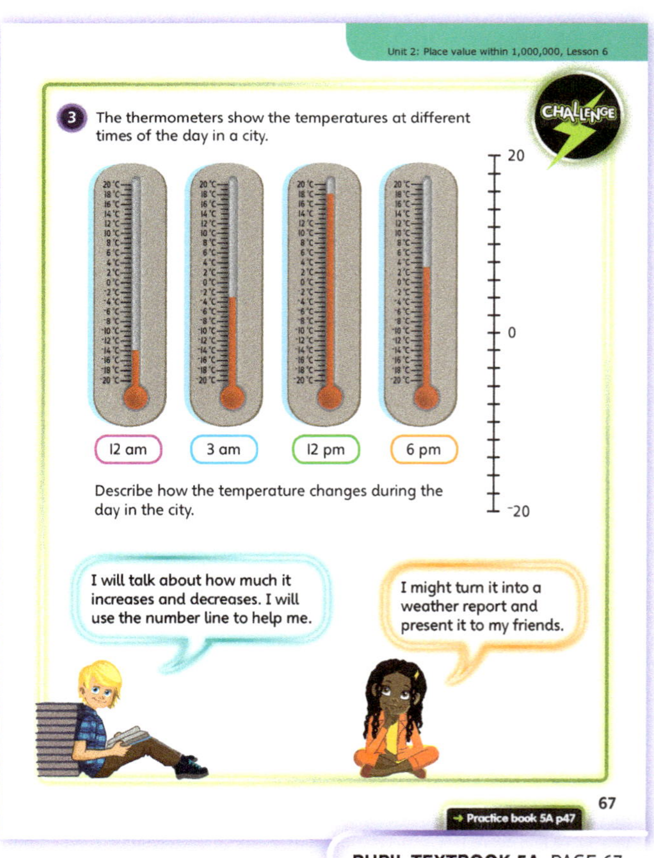

PUPIL TEXTBOOK 5A PAGE 67

Practice

WAYS OF WORKING Independent thinking

IN FOCUS Questions ❶ and ❷ will develop children's ability to independently recognise negative numbers and place them along a number line. Children will also have the opportunity to practise finding the difference between a positive and negative number, using a number line to support their thinking. Question ❸ allows children to practise finding negative numbers along a number line and to use this skill to find differences and order negative numbers.

STRENGTHEN For question ❸, provide laminated paper thermometers (from the **Discover** activity) and whiteboard pens, and ask them to colour in a thermometer to show each temperature from the question. Ask them to then place the thermometers in order, from coldest to warmest. Encourage them to discuss how they decided which thermometer was showing the coldest temperature. This will cement their understanding of how to order negative numbers.

DEEPEN Question ❻ requires children to solve a problem with multiple steps, within a different real-life context. Deepen learning by changing the digits. For example, change the iceberg to be 14 metres above sea level to work out the depth of the iceberg if it is 9 or 10 times the height of the iceberg above sea level.

THINK DIFFERENTLY Question ❺ makes children aware of another real-life context for negative numbers. This question will challenge children's thinking as the 0 point (sea level) is an imaginary line, not actually below the mountain itself but some way up it.

ASSESSMENT CHECKPOINT Check that children are able to accurately identify negative numbers on a number line and find the difference between positive and negative numbers. Can they correctly order a list of numbers that includes negative numbers? Assess whether children can add a positive number to a negative number.

ANSWERS Answers for the **Practice** part of the lesson appear in the separate **Practice and Reflect answer guide**.

Reflect

WAYS OF WORKING Pair work

IN FOCUS This **Reflect** question will give a final opportunity to assess children's fluency with negative numbers. Encourage children to compare ideas with a partner and discuss what they have concluded.

ASSESSMENT CHECKPOINT Children should be able to use the concepts they have learnt, including their ability to calculate differences and order negative numbers, to draw conclusions about the data shown.

ANSWERS Answers for the **Reflect** part of the lesson appear in the separate **Practice and Reflect answer guide**.

After the lesson ⏸

- Were children able to fluently explain, using accurate mathematical vocabulary, the key differences between positive and negative numbers?
- How confident were children in calculating with negative numbers? Was there an operation they were less confident with? How will you support this in future lessons?

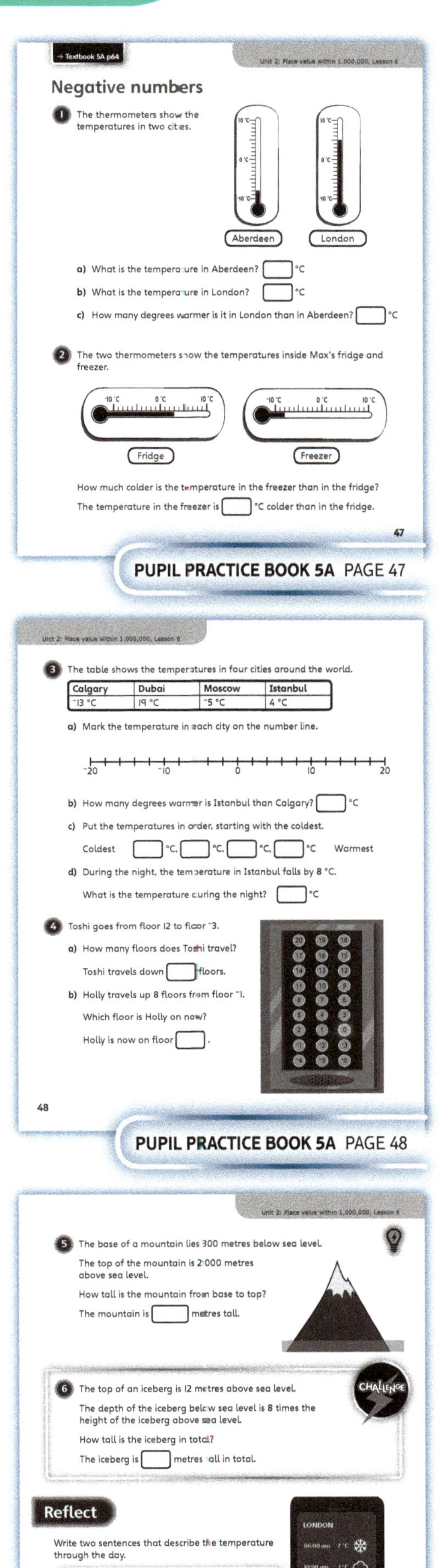

PUPIL PRACTICE BOOK 5A PAGE 47

PUPIL PRACTICE BOOK 5A PAGE 48

PUPIL PRACTICE BOOK 5A PAGE 49

Counting in 10s, 100s, 1,000s, 10,000s

Learning focus

In this lesson, children will develop their ability to count forwards and backwards in steps of 10, 100, 1,000 and 10,000.

Small steps

→ Previous step: Negative numbers
→ **This step: Counting in 10s, 100s, 1,000s and 10,000s**
→ Next step: Number sequences

NATIONAL CURRICULUM LINKS

Year 5 Number – Number and Place Value

Count forwards or backwards in steps of powers of 10 for any given number up to 1,000,000.

ASSESSING MASTERY

Children can reliably and fluently count forwards and backwards in steps of 10, 100, 1,000 and 10,000 from a given number.

COMMON MISCONCEPTIONS

Children may struggle to bridge into the next place value column when adding 10, 100, 1,000 or 10,000. For example, when adding 100 to 900 they may say 'ten hundred'. Use base 10 equipment to demonstrate how the numbers change when adding 10, 100, 1,000 and 10,000. In particular, demonstrate how 10 hundreds is the same as 1 thousand. Ask:

- *What is the correct name of the number '10 hundreds'? What number comes after 9,999? What number do you get if you add 10,000 to 90,000?*

STRENGTHENING UNDERSTANDING

Provide children with place value counters and encourage them to make up a number. Ask them to use the counters to add or subtract 10, 100, 1,000 or 10,000 to or from their number. Ask: *How did your number change? How did it stay the same? What is your new number?*

GOING DEEPER

Children could create inverse puzzles for each other. For example: *My final number was 550,000. To get to 550,000, I added 10,000, subtracted 100, added 1,000 and added another 10,000. What was my starting number?*

KEY LANGUAGE

In lesson: place value, sequence, table

Other language to be used by the teacher: ones (1s), tens (10s), hundreds (100s), thousands (1,000s), ten thousands (10,000s), hundred thousands (100,000s)

STRUCTURES AND REPRESENTATIONS

place value grid, table

RESOURCES

Mandatory: place value counters

Optional: base 10 equipment, printed place value grids

 In the eTextbook of this lesson, you will find interactive links to a selection of teaching tools.

Before you teach 

- Which of the numbers in this lesson are likely to be trickiest for some children?
- How will you scaffold their concrete understanding of those numbers?
- Can children name each column in the extended place value grid (up to 100,000s)?

Discover

WAYS OF WORKING Pair work

ASK

- Questions ❶ a) and b): *What number does Reena begin with?*
- Questions ❶ a) and b): *How will you know whether to add or subtract 10, 100, 1,000 or 10,000? What happens to a number when you add 10,000? What if both players' finishing numbers were 350,000?*

IN FOCUS This section gives children an opportunity to add and subtract 10, 100, 1,000 and 10,000 to and from a given number, within a context that they will be familiar with and find engaging. The learning is scaffolded by asking children to initially add and subtract 10,000 only (in question ❶ a), before developing this to include adding and subtracting 10, 100 and 1,000 also (in question ❶ b).

PRACTICAL TIPS It would be an engaging introduction to this lesson if children were offered the opportunity to play a similar game. Using the grid in the picture, or by creating your own on paper, ask children to investigate different routes through the maze – which way leads to the greatest or smallest total?

ANSWERS

Question ❶ a): When Isla reaches the finish her score is 120,000.

Question ❶ b): When Reena reaches the finish her score is 241,980.

Share

WAYS OF WORKING Whole class teacher led

ASK

- Question ❶ a): *How did you make sure not to lose count? How could you represent how the number changed during the game?*
- Question ❶ b): *Which numbers were easier to add or subtract? Why?*

IN FOCUS It will be important to link children's understanding of place value with the concepts covered in this part of the lesson. Encourage children to demonstrate the addition and subtraction in question ❶ b) using place value counters in a place value grid, or with base 10 equipment.

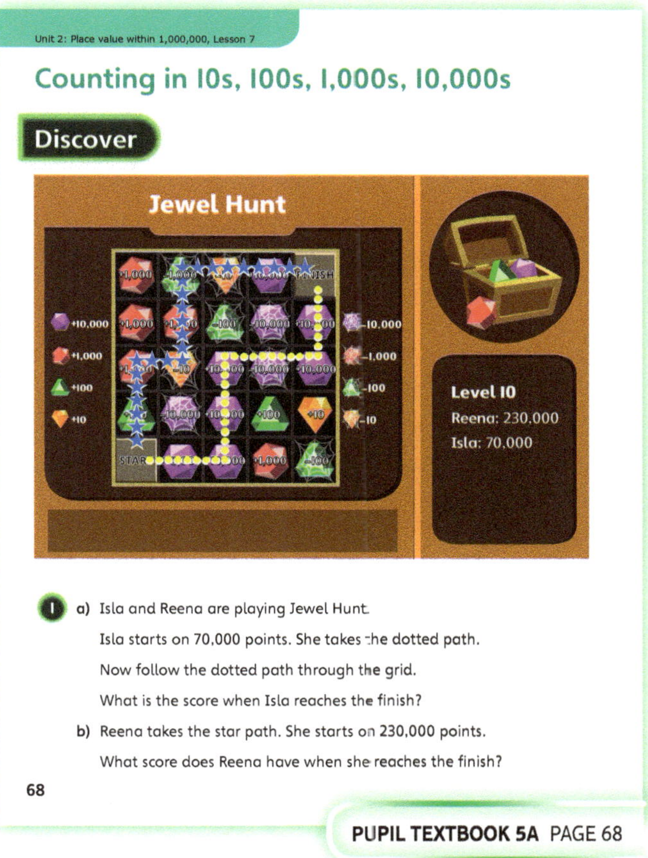

PUPIL TEXTBOOK 5A PAGE 68

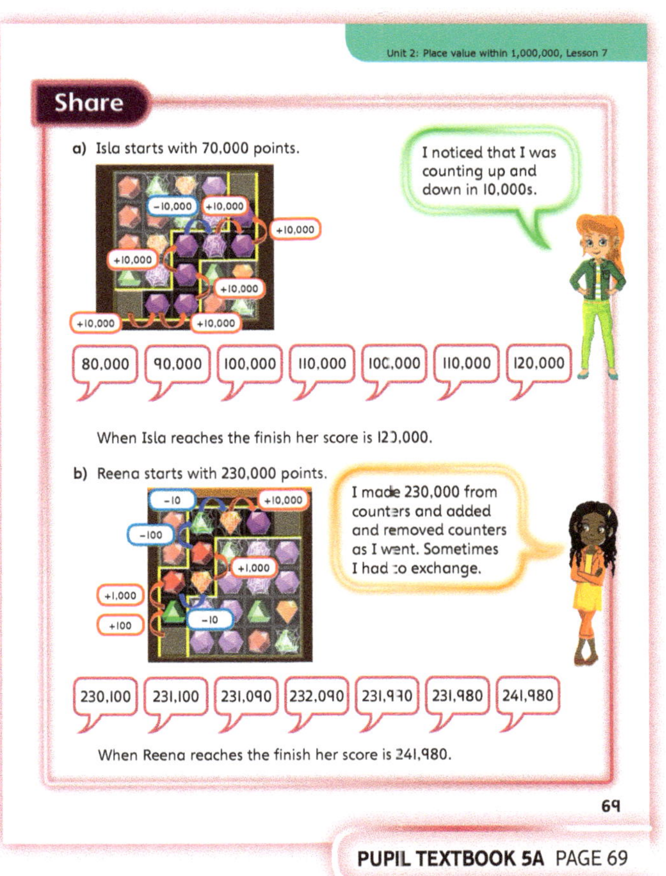

PUPIL TEXTBOOK 5A PAGE 69

Think together

WAYS OF WORKING Whole class teacher led (I do, We do, You do)

ASK

- Question **1** : *What do you notice about how the numbers change? How will this help you find the missing numbers?*
- Question **2** : *How will the place value grid help you find the solution for each calculation?*
- Question **3** b): *Is there only one solution? How many successful routes can you discover?*

IN FOCUS Question **1** will help children to identify the patterns inherent in adding and subtracting multiples of 10, enabling them to become more fluent at counting in steps of 10, 100, 1,000 and 10,000. Question **2** demonstrates how adding different multiples of 10 will affect a number differently.

STRENGTHEN Provide place value counters to help complete the table in question **2**. Ask: *Show me 100 more using place value counters. What number do you have now? Explain how you know.*

DEEPEN For children mastering the concepts of the lesson, pose another challenge once they have solved question **3**. Ask: *If my final score was ⁻6,800, what was my route?*

ASSESSMENT CHECKPOINT Assess children's recognition of patterns when counting in steps of 10, 100, 1,000 and 10,000. Can they do so fluently? Check that children recognise the link between counting in steps of 10, 100, 1,000 and 10,000 and place value.

ANSWERS

Question **1** a): 72,000, 73,000, 74,000, **75,000**, **76,000**, **77,000**

Question **1** b): 272,800, 272,900, **273,000**, **273,100**, 273,200, **273,300**

Question **1** c): 738,006, **638,006**, **538,006**, 438,006, **338,006**, **238,006**

Question **2** :

100,000 less	47,300	100,000 more	247,300
10,000 less	137,300	10,000 more	157,300
1,000 less	146,300	1,000 more	148,300
100 less	147,200	100 more	147,400
10 less	147,290	10 more	147,310

Question **3** a): Reena has a score of 400,500 by the end of the level. Children's explanation to detail how for every positive number Reena travels through, she also travels through a negative number cancelling out any points she scores.

Question **3** b): Look for children to find routes that total less than 500,000. An example is pictured below.

Think together

1 Work out the missing numbers in each sequence.

a)

| 72,000 | 73,000 | 74,000 | | | |

b)

| 272,800 | 272,900 | | | 273,200 | |

c)

| 738,006 | | | 438,006 | | |

2 Use the place value grid to complete the table below.

HTh	TTh	Th	H	T	O
1	4	7	3	0	0

100,000 less		100,000 more	
10,000 less		10,000 more	
1,000 less		1,000 more	
100 less		100 more	
10 less		10 more	

70

PUPIL TEXTBOOK 5A PAGE 70

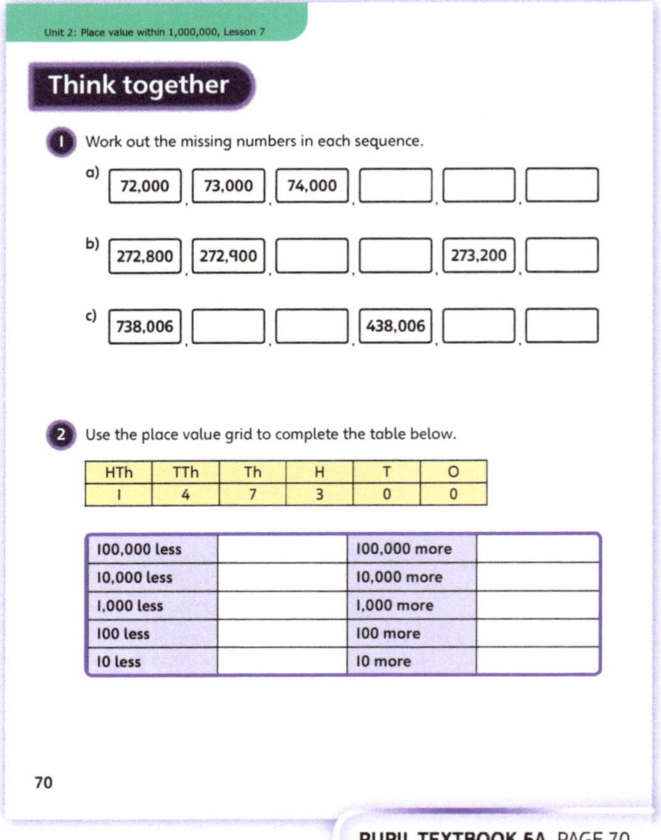

3 Isla and Reena are now on a different level of Jewel Hunt.

a) Reena starts with 400,500 points. She takes this path.

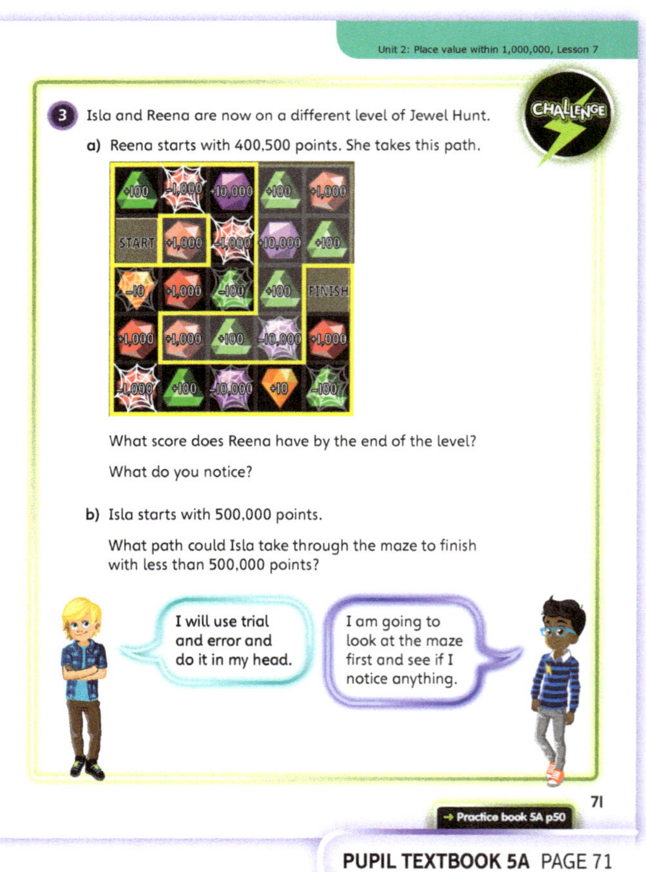

What score does Reena have by the end of the level?

What do you notice?

b) Isla starts with 500,000 points.

What path could Isla take through the maze to finish with less than 500,000 points?

→ Practice book 5A p50

71

PUPIL TEXTBOOK 5A PAGE 71

Practice

WAYS OF WORKING Independent thinking

IN FOCUS Question ❶ supports children's understanding of counting in steps of 10, 100, 1,000 and 10,000 from a given number, using the more concrete representation of place value counters in a place value grid. Children could replicate this representation using classroom resources to help them solve this question. Question ❷ offers children the opportunity to complete counts using simple numbers that are rounded to the place value that the count is in (for example, if they are counting in 1,000s, then the numbers do not have any 100s, 10s or 1s). This learning is extended in question ❸, which offers children an opportunity to continue counts with trickier numbers that are not conveniently rounded.

STRENGTHEN To strengthen understanding when counting on or back from the given numbers in questions ❻ and ❼, ask: *How could you represent the numbers in a way that would make it easier to see what happens when you count on or back?*

DEEPEN Question ❽ deepens children's reasoning and fluency with number. If children solve the given problem quickly, deepen their reasoning further by suggesting they create their own puzzle, like the one in the question, for a partner.

ASSESSMENT CHECKPOINT Check that children can count on and back in steps of 10, 100, 1,000 and 10,000, fluently from a given number.

Assess children's recognition of place value, looking for clear understanding of how this is affected by counting on and back in steps of 10, 100, 1,000 and 10,000.

ANSWERS Answers for the **Practice** part of the lesson appear in the separate **Practice and Reflect answer guide**.

Reflect

WAYS OF WORKING Independent thinking

IN FOCUS This **Reflect** question will give an opportunity to assess children's fluency and reasoning with the numbers they have been learning about. They should be able to identify that counting in 10,000s will be quicker because the steps are much larger. Encourage children to prove their ideas in a concrete or pictorial way.

ASSESSMENT CHECKPOINT Look for recognition from children that they would need to count 1,000 hundreds but only 10 ten thousands.

ANSWERS Answers for the **Reflect** part of the lesson appear in the separate **Practice and Reflect answer guide**.

After the lesson ⏸

- Were all children equally able to count in all of the different steps confidently?
- How was children's conceptual understanding of the concepts developed in this lesson?
- Were children offered concrete, pictorial and abstract opportunities to access the learning?

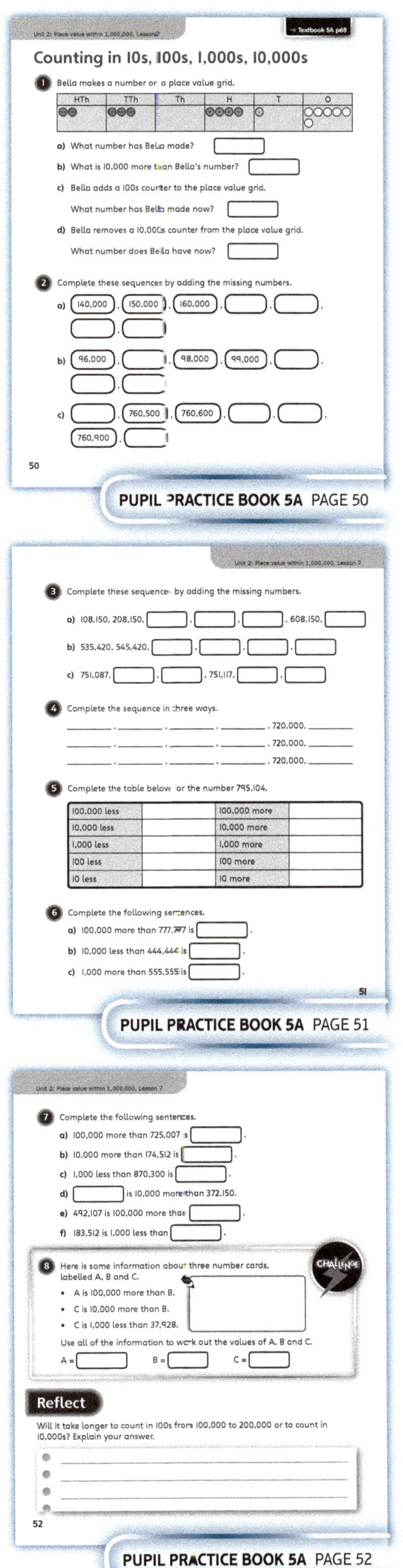

PUPIL PRACTICE BOOK 5A PAGE 50

PUPIL PRACTICE BOOK 5A PAGE 51

PUPIL PRACTICE BOOK 5A PAGE 52

Number sequences

Learning focus

In this lesson, children will use their understanding of numbers and number patterns to recognise and complete sequences. They will develop their ability to find the rule to a given number sequence.

Small steps

→ Previous step: Counting in 10s, 100s, 1,000s, 10,000s
→ **This step: Number sequences**
→ Next step: Adding whole numbers with more than 4 digits (1)

NATIONAL CURRICULUM LINKS

Year 5 Number – Number and Place Value

Solve number problems and practical problems that involve [reading, writing, ordering and comparing numbers to at least 1,000,000 and determining the value of each digit].

ASSESSING MASTERY

Children can read and recognise the patterns within a number sequence. They can use their understanding of and fluency with number to discern and communicate the rule or rules in a number sequence.

COMMON MISCONCEPTIONS

Children may only look for additions within a sequence. This may mean that, where a sequence involves multiplying, children record it as a constantly changing addition. For example, the rule for the sequence 2, 4, 8, 16 may be recorded as + 2, + 4, + 8 instead of × 2. Ask:
• *Is there a way of finding one rule that matches this pattern every time? What else do you notice about how the numbers change?*

STRENGTHENING UNDERSTANDING

Give children opportunities to count through number sequences that are familiar to them, for example their times-tables. Ask: *How could you represent these number patterns in different ways? What is the same about each step in this number sequence? What is different?*

GOING DEEPER

Encourage children to write their own number sequences for a partner to finish. For children who want to be further challenged, ask: *Can you create a number sequence where the number is changed twice before the next step in the pattern, for example + 5 × 2?*

KEY LANGUAGE

In lesson: pattern, sequence

RESOURCES

Optional: cubes, printed number lines, squared paper

 In the eTextbook of this lesson, you will find interactive links to a selection of teaching tools.

Before you teach 

• How could you link the learning in this lesson to number patterns children will have met before, for example multiplication tables?
• How can you use children's previous learning in this unit to deepen their experiences in this lesson?

Discover

 Pair work

ASK

- Question ❶ a): *How can you find out how many tiles there are in pattern 5? Do you need to know pattern 4 first? Why? What is happening to the numbers every time?*
- Question ❶ b): *What would be the quickest way of finding which pattern number has 42 tiles? Explain your answer. Design a similar question for your partner.*

IN FOCUS Question ❶ a) gives children an opportunity to begin experimenting with number sequences, including predicting how the number sequence will continue. Question ❶ b) demonstrates an inverse way of investigating a number sequence, by giving children the number of tiles in the pattern rather than the number of the pattern itself.

PRACTICAL TIPS Encourage children to make the pattern from cubes or shade in squares on squared paper. Discuss with children what changes and what stays the same each time they make a new shape. Challenge them to use what they have made to predict the next shape and the number of blocks required.

ANSWERS

Question ❶ a): Olivia and Ebo will need 27 tiles to make pattern number 5.

Question ❶ b): Pattern number 8 has 42 tiles.

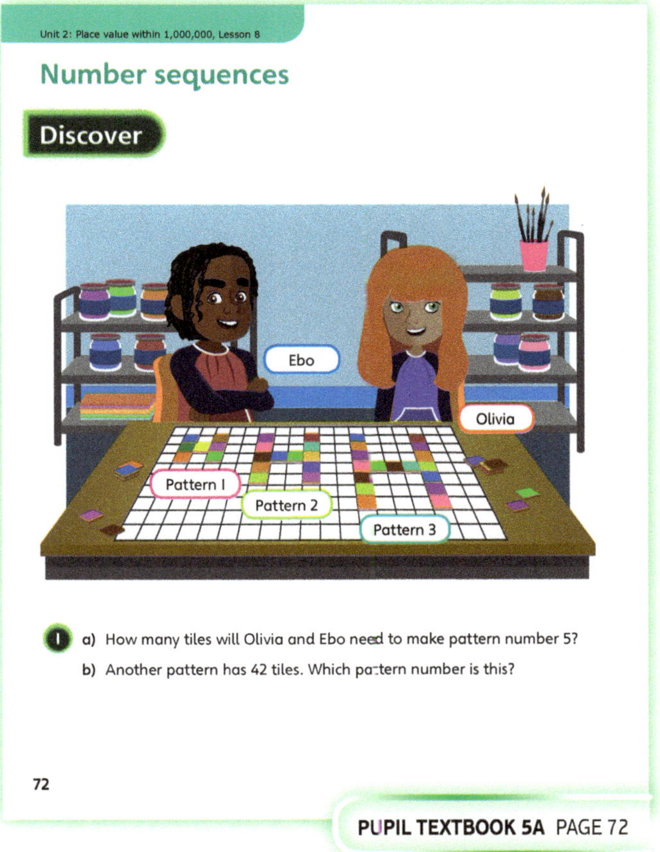

PUPIL TEXTBOOK 5A PAGE 72

Share

 Whole class teacher led

ASK

- Question ❶ a): *Why was it important to complete pattern 4 before finding pattern 5? How else could you represent the sequence?*
- Question ❶ b): *How does the picture make the sequence clear? Did you find any other ways of recording the pattern? Explain your method.*

IN FOCUS Questions ❶ a) and ❶ b) are included to ensure that children are aware of how important it is to identify what is the same and what is different about number sequences when trying to find the rule. Encourage children to recognise that their ability to spot this is essential when trying to find a general rule.

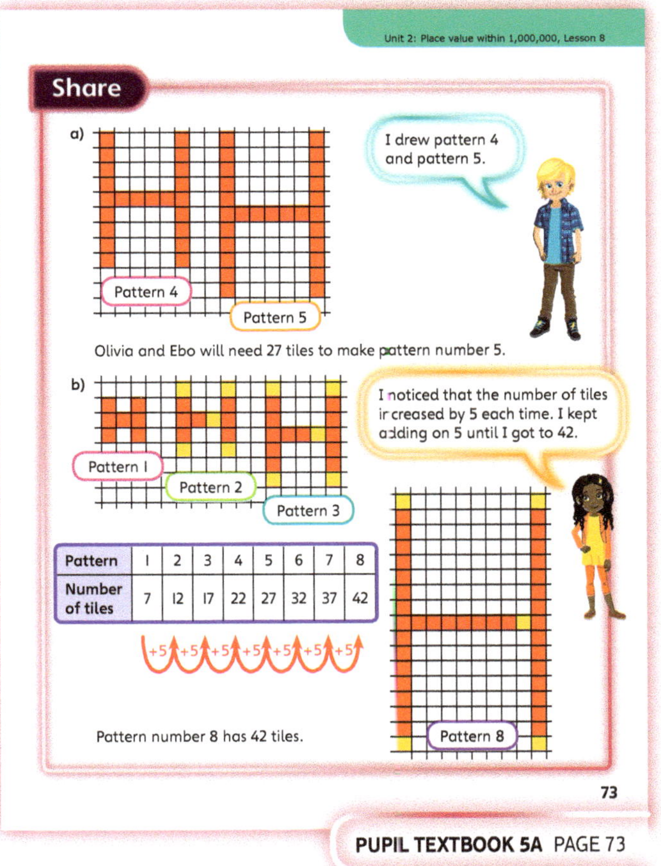

Pattern	1	2	3	4	5	6	7	8
Number of tiles	7	12	17	22	27	32	37	42

+5 +5 +5 +5 +5 +5 +5

PUPIL TEXTBOOK 5A PAGE 73

Think together

WAYS OF WORKING Whole class teacher led (I do, We do, You do)

ASK

- Question ❶ a): *What do you need to know to be able to find the fifth number?*
- Question ❶ b): *How will you know when to stop following the pattern?*
- Question ❷ : *What will help you complete the sequences? What is the same in each sequence? What is different? In what ways could you describe the sequences?*

IN FOCUS Question ❶ offers children the opportunity to work with number sequences in a concrete way. It will be important to encourage children to build the pattern with resources. This will help them to spot what is the same and what is different about each step in the sequence. Question ❷ offers children the opportunity to work with abstract number sequences, including sequences involving negative numbers.

STRENGTHEN To strengthen understanding when completing the sequences in question ❷, encourage children to build the numbers as multilink towers. This will help to emphasise the difference between each step in the pattern. For the sequences that include negative numbers, ask: *Show where these numbers are on a number line. How does the number line help you find the rule for the sequence?*

DEEPEN If children solve question ❸, use Ash's comment about forming other letters with tiles to encourage children to investigate different number patterns. Encourage children to generalise by asking: *What letters result in what numbers being added? (for example, which letters add 3 each time or 4 each time?). Why do you think this happens?*

ASSESSMENT CHECKPOINT Assess children's ability to spot number sequences in concrete representations of patterns. Are children able to find the rule that governs a number sequence and use it to find later numbers in the sequence? Check that children can complete abstract number sequences and identify the rules that govern them.

ANSWERS

Question ❶ a): 13 tiles are needed to make pattern number 5.

Question ❶ b): Pattern 6 needs 15 tiles.

Question ❷ a): 6, 10, 14, **18, 22, 26** – add 4 each time

Question ❷ b): 23, 26, 29, 32, **35, 38, 41** – add 3 each time

Question ❷ c): 800, 750, 700, 650, **600, 550, 500** – subtract 50 each time

Question ❷ d): 10, 7, 4, **1, ⁻2, ⁻5** – subtract 3 each time

Question ❷ e): ⁻11, ⁻7, ⁻3, **1, 5, 9** – add 4 each time

Question ❸ : The sequence is: 11, 16, 21, 26, 31, 36, 41, 46, 51 – add 5 each time

Pattern 9 will be the first to have more than 50 tiles.

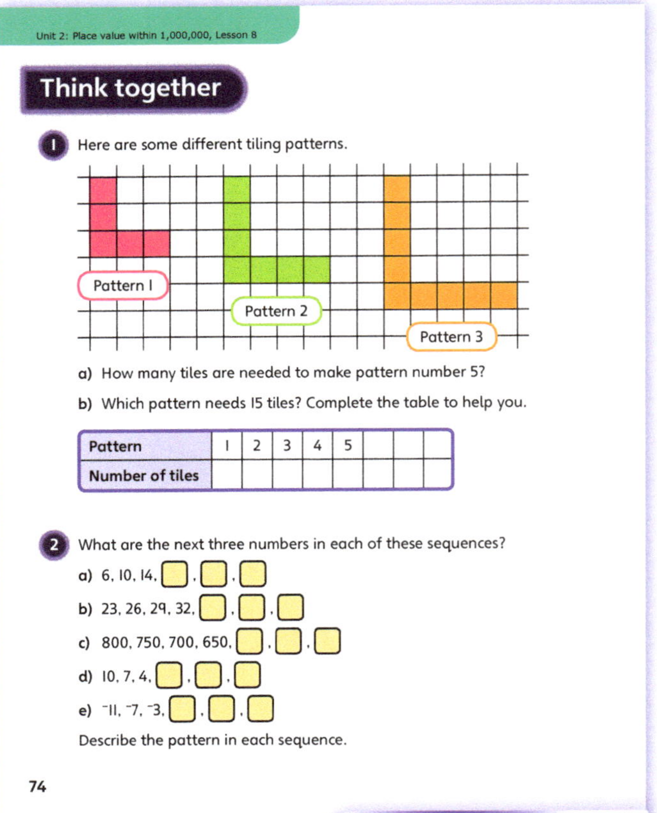

PUPIL TEXTBOOK 5A PAGE 74

PUPIL TEXTBOOK 5A PAGE 75

Practice

WAYS OF WORKING Independent thinking

IN FOCUS Question ❶ gives children an opportunity to independently investigate a number sequence that has been shown pictorially. Providing children with the resources to create these patterns will scaffold their concrete understanding. Question ❷ offers children the opportunity to show their understanding of number sequences by diagnosing and explaining the mistakes made by another person. In question ❹, children will see number sequences presented in ways they may not yet have considered. Representing the sequence as pentagons will potentially bring children out of their comfort zone, having experienced mainly rectilinear shapes so far.

STRENGTHEN Physically making the pentagons will help children identify the pattern in question ❹. This can be done with art straws or lolly sticks. Ask: *What is the same and what is different about each step in the sequence? Can you show the pattern in a chart or highlight the numbers along a number line?*

DEEPEN Question ❻ offers children the opportunity to work with a number pattern that initially looks complicated. Encourage children to record their findings in a systematic way. Ask: *How can you make sure you have followed the pattern carefully? What is the rule for this number sequence?*

THINK DIFFERENTLY Question ❺ requires children to work with larger numbers in number sequences and to work further through the sequence. To deepen children's understanding of this, ask: *How did you find the solution? Prove that you are correct, using resources or a picture.*

ASSESSMENT CHECKPOINT Look for recognition of number sequences in concrete and pictorial representations. Can children follow these rules to find other numbers in the sequence? Check that children can identify the rule of an abstract number sequence and use this to follow and complete it.

ANSWERS Answers for the **Practice** part of the lesson appear in the separate **Practice and Reflect answer guide**.

Reflect

WAYS OF WORKING Independent thinking

IN FOCUS This **Reflect** question will help you to quickly assess whether children have recognised the key feature of a number sequence – that it should change in the same way with each step.

ASSESSMENT CHECKPOINT Look for children's ability to successfully write a number sequence that works. To further help with your assessment of their ability, encourage them to try to solve their partner's sequence.

ANSWERS Answers for the **Reflect** part of the lesson appear in the separate **Practice and Reflect answer guide**.

After the lesson ⏸

- Did children work systematically to find the rule of a sequence and use it to find later numbers in the pattern?
- Can children work with patterns that use both rectilinear shapes and polygons?
- Are children able to record their findings of number sequences in a table?

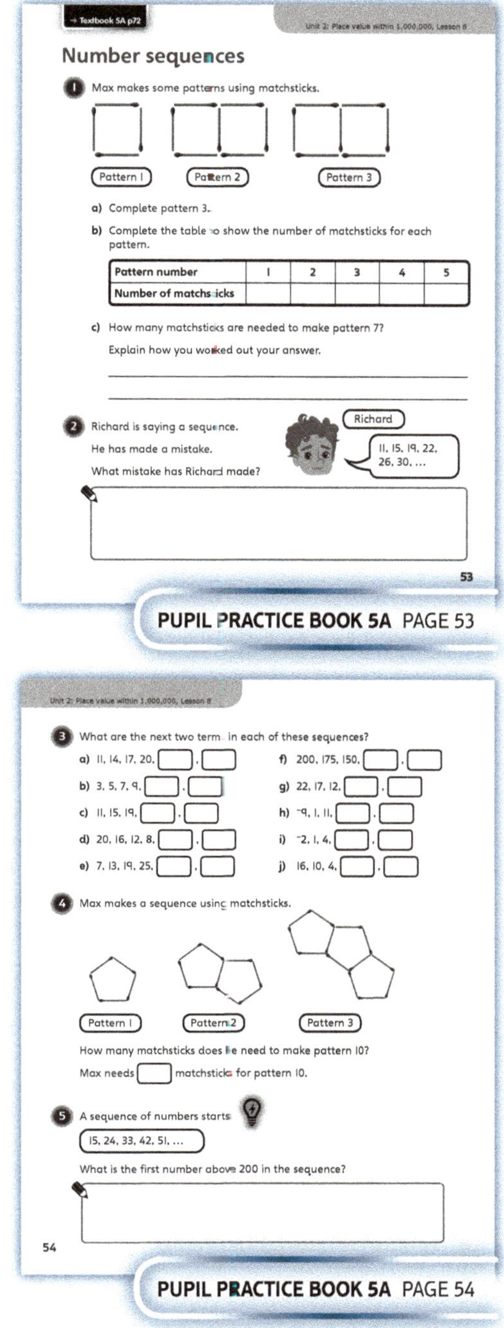

PUPIL PRACTICE BOOK 5A PAGE 53

PUPIL PRACTICE BOOK 5A PAGE 54

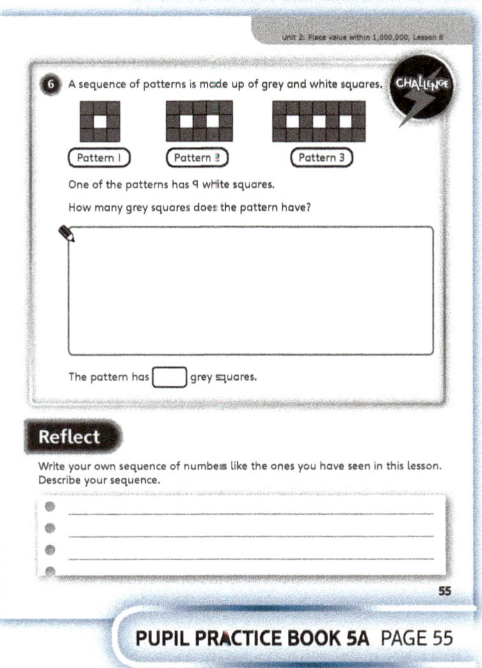

PUPIL PRACTICE BOOK 5A PAGE 55

End of unit check

Don't forget the *Power Maths* unit assessment grid on p26.

WAYS OF WORKING Group work adult led

IN FOCUS These questions are designed to draw out misconceptions or misunderstandings.

- Question **1** assesses children's recognition of the place value of digits in a number up to 1,000,000.
- Question **5** assesses children's ability to recognise numbers up to 1,000,000 along an unlabelled number line.
- Question **6** is a SATs-style question and requires children to use their understanding of the rules for rounding and how these are linked to the properties of numbers.
- Question **7** is a SATs-style question and assesses children's ability to calculate with positive and negative numbers.

ANSWERS AND COMMENTARY

Children who have mastered the concepts in this unit will be able to demonstrate fluency in place value within numbers up to 1,000,000. They will name, partition and write the names of numbers accurately and compare and order numbers up to 1,000,000 with confidence. They will complete partial number lines, using the unlabelled marks to help them calculate what each interval is worth. They will round up and down to the nearest multiple of 10 and will use negative numbers up to ⁻1,000,000. They will identify and use the rules that govern number sequences.

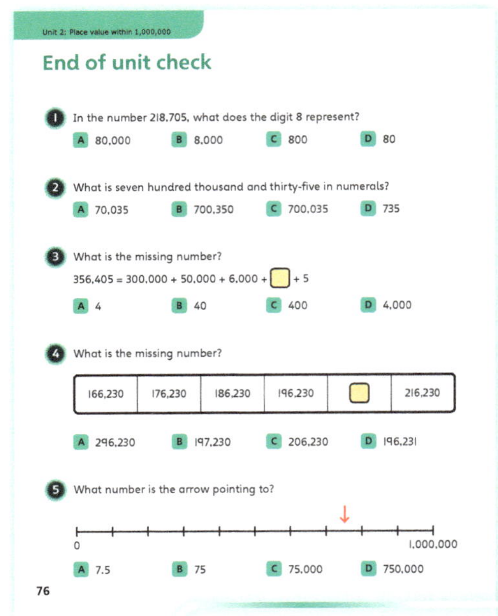

PUPIL TEXTBOOK 5A PAGE 76

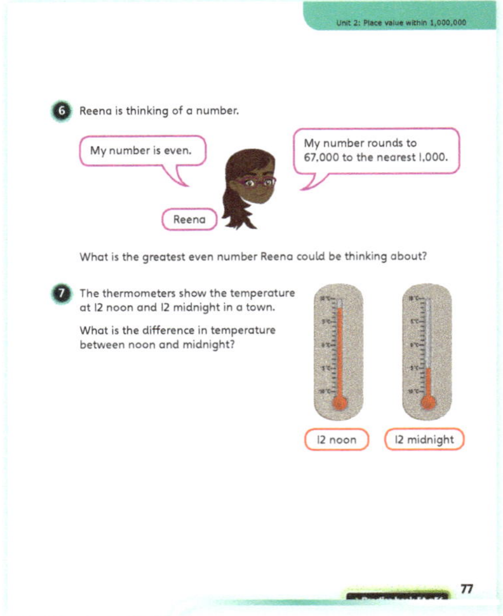

PUPIL TEXTBOOK 5A PAGE 77

Q	A	WRONG ANSWERS AND MISCONCEPTIONS	STRENGTHENING UNDERSTANDING
1	B	Incorrect choices indicate children have not mastered the unit.	In questions 1, 2 and 3, allow children to build numbers using manipulatives including place value arrow cards, grids and place value counters.
2	C	A and D suggest children are struggling to recognise and understand the written names of numbers. B suggests difficulties with place value.	
3	C	A, B or D indicate mistakes when recombining partitioned numbers.	When finding the difference between a negative number and any other number, offer children who need support the opportunity to use a number line to help them. Ask:
4	C	A and B suggest recognising the repetition of 230 without knowing the reason. D suggests counting on 1.	
5	D	Incorrect options may indicate an issue with place value.	
6		**67,498.** An incorrect answer will indicate children are not clear on the rules for rounding.	• *Can you find the two numbers on the number line?* • *Can you count the difference between them?*
7		**13 °C** An incorrect answer indicates children are miscalculating the difference between positive and negative numbers.	

My journal

WAYS OF WORKING Independent thinking

ANSWERS AND COMMENTARY

A number between 250,000 and 350,000.	Solutions such as: 315,689, 315,896, 316,589, …
A number that has a smaller number of 100s than 10,000s.	Solutions such as: 296,153, 962,153, 965,213, …
The greatest even number that can be made.	985,316
A number that rounds to 600,000 to the nearest 100,000.	Any number between 550,000 and 649,999
The smallest number that rounds to 600,000 to the nearest 100,000.	561,389
The number that is 10,000 less than 875,913.	865,913

To strengthen children's understanding, ask:
- *If the answer must be between 250,000 and 350,000, what does that mean for the 100,000s digit?*
- *Are there any clues where there is only one solution? Can you prove it?*

Power check

WAYS OF WORKING Independent thinking

ASK

- *How confident are you with the place value of numbers up to 1,000,000?*
- *Could you explain to someone how to round numbers up to 1,000,000?*
- *Are you confident to compare and order numbers to 1,000,000?*
- *What are negative numbers?*

Power puzzle

WAYS OF WORKING Pair work

IN FOCUS Use this Power puzzle to assess whether children can recognise and follow number sequences. Understanding that number sequences follow a rule will help children to see that the numbers in a sequence will change in a regular way.

ANSWERS AND COMMENTARY Can children recognise consistent and regular number sequences in the given cards? The sequences are: ¯4, 2, 8, 14, 20 and 16, 13, 10, 7, 4, 1 or 1, 4, 7, 10, 13, 16 and 20, 14, 8, 2, ¯4

If children need support, ask:
- *Which number is likely to be the first number in the sequence that gets bigger? Why?*

After the unit ⏸

- How confident are you that children have a deep conceptual understanding of the place value of 6-digit numbers?
- What one strategy or teaching approach worked particularly well in this unit? How can you apply the same strategy in other areas of your mathematics teaching?

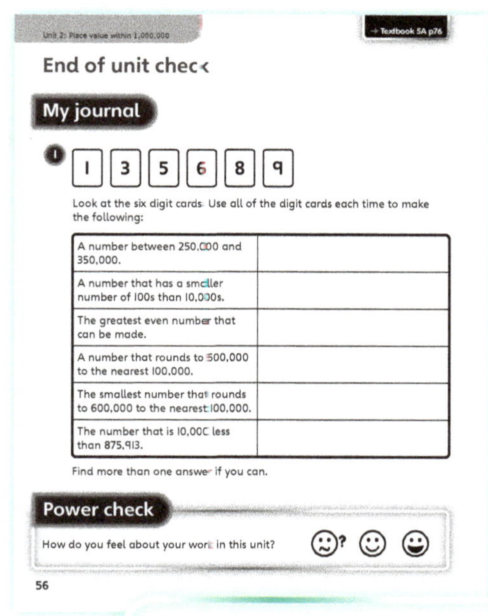

PUPIL PRACTICE BOOK 5A PAGE 56

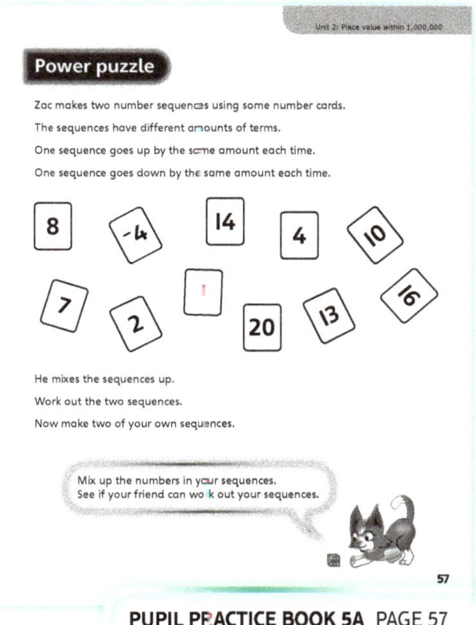

PUPIL PRACTICE BOOK 5A PAGE 57

Strengthen and **Deepen** activities for this unit can be found in the *Power Maths* online subscription.

Unit 3
Addition and subtraction

Mastery Expert tip! "To build confidence and recall with addition and subtraction methods, I encourage adding and subtracting in different areas of the curriculum throughout the year! It definitely helps children become more familiar and confident with the method."

Don't forget to watch the Unit 3 video!

WHY THIS UNIT IS IMPORTANT

This unit is important because it allows children to apply the formal methods of addition and subtraction to numbers with up to 5 digits. The range of problem solving questions involving adding and subtracting, including mentally, will develop confidence and flexibility when exploring the most efficient ways to add and subtract.

WHERE THIS UNIT FITS

→ Unit 2: Place value within 1,000,000
→ **Unit 3: Addition and subtraction**
→ Unit 4: Graphs and tables

This unit builds on children's work in Year 4, adding and subtracting 4-digit numbers. It will extend their knowledge of addition and subtraction using formal methods for numbers with up to 5 digits. It will give children the opportunity to build confidence with problem solving and to explore efficient methods for addition and subtraction calculations, including those that can be solved mentally.

Before they start this unit, it is expected that children:
- can add and subtract numbers with up to 4 digits
- are able to solve sum and difference word problems
- have experience of using inverse operations to check calculations
- are able to round numbers to the nearest 10, 100, 1,000 and 10,000.

ASSESSING MASTERY

Children who have mastered this unit can add and subtract numbers with up to 5 digits using a variety of methods, including formal written methods and mental methods. They can confidently apply their knowledge of addition and subtraction to solving word problems.

COMMON MISCONCEPTIONS	STRENGTHENING UNDERSTANDING	GOING DEEPER
When adding or subtracting numbers with a different number of digits, children may line the numbers up from the left, not the right.	Provide place value grids and counters to help children visualise when they will need to exchange.	Compare calculations laid out correctly and incorrectly. What is the difference between the correct and incorrect answer? Why?
When solving a problem, children may choose the wrong calculation. For example, adding when they should subtract.	Use a bar model to help children understand whether a calculation requires addition or subtraction.	Ask increasingly difficult word problems, including ones that bring in other areas of the mathematics curriculum.
Children may think they need to rely on written methods, even when a mental method might be more efficient.	Ensure that children know the bonds to any number within 10 to help them mentally add and subtract quickly.	Ask children to justify the most efficient way of solving addition and subtraction problems. This is not always the column method.

Unit 3: Addition and subtraction

WAYS OF WORKING

Go through the unit starter pages of the textbook with the whole class. Talk through the key learning points that the characters mention and the key vocabulary. Discuss Dexter's comment about the importance of setting out column addition neatly.

STRUCTURES AND REPRESENTATIONS

Place value grid: This model uses counters to show the value of each column, which supports the column method layout.

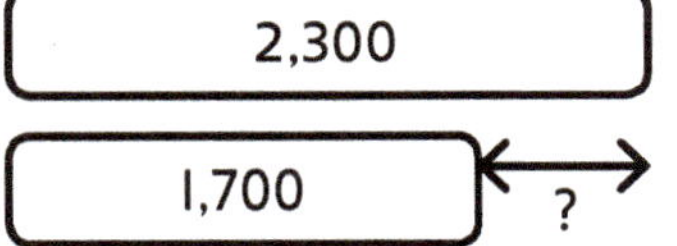

Column method addition and column method subtraction: This model demonstrates the place value of each digit in addition and subtraction calculations and shows exchanges between columns.

```
 TTh Th  H   T   O        TTh Th  H   T   O
   1   6   9   9   8        ⁷8̸  ¹2  ⁶7̸  ¹0   6
 +     2   1   5   6      -   3   9   4   1   5
 ─────────────────        ─────────────────
   1   9   1   5   4          4   3   2   9   1
       |   |   |
```

Bar model: This model can be used to represent the situation in some addition and subtraction word problems.

```
┌─────────────────────┐
│       2,300         │
└─────────────────────┘
┌──────────────┐
│    1,700     │  ←──→
└──────────────┘    ?
```

KEY LANGUAGE

There is some key language that children will need to know as part of the learning in this unit:

➜ add, subtract
➜ 1s (ones), 10s (tens), 100s (hundreds), 1,000s (thousands), 10,000s (ten thousands)
➜ total
➜ difference
➜ inverse
➜ round
➜ mentally
➜ estimate

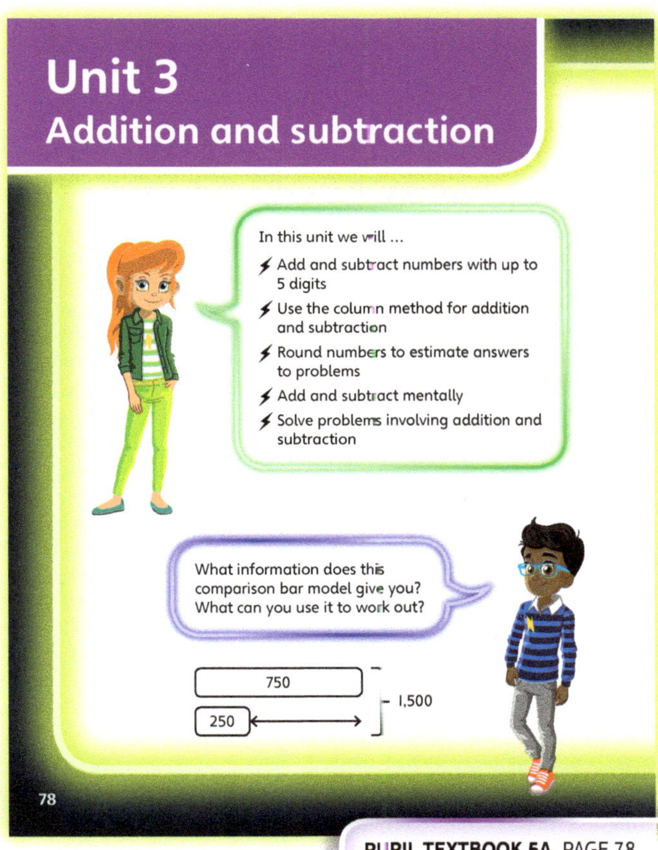

PUPIL TEXTBOOK 5A PAGE 78

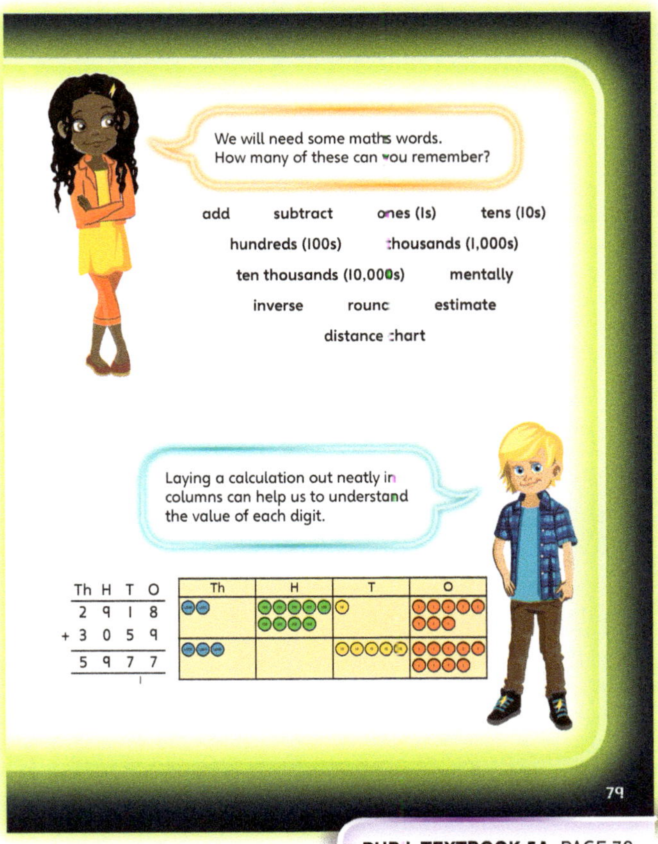

PUPIL TEXTBOOK 5A PAGE 79

Adding whole numbers with more than 4 digits ❶

Learning focus

In this lesson, children will use the written column method to add whole numbers with more than 4 digits, recognising the importance of place value.

Small steps

→ Previous step: Number sequences

→ **This step: Adding whole numbers with more than 4 digits (1)**

→ Next step: Adding whole numbers with more than 4 digits (2)

NATIONAL CURRICULUM LINKS

Year 5 Number – Addition and Subtraction
Add and subtract whole numbers with more than 4 digits, including using formal written methods (columnar addition and subtraction).

ASSESSING MASTERY

Children can use the written method of column addition to add whole numbers with more than 4 digits.

COMMON MISCONCEPTIONS

Children may not know which place value column to start with when adding two whole numbers together. Ask:
• *Why do you always start by adding the column that has the smallest place value in this method?*

Children may not understand the concept of exchanging between columns. Ask:
• *How might a place value grid and counters help you to see what is happening?*

STRENGTHENING UNDERSTANDING

Children should first practise adding whole numbers with 2 or 3 digits before moving on to whole numbers with more than 4 digits. Encourage them to clearly describe the place value of each column, hundreds, tens and ones, and make sure they understand the importance of this, in particular when making an exchange.

GOING DEEPER

Give children a total and ask how many different ways they can make the total, for example _ + _ = 8,876. This could also be represented on a part-whole model to help children to see the link between adding and subtracting.

KEY LANGUAGE

In lesson: add, total, digit, column, place value

Other language to be used by the teacher: exchange, ones, tens, hundreds, thousands, ten thousands

STRUCTURES AND REPRESENTATIONS

place value grid, column method addition

RESOURCES

Mandatory: place value counters

 In the eTextbook of this lesson, you will find interactive links to a selection of teaching tools.

Before you teach ⏸

• Do children know how to add 2- and 3-digit numbers?
• Do children know how to make an exchange when using the column addition method?

Discover

WAYS OF WORKING Pair work

ASK

- Question ❶ a): *How many views are there on Tuesday? How many views are there on Wednesday? What does total mean? How could you add these amounts together?*
- Question ❶ b): *What method could you use to find which two days make this total?*

IN FOCUS Question ❶ a) requires children to identify numbers and find the total of them. This calculation requires children to make one exchange when using the column addition method. Question ❶ b) gives children a total and asks them to work out which two numbers make this total.

PRACTICAL TIPS Make sure children understand that Richard is watching videos and can see the amount of views for each video. Show a video on a real video sharing website and ask children to point out the number of views.

ANSWERS

Question ❶ a): The total number of video views for Tuesday and Wednesday is 39,328.

Question ❶ b): Wednesday and Friday have the total views of 37,592.

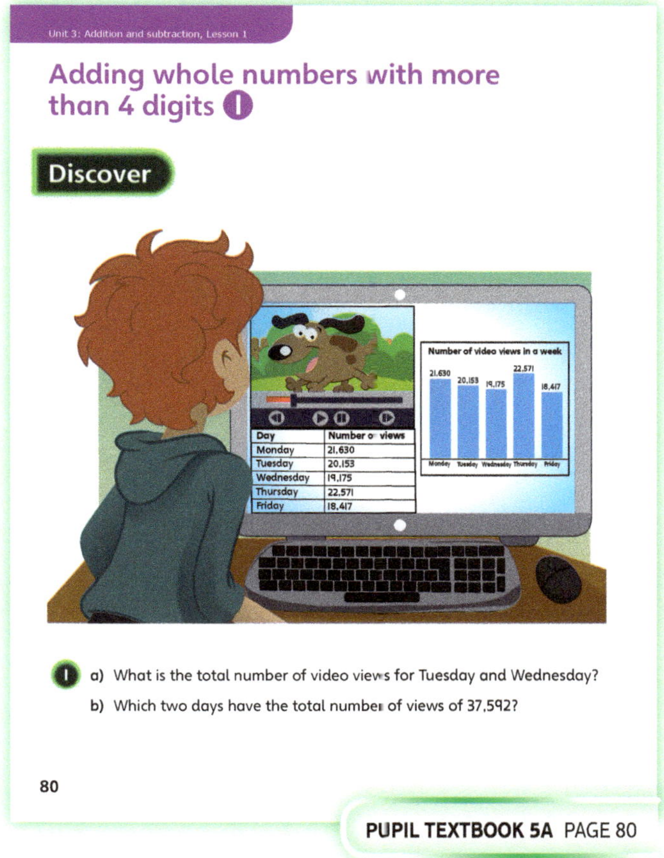

PUPIL TEXTBOOK 5A PAGE 80

Share

WAYS OF WORKING Whole class teacher led

ASK

- Question ❶ a): *What method can you use to add the two numbers together?*
- Question ❶ a): *Which place value column do you need to start with? Do you need to make an exchange?*
- Question ❶ b): *What method could you use to work out which two numbers make that total? Is there a quicker way?*

IN FOCUS For question ❶ a), take the opportunity to discuss how the word 'total' leads us to carry out an addition for this question and make sure children know which place value we begin with when adding. Check that children are able to identify and say the larger numbers correctly. The place value grids can be used to reinforce the place value of each digit when carrying out the calculation, for example: *What is 3 ones add 5 ones? What is 5 tens add 7 tens?* Demonstrate why this is important when children are required to carry out one exchange, of 10 tens for 1 hundred.

Discuss the use of the trial and improvement method in question ❶ b), emphasising the need to not miss any calculations out. Draw out that this will be time-consuming with such big numbers and encourage children to think flexibly about using a different strategy. For example, Flo's method of adding just the ones of each number instead to see which gives a total ending with 2.

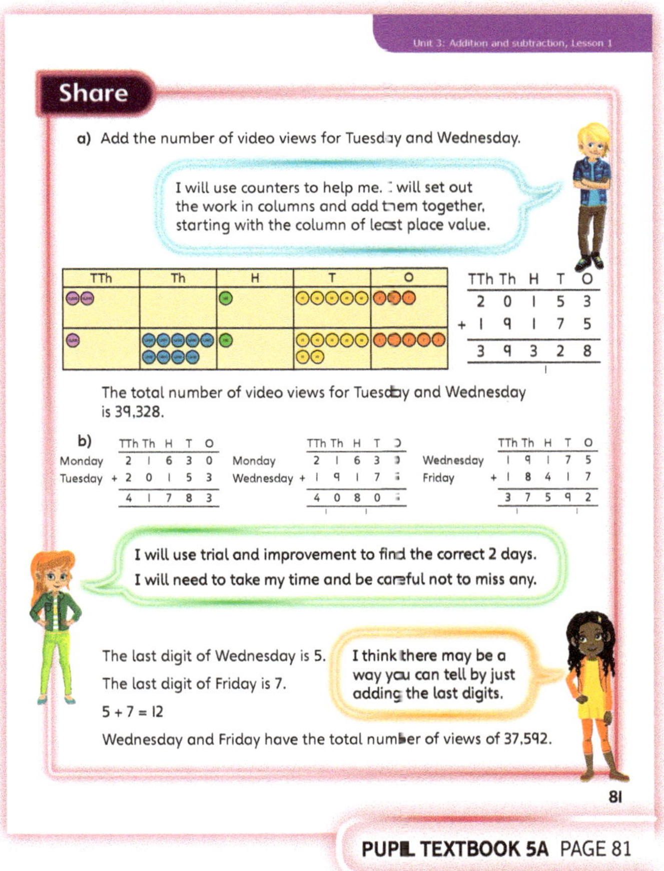

PUPIL TEXTBOOK 5A PAGE 81

Think together

WAYS OF WORKING Whole class teacher led (I do, We do, You do)

ASK

- Question **1** : *How many views are there on Saturday? How many views are there on Friday? What method can you use to add these amounts together?*
- Question **2** : *Do you need to make any exchanges for this addition?*
- Question **3** : *How do you set numbers out in columns when they have different amounts of digits?*

IN FOCUS In question **3** children may choose to add two numbers that do not have the same amount of digits. Make sure they lay the column addition out correctly. When working out which two numbers have made a given total, encourage children to look at the last digit in each number instead of carrying out the full calculation.

STRENGTHEN Support understanding in question **3** by representing calculations using counters on a place value grid.

DEEPEN For question **3** ask children to work out the total for other combinations of numbers. Encourage them to add numbers that have a different amount of digits.

ASSESSMENT CHECKPOINT Can children use the written column method to add whole numbers with 4 or more digits where one or more exchanges are required? Make sure they pay attention to laying it out neatly and accurately and identifying the importance of the place value of each column.

ANSWERS

Question **1** : 22,571 + 18,417 = 40,988

The total number of views is 40,988.

Question **2** : 1,564 + 18,417 = 19,981

The total number of views is 19,981.

Question **3** : Children should work out any two from the following:

34,171 + 61,426 = 95,597

34,171 + 5,458 = 39,629

34,171 + 1,023 = 35,194

61,426 + 5,458 = 66,884

61,426 + 1,023 = 62,449

5,458 + 1,023 = 6,481

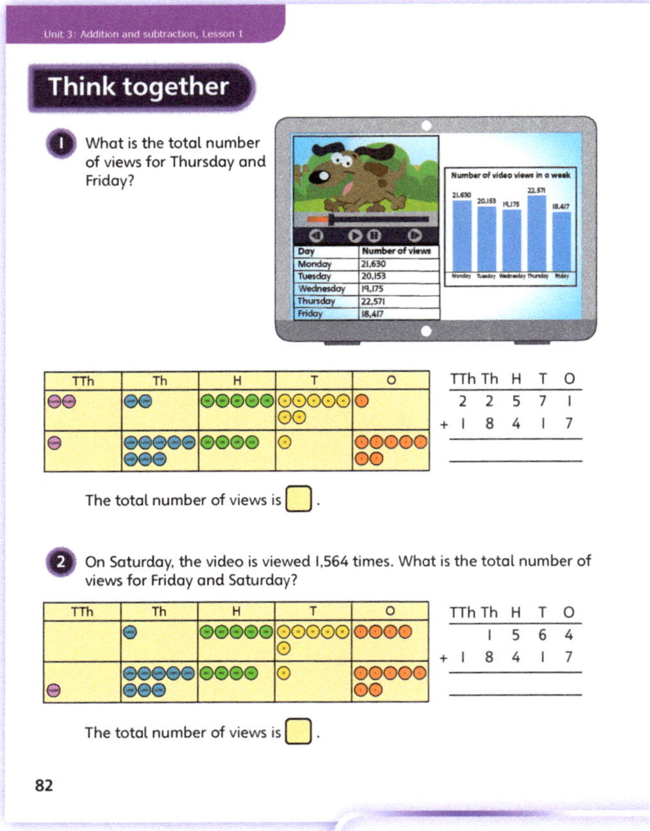

PUPIL TEXTBOOK 5A PAGE 82

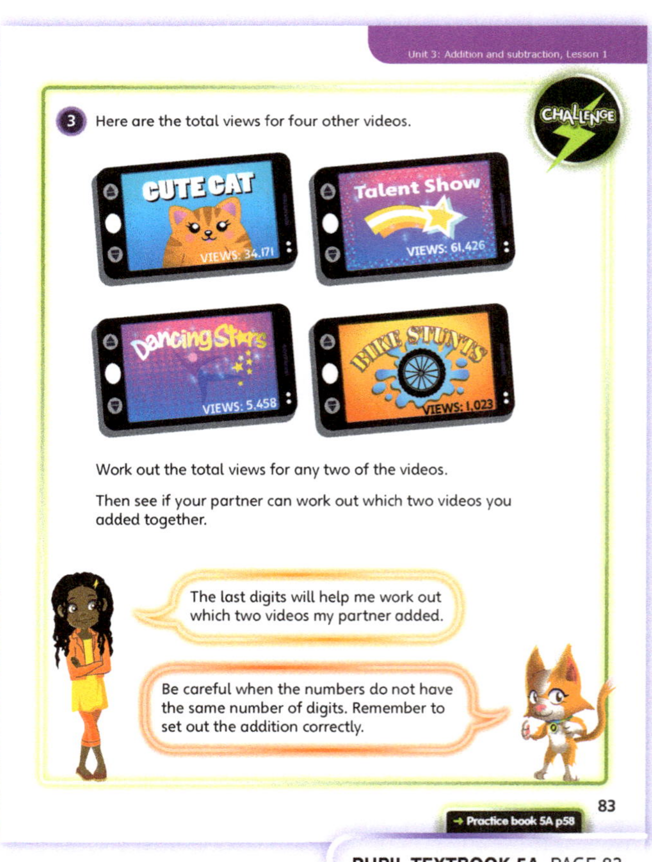

PUPIL TEXTBOOK 5A PAGE 83

Practice

WAYS OF WORKING Independent thinking

IN FOCUS Questions **1** to **3** consolidate understanding of adding two whole numbers using the column method where the information is represented with counters on a place value grid, in a column and abstractly. Question **4** asks children to problem solve and work out missing digits in addition calculations while linking to subtraction.

Question **5** introduces a context for adding two whole numbers that have different amounts of digits. In question **6** numbers are given in words and children need to write these as numbers in order to work out the total. Some children may be able to work out the answers without using the column method.

STRENGTHEN Encourage children to use counters on a place value grid to support understanding and when the calculation is not given in a column layout, encourage them to write it in columns.

DEEPEN Explore question **4** in more depth by giving other missing number problems. Question **6** can be explored further by saying numbers for children to add together mentally, rather than using a written method.

ASSESSMENT CHECKPOINT Children are confident in using the column layout to add whole numbers with 4 or more digits.

ANSWERS Answers for the **Practice** part of the lesson appear in the separate **Practice and Reflect answer guide**.

Reflect

WAYS OF WORKING Pair work

IN FOCUS This **Reflect** activity checks understanding of adding two whole numbers with 4 or more digits. Encourage children to explain how they would carry out the calculation as well as actually answering it. Look for children who are able to do this without any support.

ASSESSMENT CHECKPOINT Assess if children can correctly explain how to find the total of two whole numbers, emphasising the importance of the place value of each column, and identifying the need to make exchanges.

ANSWERS Answers for the **Reflect** part of the lesson appear in the separate **Practice and Reflect answer guide**.

After the lesson ⏸

- Can children show how to use the column method to add whole numbers with 4 or more digits?
- Do children understand the importance of neat and accurate layout for this method?
- Which children needed to use counters on a place value grid for support?

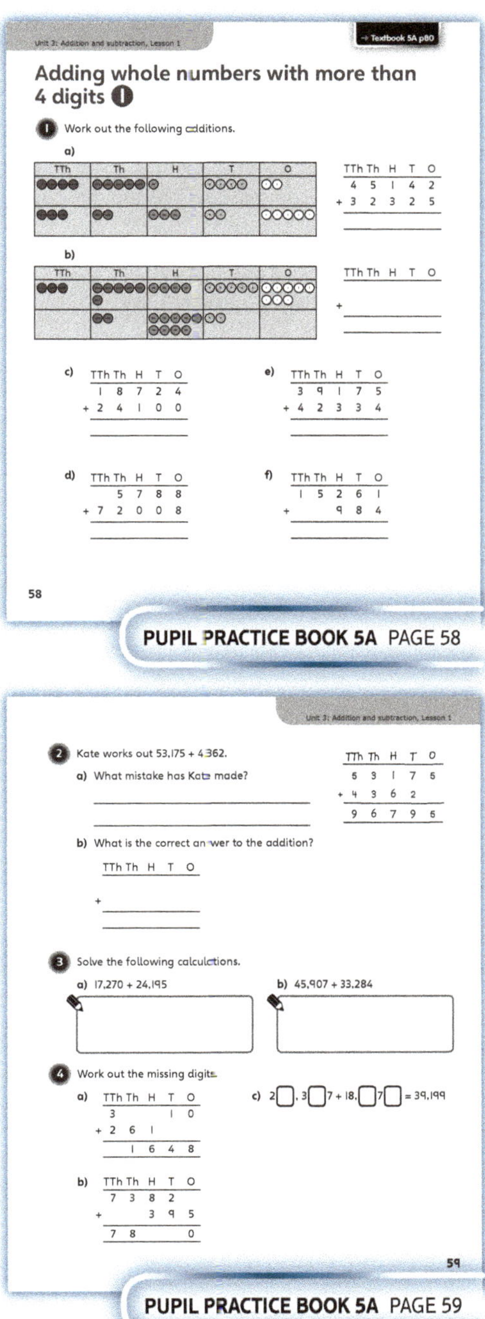

PUPIL PRACTICE BOOK 5A PAGE 58

PUPIL PRACTICE BOOK 5A PAGE 59

PUPIL PRACTICE BOOK 5A PAGE 60

Adding whole numbers with more than 4 digits ②

Learning focus

In this lesson, children will identify large numbers in the context of distance and will use the written column method to add two or more whole numbers with more than 4 digits.

Small steps

→ Previous step: Adding whole numbers with more than 4 digits (1)
→ **This step: Adding whole numbers with more than 4 digits (2)**
→ Next step: Subtracting whole numbers with more than 4 digits (1)

NATIONAL CURRICULUM LINKS

Year 5 Number – Addition and Subtraction
Add and subtract whole numbers with more than 4 digits, including using formal written methods (columnar addition and subtraction).

ASSESSING MASTERY

Children can identify, compare and add two or more whole numbers with more than 4 digits using the written column method, in a real-life context.

COMMON MISCONCEPTIONS

Children may not set the column method out correctly when adding numbers with different amounts of digits. Ask:
• *What is the value of each digit? Are you sure each digit in the same column has the same place value?*

STRENGTHENING UNDERSTANDING

Encourage children to use counters on a place value grid to help them see and understand things in a concrete way. This may be done alongside the column method so children see how the concrete links to the abstract method. Children are encouraged more to write out their own column additions to increase familiarity with the place value columns.

GOING DEEPER

Give children information in different forms such as tables and charts where they must first read and extract the information before adding numbers together. Provide missing number problems, for example 15,239 + _ = 29,034, and encourage children to make the link between addition and subtraction when solving this type of calculation.

KEY LANGUAGE

In lesson: add, total, distance, kilometres (km), metres (m), column, shortest, digit, **distance chart**

Other language to be used by the teacher: place value, ones, tens, hundreds, thousands, ten thousands

STRUCTURES AND REPRESENTATIONS

column method addition, distance chart

RESOURCES

Mandatory: place value counters

 In the eTextbook of this lesson, you will find interactive links to a selection of teaching tools.

Before you teach 

• Do children know how to make an exchange when using the column addition method?
• Can children set out a column addition where the two numbers have different amounts of digits?

Discover

WAYS OF WORKING Pair work

ASK

- Question **1** a): *What information does the pilot's chart show us?*
- Question **1** a): *Which column represents London? Which row represents Sydney? Where do they meet? Can you find 9,385 km on the table?*
- Question **1** b): *What is the distance between London and Auckland? What is the distance between Auckland and Dubai?*

IN FOCUS Question **1** a) helps to ensure that children can use a distance chart to find the relevant information. Question **1** b) asks children to identify numbers and find the total. This calculation requires children to make one exchange.

PRACTICAL TIPS Make sure children are confident in using a distance chart to find and use information. If necessary, provide similar charts with fewer smaller numbers, so that children can practise looking across the rows and down the columns to identify the numbers they need.

ANSWERS

Question **1** a): The distance between London and Sydney is 16,998 km.

Shanghai and Auckland are 9,385 km apart.

Question **1** b): 18,360 + 14,212 = 32,572 km

Holly flies 32,572 km in total.

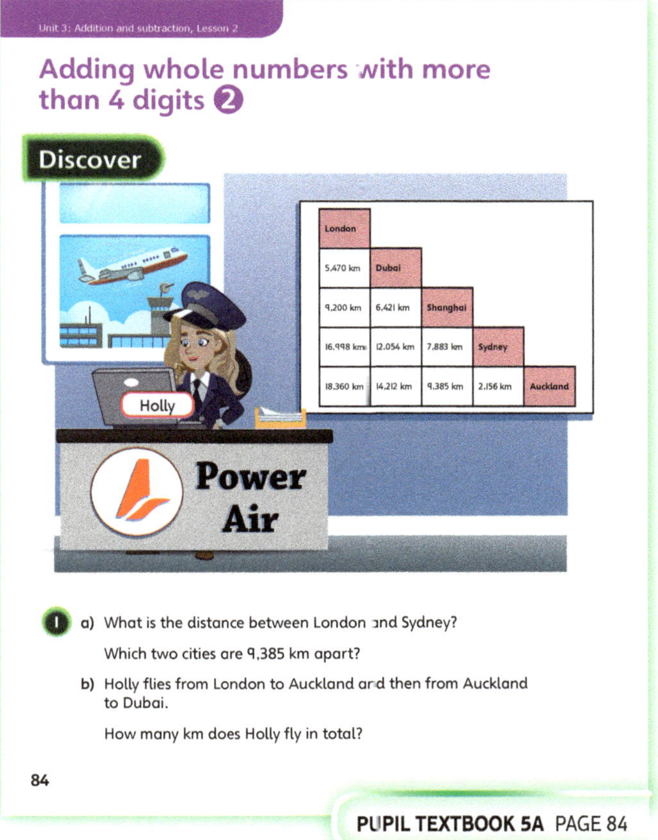

PUPIL TEXTBOOK 5A PAGE 84

Share

WAYS OF WORKING Whole class teacher led

ASK

- Question **1** a): *Look down the column for London, what number do you need to look at so you are in the row for Sydney? Which row and which column is 9,385 km in? Which two cities does this represent?*
- Question **1** b): *What method will you use to work out the total distance? Do you need to make an exchange?*

IN FOCUS Question **1** a) presents information in a real-life context. Children will need to understand the significance of each column and row in the table and use this to draw out the relevant information. Can children explain how they know that the distance between the two cities is 16,998 km, and which two cities are 9,385 km apart?

For question **1** b), show children the column method of addition. They should be able to explain which place value column we start with when adding numbers together and why. Use this question to reinforce the place value of each digit when carrying out the calculation and discuss the exchange that needs to take place, of 10 thousands for 1 ten thousand. Emphasise the use of the correct units of measure (km) in this context.

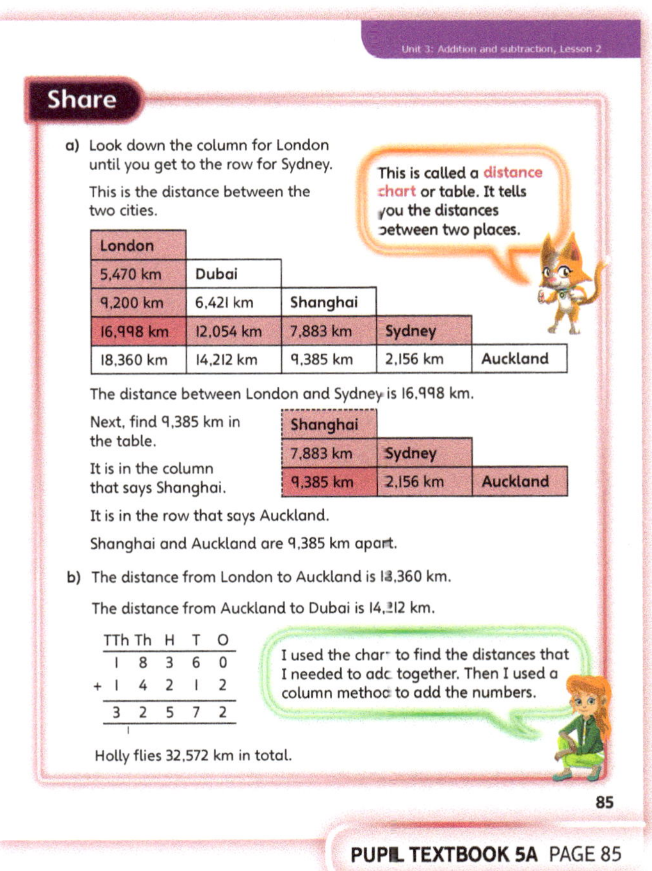

PUPIL TEXTBOOK 5A PAGE 85

Think together

WAYS OF WORKING Whole class teacher led (I do, We do, You do)

ASK

- Question **1** : *What is the distance from London to Sydney? What is the distance from Sydney to Auckland? What method will you use to add these distances together?*
- Question **2** : *How will you find the total distance from London to Auckland for each route? How do you know which route is the shortest distance?*
- Question **3** : *Can you add all three distances using the column method? If you add the first two distances then add on the final distance afterwards, do you still get the same answer?*

IN FOCUS Question **1** looks at adding together whole numbers that require exchanges. Encourage children to use the column method and ensure they lay it out correctly as the two numbers have different amounts of digits. Question **2** requires children to add two whole numbers together involving more than one exchange, and then to compare their answers to say which is smaller. In question **3** children can explore adding more than two whole numbers together. Children may choose to add all three numbers in one go or they may add two together and then add on the final number to get the total. Encourage children to try both ways ensuring they lay out the column addition correctly. Discuss which method children prefer.

STRENGTHEN To support understanding, represent each calculation using counters on a place value grid and display this alongside the abstract calculations.

DEEPEN For question **3**, add in one more city after London and ask children to work out the total distance now. Encourage children to find the answer using more than one method.

ASSESSMENT CHECKPOINT Can children explore the best method to use to add two or more whole numbers with 4 or more digits, including those with more than one exchange? Can they compare their answers to explain which is bigger and which is smaller?

ANSWERS

Question **1** : Mo flies 19,154 km in total.

Question **2** : The total distance of Route 1 is 19,682 km.

The total distance of Route 2 is 18,585 km.

David should choose Route 2.

Question **3** a): Ebo travels 34,166 km in total.

Question **3** b): Both methods give the same answer.

Question **3** c): This will be each child's personal choice.

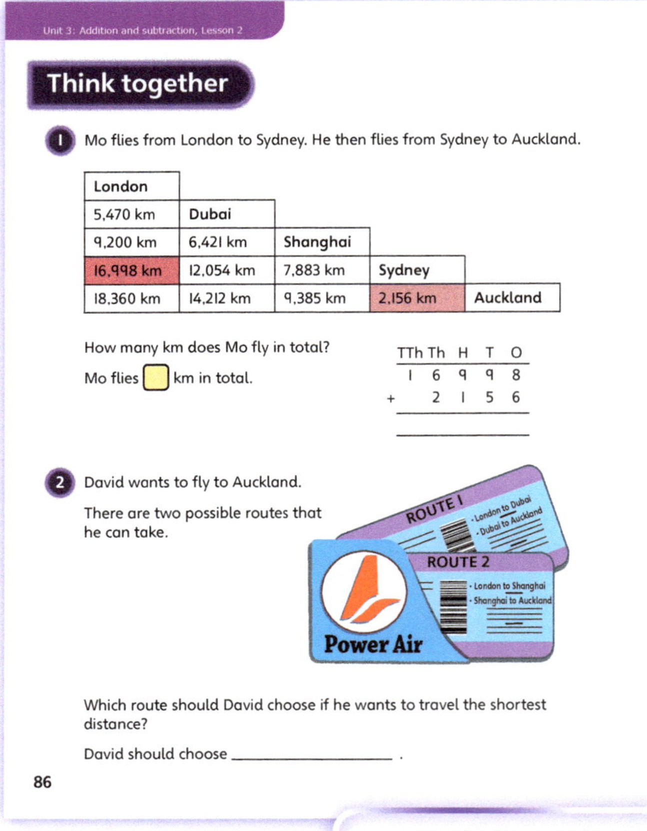

PUPIL TEXTBOOK 5A PAGE 86

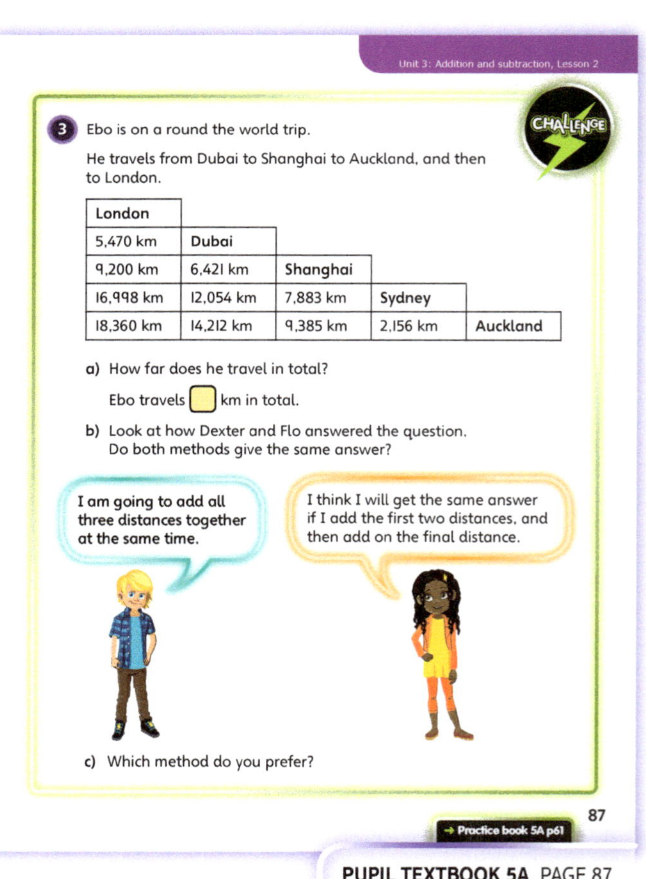

PUPIL TEXTBOOK 5A PAGE 87

Practice

WAYS OF WORKING Independent thinking

IN FOCUS Questions ❶ and ❷ aim to consolidate children's understanding of adding two whole numbers together using the column method where the information is represented with counters on a place value grid, in a column and abstractly. In question ❸, children will need to use their preferred method to add together three whole numbers, and then decide if the total is less than or more than 6,000 metres.

STRENGTHEN Encourage children to use counters on a place value grid to support their understanding. They should write each calculation in columns, paying close attention to laying this out carefully and accurately. They can use this to help write out their own column additions.

DEEPEN Question ❻ is more open-ended and allows children to look at different ways to make totals. Explore this further by giving other totals, but where the digit cards may be used more than once. Encourage children to start with the ones column and think about which digits they could put there to make the total of the ones column.

THINK DIFFERENTLY Question ❺ encourages children to problem solve and work out missing digits in addition calculations. They will need to make the link between addition and subtraction to solve them.

ASSESSMENT CHECKPOINT Are children confident in using the column addition method to add two or more whole numbers with 4 or more digits, including those with exchanges? For questions ❺ and ❻, do they understand how they can use the link between addition and subtraction to find missing numbers?

ANSWERS Answers for the **Practice** part of the lesson appear in the separate **Practice and Reflect answer guide**.

Reflect

WAYS OF WORKING Independent thinking

IN FOCUS This **Reflect** activity checks understanding of adding two whole numbers that both have 5 digits and that require two exchanges. Encourage children to explain how they know that their calculation will have exactly two exchanges and ask them to also work out the answer to the calculation. Identify children who still need support to do this.

ASSESSMENT CHECKPOINT Can children work out what information they will need to know to make sure they are writing an addition question that requires exactly two exchanges? Can they explain why the place value of each column is important for this? Can they lay their calculation out neatly and accurately and correctly find the answer to it?

ANSWERS Answers for the **Reflect** part of the lesson appear in the separate **Practice and Reflect answer guide**.

After the lesson ⏸

- Can children confidently use the column method to add two or more whole numbers with 4 or more digits?
- Can children identify calculations that require exchanges and correctly solve these?
- Are any children still overlooking the importance of place value in these calculations?

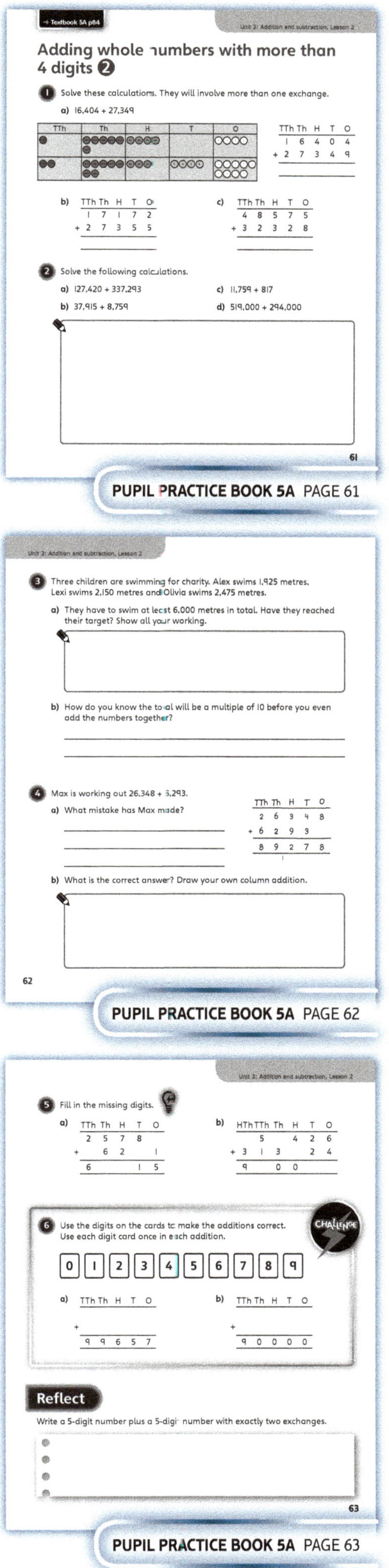

PUPIL PRACTICE BOOK 5A PAGE 61

PUPIL PRACTICE BOOK 5A PAGE 62

PUPIL PRACTICE BOOK 5A PAGE 63

Subtracting whole numbers with more than 4 digits ①

Learning focus

In this lesson, children will use the column method to subtract whole numbers with more than 4 digits, in the context of taking away and of finding a difference. This includes examples where an exchange is required.

Small steps

→ Previous step: Adding whole numbers with more than 4 digits (2)

→ **This step: Subtracting whole numbers with more than 4 digits (1)**

→ Next step: Subtracting whole numbers with more than 4 digits (2)

NATIONAL CURRICULUM LINKS

Year 5 Number – Addition and Subtraction

Add and subtract whole numbers with more than 4 digits, including using formal written methods (columnar addition and subtraction).

ASSESSING MASTERY

Children can recognise why exchanges are needed when subtracting whole numbers with more than 4 digits and can use the column method to find the answer. They can use comparison bar models to express a problem and show what they need to work out.

COMMON MISCONCEPTIONS

Children may just subtract the smaller digit from the bigger digit. For example, when working out 38,792 – 17,345 they may write 3 in the ones column as they have worked out 5 – 2 and not made an exchange. Ask:
• *Which digit do you need to subtract? What will you do if the digit you are subtracting is bigger?*

Children may not know how to correctly set out the column method when subtracting two numbers that have a different number of digits. Ask:
• *What is the place value of each digit in these numbers? Why is it important to lay this out accurately?*

STRENGTHENING UNDERSTANDING

Children should first practise subtracting whole numbers with 2 or 3 digits before moving on to subtracting whole numbers with more than 4 digits.

GOING DEEPER

Give children an answer and ask how many different ways they can make this number using subtraction of a 4-digit number from a 5-digit number. For example, ? + ? = 34,728.

KEY LANGUAGE

In lesson: subtract, difference, exchange, greater than, more, less, digit, capacity

Other language to be used by the teacher: column, place value, ones, tens, hundreds, thousands, ten thousands

STRUCTURES AND REPRESENTATIONS

place value grid, column method subtraction, bar model

RESOURCES

Mandatory: counters

 In the eTextbook of this lesson, you will find interactive links to a selection of teaching tools.

Before you teach 

• Can children subtract with 2- and 3-digit numbers?
• Do children understand key vocabulary, such as 'greater than', 'less than' and 'difference'?

Discover

 Pair work

ASK

- Question **1** a): *What calculation do you need to do to work out how much greater the capacity of the velodrome is to the archery field?*
- Question **1** b): *What is the capacity of the athletics stadium? How will you work out how many seats were empty?*

IN FOCUS Question **1** a) presents subtraction in the context of a 'how much greater than' question and requires children to identify and subtract a 4-digit number from a 5-digit number using the column method and making one exchange. Question **1** b) requires children to work out the difference between a given number and the total, by subtracting a 5-digit number from a 5-digit number with one exchange.

PRACTICAL TIPS Explain that the word 'capacity' means how many people will fit in each area of the sports park. Find images and capacity information about local and national sports centres that children may be familiar with to help them visualise and understand the large numbers and the difference in capacity between these.

ANSWERS

Question **1** a): 15,735 − 2,582 = 13,153

> The velodrome capacity is 13,153 greater than the archery field capacity.

Question **1** b): 75,450 − 52,700 = 22,750

> 22,750 seats were empty in the athletics stadium.

Share

 Whole class teacher led

ASK

- Question **1** a): *What calculation will you do if the question is 'how much greater than'? What method will you use? Which place value column do you start with? Do you need to make an exchange?*
- Question **1** b): *What type of calculation will this be? How do you lay this out as a written method?*

IN FOCUS For question **1** a) make sure children are able to correctly identify and say the given capacity of the velodrome and the archery field. Discuss how the words 'how much greater' lead us to carry out a subtraction. Show children the column subtraction, and ensure they are confident with using this layout. Use this to reinforce the place value of each digit. Explore why it is not possible to subtract 8 tens from 3 tens, and discuss the need to exchange 1 hundred for 10 tens. For question **1** b), discuss which calculation we need to use and show the bar model so children can see why we are subtracting. Emphasise the place value of each digit and the exchange of 1 thousand for 10 hundreds so we can easily subtract the 7 hundreds.

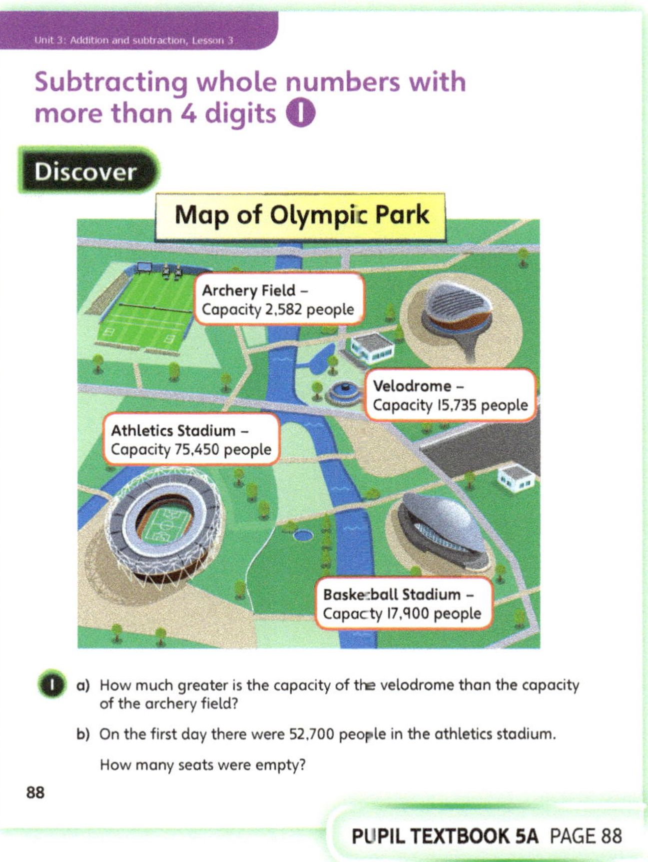

PUPIL TEXTBOOK 5A PAGE 88

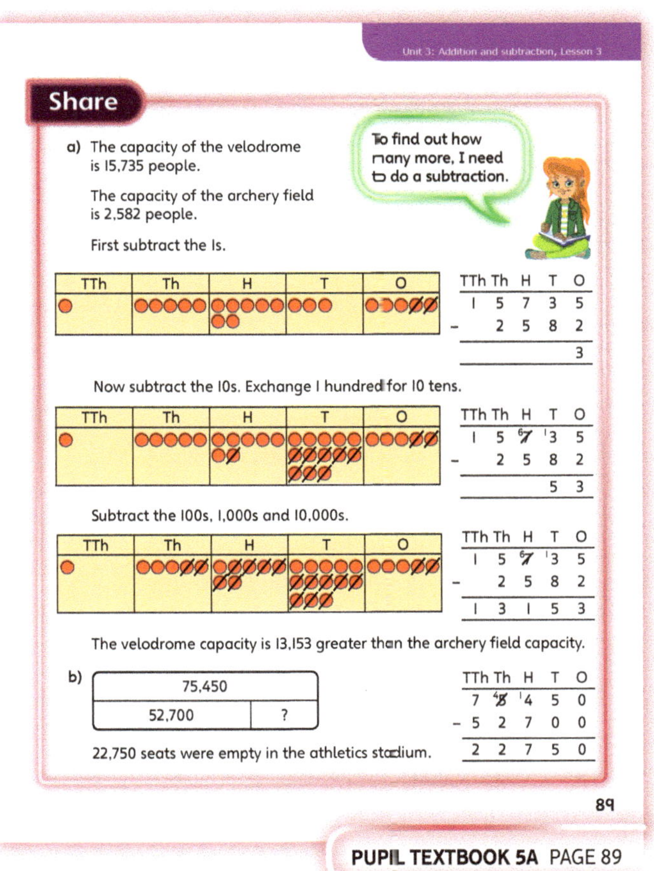

PUPIL TEXTBOOK 5A PAGE 89

Think together

Unit 3: Addition and subtraction, Lesson 3

Think together

WAYS OF WORKING Whole class teacher led (I do, We do, You do)

ASK

- Question ❶ : *If you are taking away the number of people who leave, what calculation will you need to do? What method will you use?*
- Question ❷ : *What is the capacity of the basketball stadium? What calculation will you do to work out how many more people could have watched the game?*
- Question ❸ : *What information do you have? How does the model show this? How will you find the capacity of the hockey centre? What calculation will you do to work out how much greater this is than the velodrome?*

IN FOCUS Question ❶ involves subtracting whole numbers without any exchange. Encourage children to use the column subtraction method. In question ❷ children will subtract whole numbers with one exchange. Children can use the bar model to help them decide which calculation they need to do. To solve question ❸, children will need to problem solve and explore a two-step subtraction. Encourage children to look at the comparison bar model to help them decide which calculation and numbers to use.

STRENGTHEN To support understanding children should represent each calculation using counters on a place value grid alongside the abstract calculations. Where there are exchanges, encourage them to make and move each number and to explain what is happening and why.

DEEPEN Extend question ❸ by asking children to use subtraction to compare the capacities of other structures in the sports parks. For example, how much greater is the capacity of the hockey centre than the archery field?

ASSESSMENT CHECKPOINT Can children identify where a subtraction is needed to find an answer and relate this to finding a difference or taking away? Can they use the column method to subtract whole numbers with 4 or more digits, including those where an exchange is required?

ANSWERS

Question ❶ : 15,735 – 3,620 = 12,115

There are 12,115 people left in the velodrome.

Question ❷ : 17,900 – 10,840 = 7,060

7,060 more people could have watched the game.

Question ❸ : 72,450 – 42,300 = 33,150

The capacity of the hockey centre is 33,150.

33,150 – 15,735 = 17,415

The capacity of the hockey centre is 17,415 greater than the capacity of the velodrome.

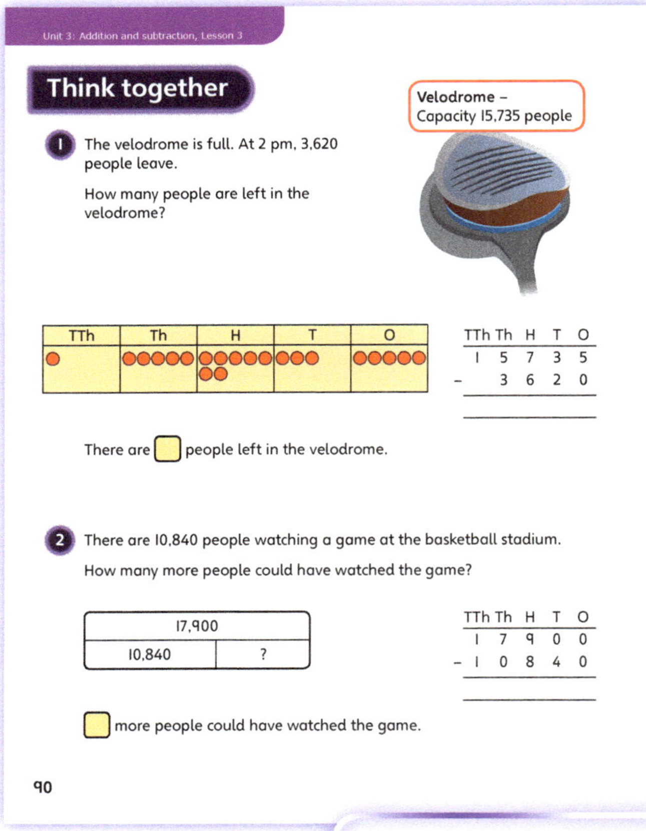

PUPIL TEXTBOOK 5A PAGE 90

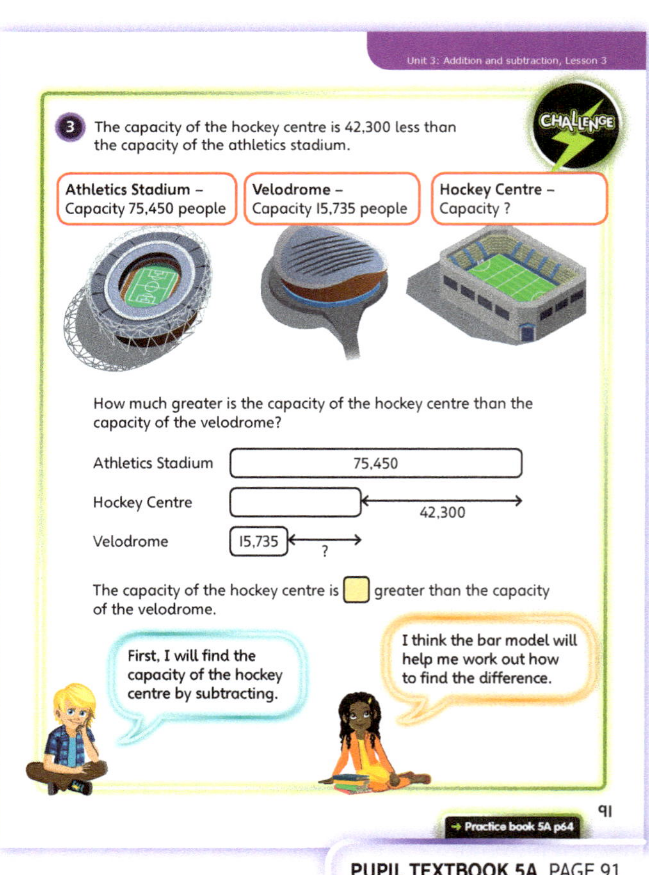

PUPIL TEXTBOOK 5A PAGE 91

Practice

WAYS OF WORKING Independent thinking

IN FOCUS Questions ① and ② aim to consolidate children's understanding of subtracting two whole numbers using the column method where the information is represented with counters on a place value grid, in columns and as a bar model. Question ③ uses the context of prices; children will need to identify that a subtraction is required. Encourage them to lay out the column method correctly, paying attention to place value.

STRENGTHEN Encourage children to use counters on a place value grid to support their understanding. When the calculation is not given in columns, encourage them to lay it out carefully in columns, describing the place value of each column. Support children in drawing given numbers in a comparison bar model.

DEEPEN To extend question ⑤, give examples of other two-step problems that require more than one calculation. Ask a child to present some information in the form of a comparison bar model and then ask the other children to write a word problem to match it that they can solve using column subtraction.

THINK DIFFERENTLY Question ④ encourages children to problem solve and work out missing digits in subtraction calculations by linking them to addition. This can be explored further by giving children other missing number problems, such as $13{,}565 - ? = 4{,}099$.

ASSESSMENT CHECKPOINT Are children confident in laying out the column method for subtraction and using it to subtract whole numbers with 4 or more digits, including where an exchange is needed?

ANSWERS Answers for the **Practice** part of the lesson appear in the separate **Practice and Reflect answer guide**.

Reflect

WAYS OF WORKING Pair work

IN FOCUS This **Reflect** activity checks children's understanding of the column method of subtraction by asking them to explain how to use it to subtract two whole numbers that both have 5 digits. Encourage children to actually answer the calculation as well as explaining the steps. Look out for children who do not address the importance of place value or who do not identify how to correctly carry out the exchange that is needed.

ASSESSMENT CHECKPOINT Can children explain how to use the column method to subtract two 5-digit whole numbers with one exchange? Have children clearly explained place value and the process of carrying out one exchange?

ANSWERS Answers for the **Reflect** part of the lesson appear in the separate **Practice and Reflect answer guide**.

After the lesson ⏸

- Can children layout the column method of subtraction neatly and accurately?
- Can children identify which information in a problem shows that a subtraction is needed?
- Do children understand how comparison bar models can help them to work out missing information?

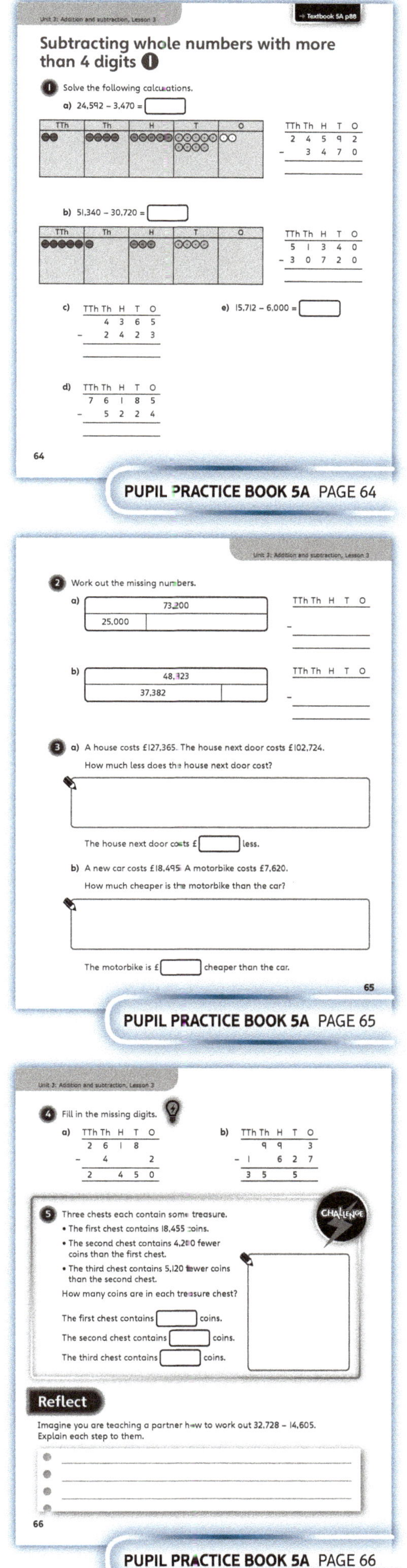

PUPIL PRACTICE BOOK 5A PAGE 64

PUPIL PRACTICE BOOK 5A PAGE 65

PUPIL PRACTICE BOOK 5A PAGE 66

Subtracting whole numbers with more than 4 digits ②

Learning focus

In this lesson, children will explore how and why exchanges can occur in subtractions. They use the column method to subtract whole numbers with more than 4 digits, including where exchanges are needed in some or all columns.

Small steps

→ Previous step: Subtracting whole numbers with more than 4 digits (1)
→ **This step: Subtracting whole numbers with more than 4 digits (2)**
→ Next step: Using rounding to estimate and check answers

NATIONAL CURRICULUM LINKS

Year 5 Number – Addition and Subtraction
Add and subtract whole numbers with more than 4 digits, including using formal written methods (columnar addition and subtraction).

ASSESSING MASTERY

Children can lay the column method out neatly and accurately and can use it to subtract whole numbers with more than 4 digits, including needing to make multiple exchanges.

COMMON MISCONCEPTIONS

Children may make mistakes where an exchange is needed because there are zeros in the bigger number, for example, for 28,300 – 18,976, exchanging 1 hundred for 10 ones, not 1 hundred for 10 tens, and then 1 ten for 10 ones. Ask:
- *Can you make an exchange from this column? Why not? Which column can you exchange from? How much will you be exchanging?*

STRENGTHENING UNDERSTANDING

Encourage children to use counters on a place value grid to help them see the calculation in a concrete way. Do this alongside the column method so that children can see how the concrete links to the abstract method. If children do not understand the concept of exchanges, focus on calculations that require just one exchange before moving on to calculations with more than one exchange.

GOING DEEPER

Give children two 5-digit numbers, for example 89,343 and 17,824, and ask how many times they can subtract the smaller number from the larger number without going into negative numbers.

KEY LANGUAGE

In lesson: subtract, exchange, digit, column, place value

Other language to be used by the teacher: ones, tens, hundreds, thousands, ten thousands, less than

STRUCTURES AND REPRESENTATIONS

column method subtraction

RESOURCES

Mandatory: digit cards 0–9, counters

 In the eTextbook of this lesson, you will find interactive links to a selection of teaching tools.

Before you teach 

- Do children know how to accurately lay out a column subtraction?
- Do children know how to make an exchange when using the column subtraction method?

Discover

WAYS OF WORKING Pair work

ASK

- Question **1** a): *What is the new calculation if you swap the 2 and the 8 around? What do you need to look at to decide if an exchange is needed?*
- Question **1** b): *What has changed in each calculation? Can you see any new exchanges? How do you know?*

IN FOCUS Question **1** a) requires children to identify when an exchange is needed and to use the column method to work out the answer to a subtraction with one exchange. Question **1** b) introduces more subtractions where the same digits are rearranged so that each question has a different number of exchanges in different place value columns.

PRACTICAL TIPS Ask children to arrange real digit cards 0–9 to make Max and Jamilla's subtraction from the picture. These can then be used to create subtractions throughout the lesson.

ANSWERS

Question **1** a): One exchange is needed now, as there are not enough thousands to subtract from.

The answer to the new subtraction is 44,563.

Question **1** b): 62,097 − 18,534 = 43,563

62,037 − 18,594 = 43,443

62,034 − 18,597 = 43,437

PUPIL TEXTBOOK 5A PAGE 92

Share

WAYS OF WORKING Whole class teacher led

ASK

- Question **1** a): *In which column do you need to make an exchange? How do you know?*
- Question **1** b): *Do you need to make more than one exchange in each of these calculations?*

IN FOCUS For question **1** a) encourage children to write out the new calculation, or make it with digit cards, and to say the numbers correctly. Show the column subtraction and ensure children are confident in using this layout. Check that children can explain how we know that we need to make an exchange in the thousands column. Use the column method model to reinforce the place value of each digit when carrying out the calculation and explain that we always start by subtracting in the ones column in case we need to make an exchange. For question **1** b), show children the new column subtractions and reinforce the place value of each digit. Ensure children do not just subtract the smaller digit from the larger, by highlighting and describing any exchange.

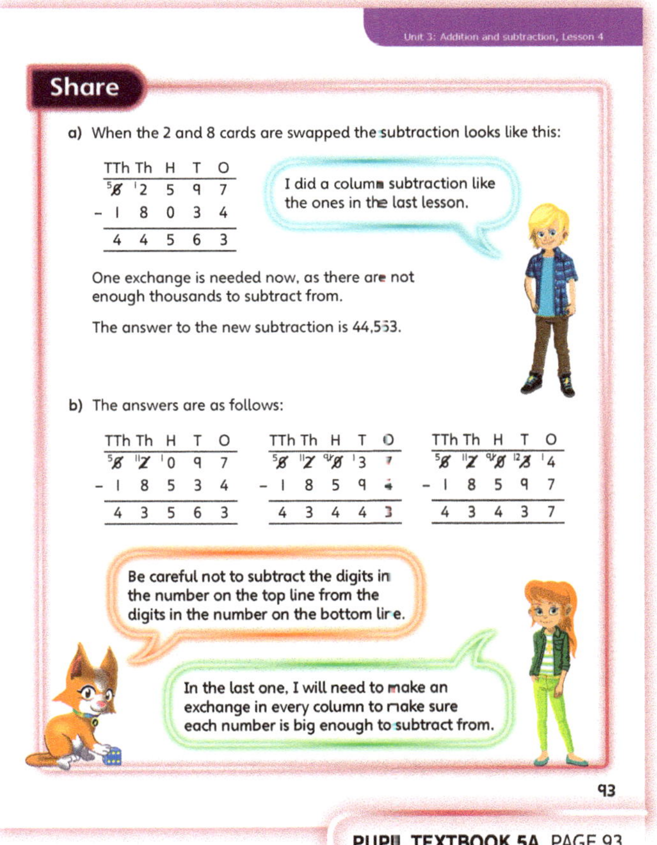

PUPIL TEXTBOOK 5A PAGE 93

Think together

WAYS OF WORKING Whole class teacher led (I do, We do, You do)

ASK

- Question ❶ : *Which place value column should you start with? What is 6 ones subtract 5 ones? Do you need to make any exchanges? How do you know?*
- Question ❷ : *What do you subtract from 3 ones to get 1 one? What is the missing digit in the ones column? Can you subtract from 0 tens to get 1 ten? What must have happened here?*
- Question ❸ : *What is the same in the first and second question, and what is different? What will be different about the answer? Is there an easier way?*

IN FOCUS Question ❷ introduces problem solving to find missing digits. Encourage children to start with the ones column and to think about how to identify if an exchange has taken place. Encourage them to check that their calculation is correct by working it out and checking they get an answer of 61,611. In question ❸, encourage children to use the column method to begin with but to think about what is the same in each question (the 15,462) and what is different (such as, 27,900 is 10 less than 27,910, so the answer should be 10 less). Where there are more zeroes in the bigger numbers, there will also be more exchanges. Use 20,000 – 15,462 to demonstrate how to get rid of the zeroes and the need for exchanges by subtracting 1 from both numbers. The answer to 19,999 – 15,461 will still be the same. Children can still carry out the original column subtraction to check this.

STRENGTHEN To support understanding children should represent the calculations using counters on a place value grid alongside the abstract calculations. For question ❸ ask children to make all four calculations using counters so they can easily see what is different in each question.

DEEPEN For question ❸ ask children to work out the answers to other calculations where they are always subtracting 15,462. Children could think of their own questions for this and then answer them.

ASSESSMENT CHECKPOINT Can children identify where exchanges are needed in subtractions and use column subtraction to subtract whole numbers with 4 or more digits that involve exchanges?

ANSWERS

Question ❶ : 82,706 – 39,415 = 43,291

Question ❷ : 7**6**,503 – **14**,**892** = 61,611

Question ❸ a): 27,910 – 15,462 = 12,448

Question ❸ b): 27,900 – 15,462 = 12,438

Question ❸ c): 27,000 – 15,462 = 11,538

Question ❸ d): 20,000 – 15,462 = 4,538

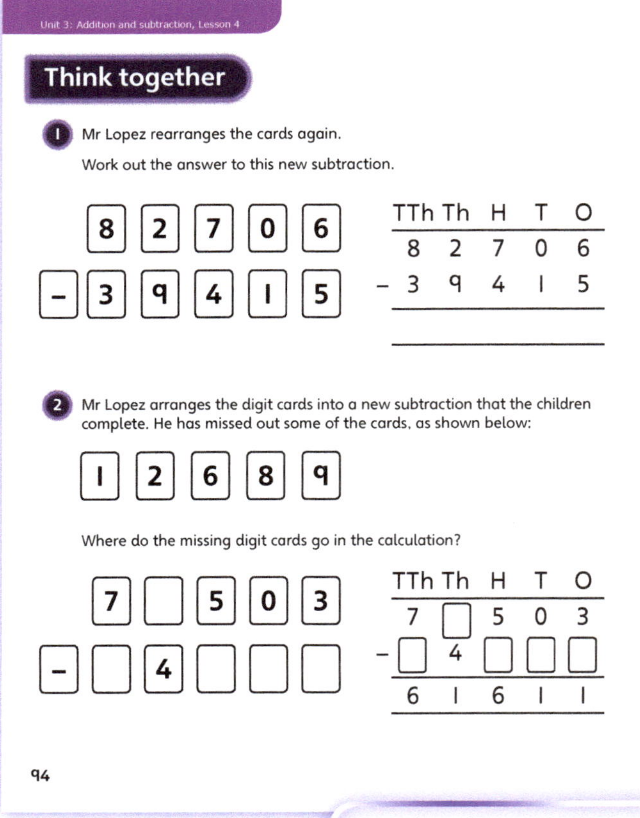

PUPIL TEXTBOOK 5A PAGE 94

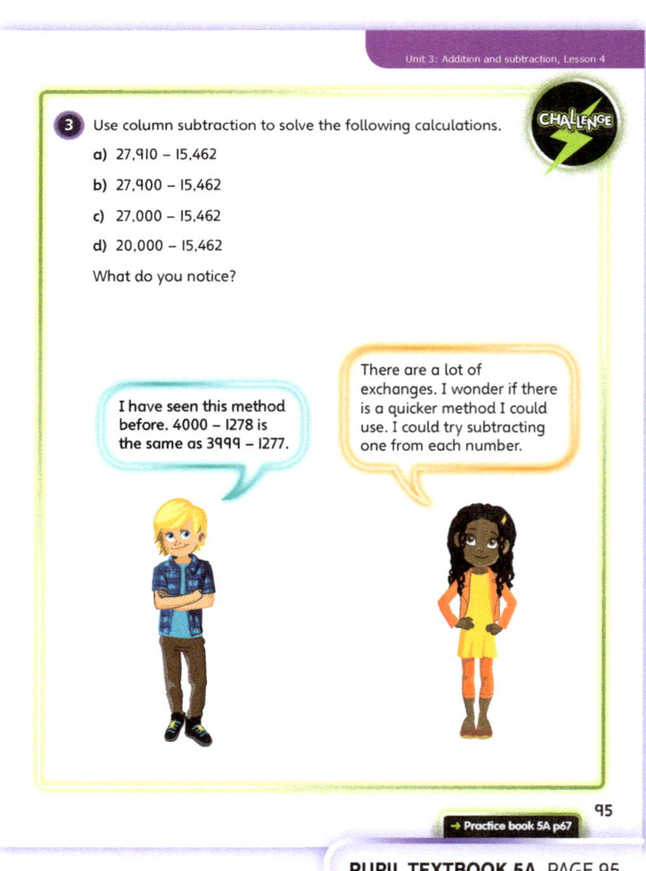

PUPIL TEXTBOOK 5A PAGE 95

Practice

WAYS OF WORKING Independent thinking

IN FOCUS Questions ❶ and ❷ aim to consolidate children's understanding of subtracting two whole numbers using the column method where the information is represented with counters on a place value grid, in a column, as a calculation and in words. Encourage children to choose their preferred method for question ❸. They may subtract 18,926 once to get an answer and then subtract 18,926 from this, and so on. Alternatively, they may choose to work out 18,926 + 18,926 + 18,926 and then subtract this answer from 76,350. Question ❺ involves a lot of 0 digits so will require multiple exchanges. Can children recall a strategy to avoid this?

STRENGTHEN Encourage children to use counters on a place value grid to support their understanding, and when a calculation is not given in columns, encourage them to write it out in columns. Children may find it useful to draw a bar model for question ❻, to help them understand which calculations they need to do.

DEEPEN Extend the context in question ❸ by asking children after how many hours the ball pool will only have 646 balls left in it.

ASSESSMENT CHECKPOINT Are children confident in laying the column method out neatly and accurately and using it to subtract whole numbers with 4 or more digits, including those with multiple exchanges? Do they use the link with addition, think flexibly to employ a variety of strategies and use models to help them work out what calculation is needed?

ANSWERS Answers for the **Practice** part of the lesson appear in the separate **Practice and Reflect answer guide**.

Reflect

WAYS OF WORKING Independent thinking

IN FOCUS This **Reflect** activity checks children's understanding of using the column method to subtract two 5-digit whole numbers with exactly two exchanges. Encourage children to explain how they know their subtraction has exactly two exchanges and encourage them to focus on using key vocabulary and the correct place value terms. Also, ask children to find the answer to their calculation.

ASSESSMENT CHECKPOINT Can children explain how they know that a calculation will have exactly two exchanges? Can they accurately use the column method to subtract the two whole numbers and find the correct answer to their calculation?

ANSWERS Answers for the **Reflect** part of the lesson appear in the separate **Practice and Reflect answer guide**.

After the lesson ⏸

- Can children confidently use the column method to subtract two whole numbers with 4 or more digits where there are multiple exchanges?
- Can children identify where exchanges will occur in subtractions?
- Do children employ a variety of strategies to solve problems?

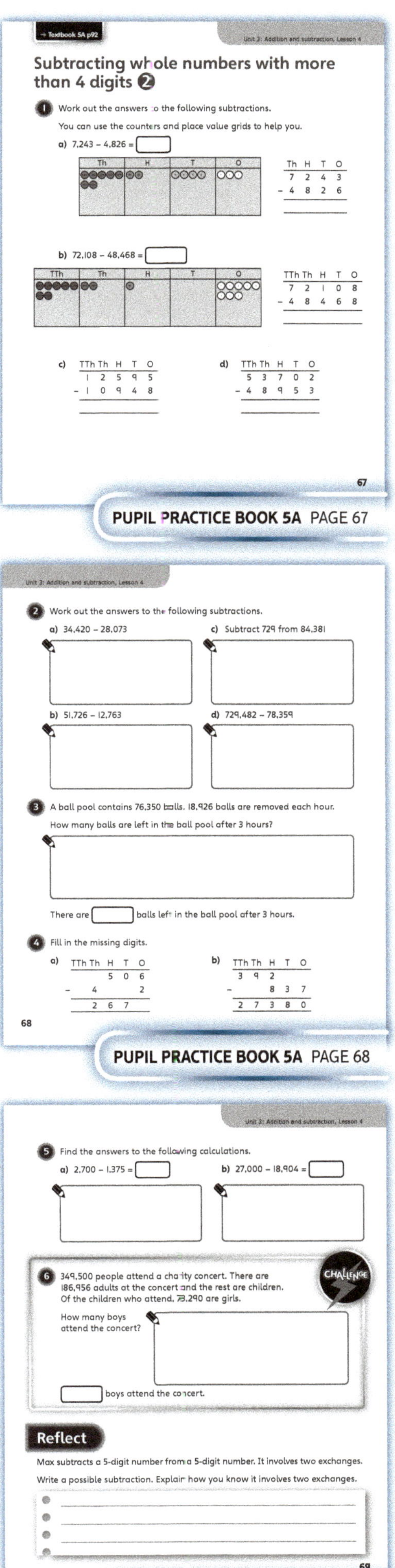

PUPIL PRACTICE BOOK 5A PAGE 67

PUPIL PRACTICE BOOK 5A PAGE 68

PUPIL PRACTICE BOOK 5A PAGE 69

Using rounding to estimate and check answers

Learning focus

In this lesson, children will learn how to use rounding numbers to help make estimates, identify sensible answers, find mistakes and check answers to calculations.

Small steps

→ Previous step: Subtracting whole numbers with more than 4 digits (2)
→ **This step: Using rounding to estimate and check answers**
→ Next step: Mental addition and subtraction (1)

NATIONAL CURRICULUM LINKS

Year 5 Number – Addition and Subtraction

Use rounding to check answers to calculations and determine, in the context of a problem, levels of accuracy.

ASSESSING MASTERY

Children can round numbers and use this to make estimates, find mistakes and check answers.

COMMON MISCONCEPTIONS

Children may not know what to round each number to. Ask:
- *Can you find the number on a number line? What number is the closest to round to?*

Some children may not round numbers appropriately. For example, rounding 39,921 + 15,839 to 39,920 + 15,840 which is still very close to the original. Ask:
- *Will that help you to easily check your answer? What would be a simpler estimate to calculate with?*

STRENGTHENING UNDERSTANDING

Encourage children to use number lines to help them identify what to sensibly round each number to before using these to make an estimate.

GOING DEEPER

Give children a rounded answer and ask what the question could have been. For example: *Two numbers are added together and an estimated answer is 12,000. What could the calculation have been?* For example, it could have been 7,136 + 4,834 which would round to 7,000 + 5,000.

KEY LANGUAGE

In lesson: round, estimate, check, difference, total, close to, sensible, reasonable

Other language to be used by the teacher: addition, subtraction, exchange, ones, tens, hundreds, thousands, ten thousands

STRUCTURES AND REPRESENTATIONS

column method addition, column method subtraction, number line

 In the eTextbook of this lesson, you will find interactive links to a selection of teaching tools.

Before you teach 

- Do children know how to round to the nearest ten thousand, thousand, hundred and ten?
- Are children confident in using the column method for addition and subtraction?

Discover

 Pair work

ASK

- Question ① a): *What could you round each number to, to make a sensible estimate?*
- Question ① b): *Do you think Bella has laid the calculation out carefully? What do you need to think about when laying out a column addition?*

IN FOCUS Question ① a) is used to help children realise the importance of making estimates by rounding numbers to check if their answer is logical. Question ① b) requires children to focus on the layout of a column addition in order to identify the mistake that has been made.

PRACTICAL TIPS Consider showing children some sample test papers similar to the one that Bella is doing. Explain the importance of checking your answers when doing a test.

ANSWERS

Question ① a): 18,000 + 4,000 = 22,000. Bella's answer should be close to 22,000.

Question ① b): Bella has lined up the numbers incorrectly in the column addition.

The thousands need to be lined up underneath the thousands, and so on.

The correct answer is 21,889.

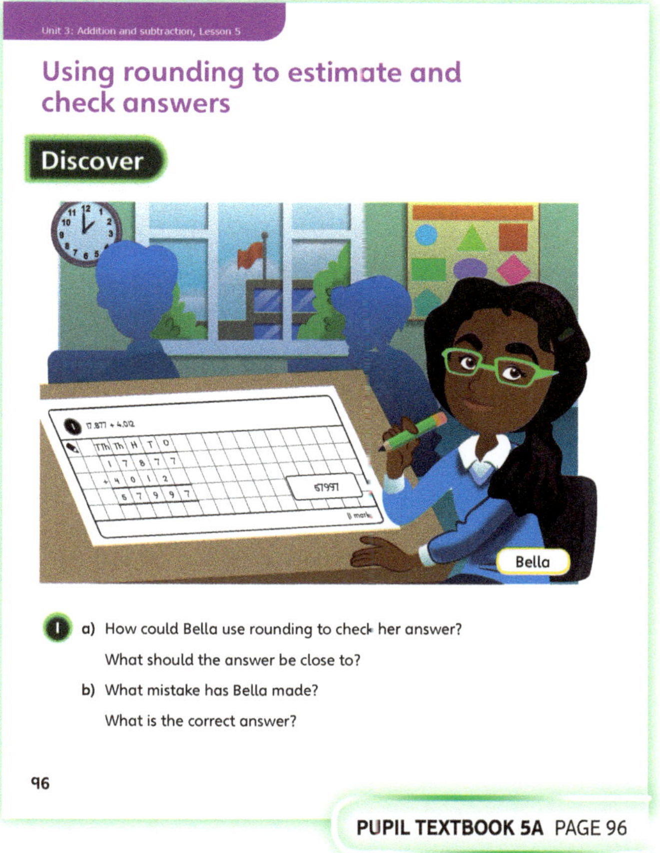

PUPIL TEXTBOOK 5A PAGE 96

Share

 Whole class teacher led

ASK

- Question ① a): *What number is 17,877 close to? What number is 4,012 close to? Can you work out 18,000 add 4,000 mentally?*
- Question ① b): *How do you know Bella has lined the numbers up incorrectly? Why is place value important?*

IN FOCUS In question ① a) the number line can be used to prompt a discussion about which numbers are closest to 17,877 and 4,012. Encourage children to say the numbers they have rounded to correctly then add them together. Can children explain why the answer should be close to 22,000? Compare Bella's column addition to the correct column addition in question ① b). Can children explain why Bella's answer must be incorrect? Emphasise the importance of lining the ones up under the ones, and so on. Discuss the exchange that needs to take place. Can children explain why the correct answer is 21,889? Discuss how 21,889 and 22,000 are close to each other to emphasise why rounding is useful for checking answers.

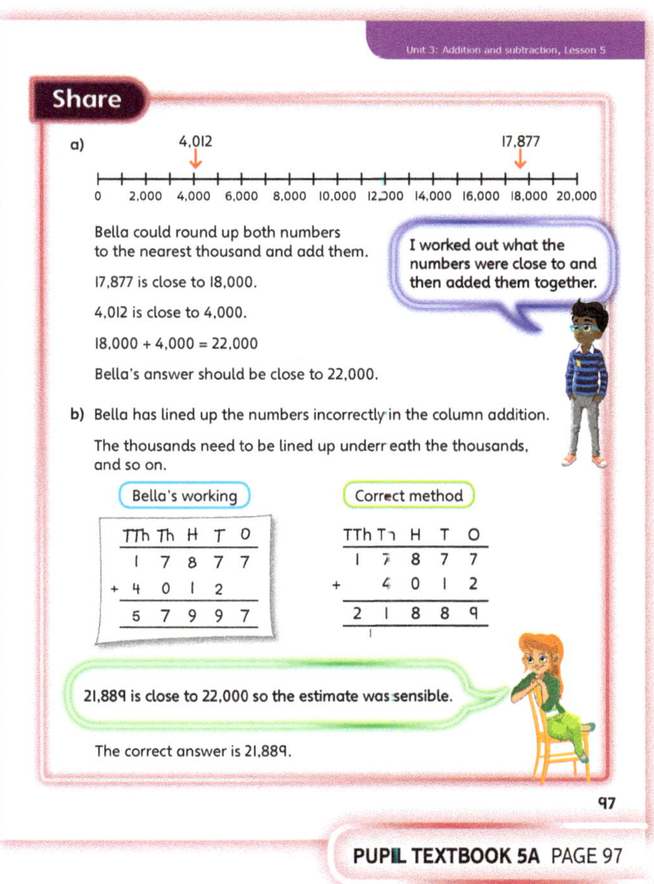

PUPIL TEXTBOOK 5A PAGE 97

Think together

WAYS OF WORKING Whole class teacher led (I do, We do, You do)

ASK

- Question ❶ : *What will you round 4,935 to? What will you round 322 to? Can you subtract mentally?*
- Question ❷ : *How do you know what to round each number to? Can you tell if your estimates are sensible?*
- Question ❸ : *What methods will you use to find the total and the difference? How will you check your answers?*

IN FOCUS Question ❶ requires children to use rounding to check if an answer is correct and to explain the mistake that has been made. Encourage children to round each number to a suitable degree of accuracy. Children may round 4,935 to 5,000 or to 4,900. Discuss why 4,900 will give a more accurate answer. In question ❷, children use rounding to estimate answers and compare them to the correct answer. Encourage children to explain how they know if their estimate is sensible. Question ❸ presents calculations in the context of prices and involving more than two numbers. Although the question does not ask children to use rounding, discuss the different strategies children could use to check their answers by rounding.

STRENGTHEN Children can use a number line to help them to decide what to round each number to. If necessary, encourage children to use the column method to work out their estimations rather than doing it mentally.

DEEPEN For question ❸, ask children to find the difference between the price of the car and the TV, and the difference between the price of the laptop and the TV, giving their answers both as an estimate and as an exact answer.

ASSESSMENT CHECKPOINT Can children round numbers to an appropriate degree of accuracy and use this to make sensible estimates, identify mistakes, and check answers when adding and subtracting?

ANSWERS

Question ❶ : 4,935 is close to 4,900 or 5,000
322 is close to 300
4,900 − 300 = 4,600 or 5,000 − 300 = 4,700
Bella has laid the calculation out incorrectly. She has not lined the ones up under the ones, and so on.

Question ❷ : 17,240 rounds to 17,000
28,385 rounds to 28,000
17,000 + 28,000 = 45,000
17,240 + 28,385 = 45,625
7,010 rounds to 7,000
3,997 rounds to 4,000
7,000 − 4,000 = 3,000
7,010 − 3,997 = 3,013
The estimates were sensible.

Question ❸ : 12,795 + 1,199 + 298 = £14,292
The items cost £14,292 in total.
12,795 − 1,199 = £11,596
The difference in price between the car and the laptop is £11,596.

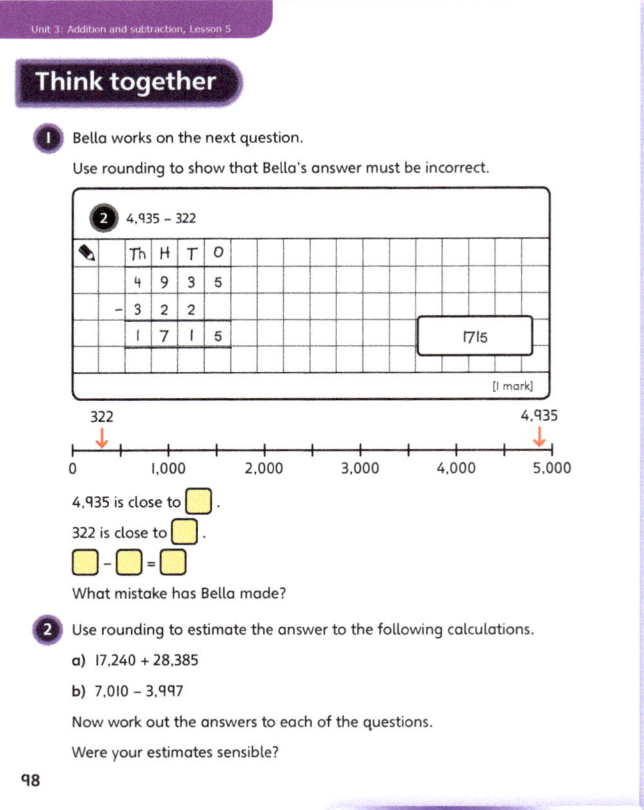

PUPIL TEXTBOOK 5A PAGE 98

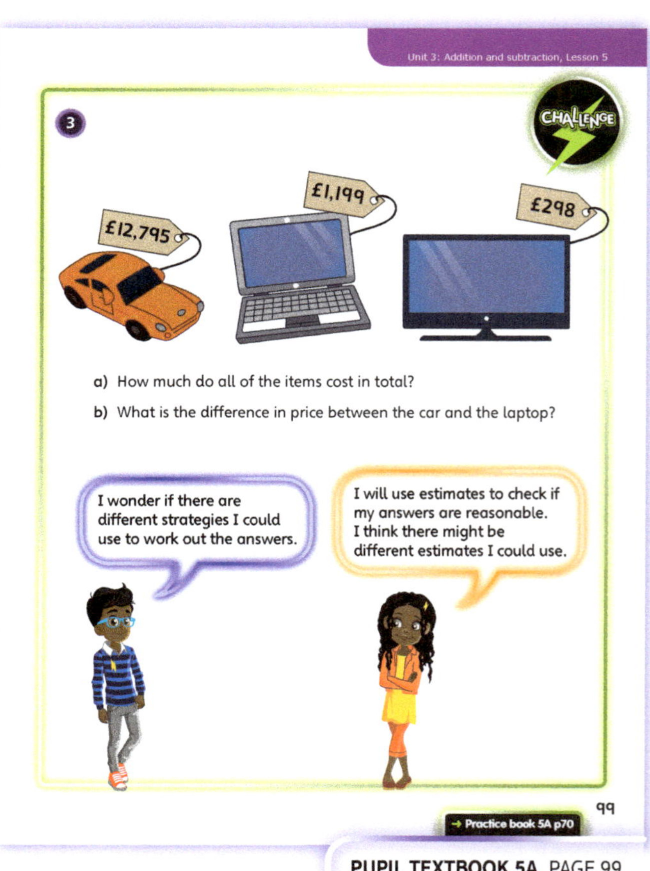

PUPIL TEXTBOOK 5A PAGE 99

Practice

WAYS OF WORKING Independent thinking

IN FOCUS Questions **1**, **2** and **3** aim to consolidate children's understanding of using rounding to check answers and to compare actual answers to estimates. Question **5** is a two-step problem involving both addition and subtraction. Discuss the different ways in which this can be calculated. Children may add the first two numbers first and then subtract. Encourage children to compare their estimate to the actual answer and to decide if it is a sensible estimate.

STRENGTHEN In question **4**, support children in realising that it can be possible to round to different numbers and that this will affect the estimate. Encourage children to use a number line with hundreds and thousands marked so they can see why 2,187 can round to both 2,000 and 2,200. Discuss which estimate is more accurate and why.

DEEPEN Ask children to come up with at least two different estimates for question **1**, question **2** and question **3**.

ASSESSMENT CHECKPOINT Are children confident in using rounding to make sensible estimates and use these to check and correct answers? Can they identify the mistakes made and explain what should be changed?

ANSWERS Answers for the **Practice** part of the lesson appear in the separate **Practice and Reflect answer guide**.

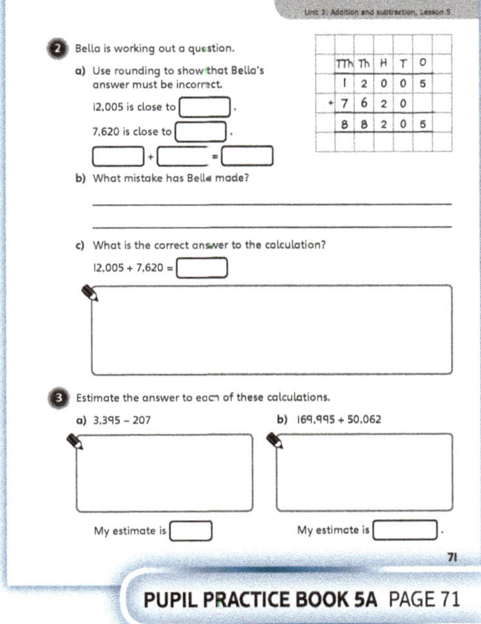

PUPIL PRACTICE BOOK 5A PAGE 70

PUPIL PRACTICE BOOK 5A PAGE 71

Reflect

WAYS OF WORKING Independent thinking

IN FOCUS This **Reflect** activity checks if children understand the importance of using rounding to make estimates. Children should be able to explain that rounding helps to find mistakes and to check if the answer to a calculation looks sensible and is likely to be correct.

ASSESSMENT CHECKPOINT Can children explain why rounding can be used to make estimates and check answers?

ANSWERS Answers for the **Reflect** part of the lesson appear in the separate **Practice and Reflect answer guide**.

After the lesson ⏸

- Can children round numbers to an appropriate degree of accuracy?
- Can children make sensible estimates and use these to find mistakes?
- Do children understand the importance of checking answers?

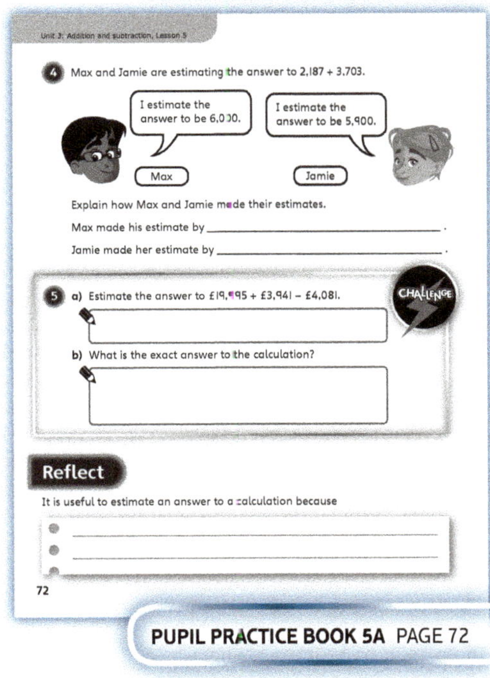

PUPIL PRACTICE BOOK 5A PAGE 72

Mental addition and subtraction 

Learning focus

In this lesson, children will learn how to mentally add whole numbers by choosing the most efficient method from a variety of strategies.

Small steps

→ Previous step: Using rounding to estimate and check answers

→ **This step: Mental addition and subtraction (1)**

→ Next step: Mental addition and subtraction (2)

NATIONAL CURRICULUM LINKS

Year 5 Number – Addition and Subtraction

Add and subtract numbers mentally with increasingly large numbers.

ASSESSING MASTERY

Children can understand which mental methods can be used to efficiently add whole numbers.

COMMON MISCONCEPTIONS

Children may miscalculate when working out in their head. For example, think the answer to 53 + _ = 80 is 37. Ask:
- *What method are you using? What will the next ten be in this calculation? Are you sure?*

Children may ignore an exchange that is needed because they are working out mentally. Ask:
- *What is the place value of each digit in your calculation? Do you need to do an exchange here?*

STRENGTHENING UNDERSTANDING

Children should start by adding 1-digit and then 2-digit numbers with no exchanges. Build up to calculations that require an exchange and calculations with 3-digit numbers. Encourage children to use a number line to envisage the calculation in a pictorial way.

GOING DEEPER

Ask children to mentally add more than two whole numbers together. Represent the calculations in different ways, for example on a part-whole model or bar model and ask them to work out the totals or missing parts mentally.

KEY LANGUAGE

In lesson: method, add, subtract

Other language to be used by the teacher: total, round, exchange, hundreds, tens, ones

STRUCTURES AND REPRESENTATIONS

digit cards, number lines, part-whole models

RESOURCES

Optional: place value counters

 In the eTextbook of this lesson, you will find interactive links to a selection of teaching tools.

Before you teach 

- Do children know how to partition a number into ones, tens, hundreds and so on?
- Can they add multiples of 10, 100, 1,000 mentally?
- Can children round to the nearest 10, 100, 1,000?

Discover

 Pair work

ASK

- Question **1** a): *What do you need to think about if you are doing this mentally? What is the same and what is different in this pair?*
- Question **1** b): *Can you partition these numbers to help you? What is the same and what is different in this pair?*

IN FOCUS Question **1** a) requires children to mentally work out how much needs to be added to a number to make another number. Also, in question **1** b) children are given two numbers and asked to work out the total mentally.

PRACTICAL TIPS Make flash cards of the calculations in the **Discover** picture. Ask children to read each calculation aloud and arrange them in pairs, as in the picture. Can they explain why the calculations have been paired like this?

ANSWERS

Question **1** a): 63 + 27 = 90

263 + 27 = 290

Question **1** b): 45 + 23 = 68

450 + 230 = 680

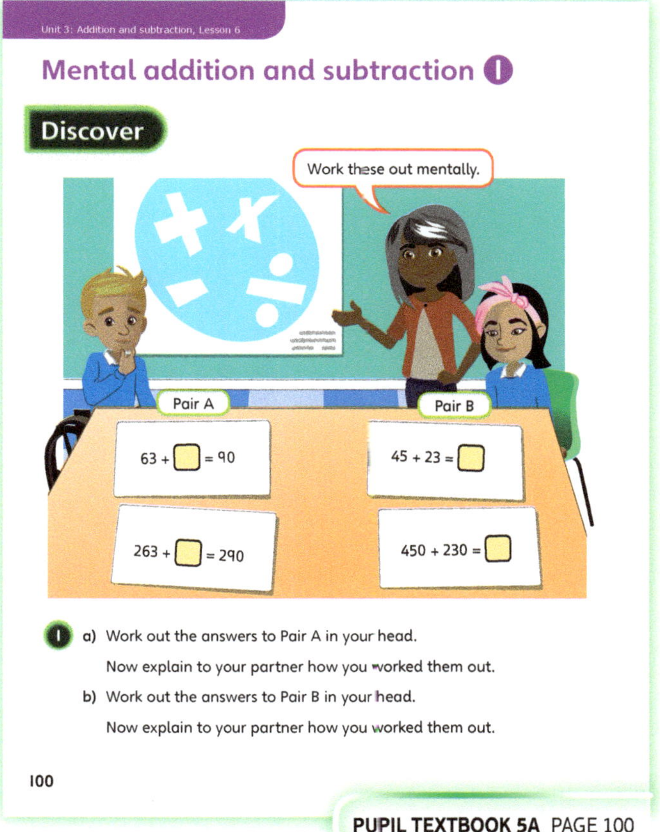

PUPIL TEXTBOOK 5A PAGE 100

Share

 Whole class teacher led

ASK

- Question **1** a): *What is the next ten after 63? How many do you need to add to get to the next ten and then how many to get to 90? What do you do with these two answers? How will the second calculation be different?*
- Question **1** b): *What is 45 made up of? What is 23 made up of? Can you add the 10s and the 1s separately? What should you do with these two answers?*

IN FOCUS For question **1** a) show children the number line from 63 to 70 and 90. Can children explain why we first need to add on 7 ones and then add on 20? Discuss with children how overall we have 7 + 20 = 27. Show the number line from 263 to 270 and 290 and encourage children to describe what is the same and what is different between 63 + _ = 90 and 263 + _ = 290 and why we add 27 on for both questions.

In question **1** b), discuss how 45, 23, 450 and 230 are made up of hundreds, tens and ones, for example 45 is 4 tens and 5 ones so 45 = 40 + 5. Show the part-whole models to reinforce this. Encourage children to add the hundreds, tens and the ones separately then combine their answers. Also, encourage children to compare 45 + 23 and 450 + 230. For example, if we know 45 + 23 = 68 then 45 tens + 23 tens = 68 tens so 450 + 230 = 680.

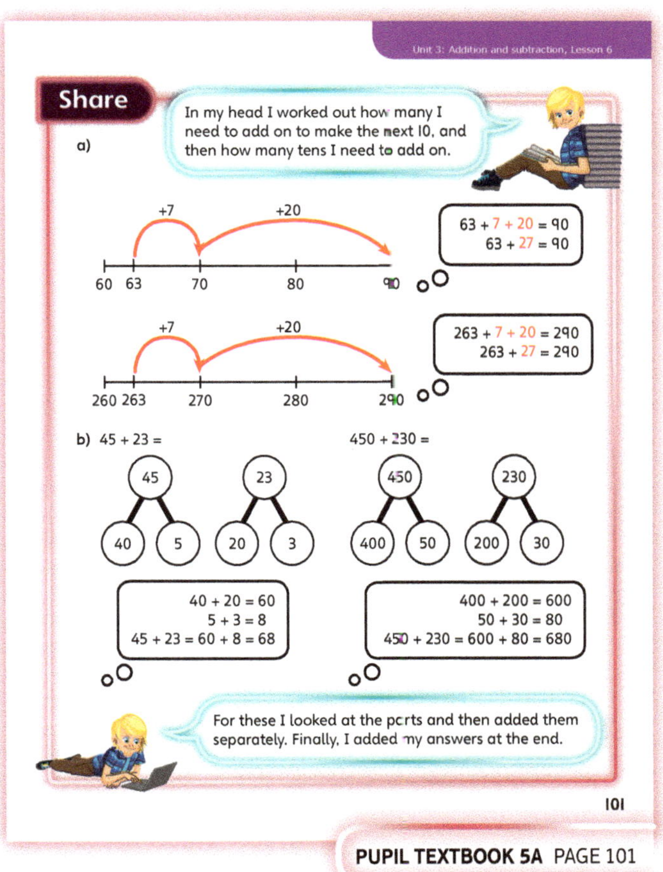

PUPIL TEXTBOOK 5A PAGE 101

Think together

WAYS OF WORKING Whole class teacher led (I do, We do, You do)

ASK

- Question **1** : *What is each number made up of? How can you add these mentally? What do you need to do with your answers?*
- Question **2** : *What is the same and what is different? Can you add each place value separately?*
- Question **3** : *How much more has Andy added on each time? How will this affect the answer? What will you do now to get to the correct answer?*

IN FOCUS Questions **1** and **2** look at adding whole numbers together mentally by adding each place value separately and combining the answers. Some of the questions require an exchange, which children can make mentally. Encourage children to discuss and explain their methods. Question **3** allows children to explore different strategies for doing mental calculations. Encourage children to realise the benefit of this mental strategy when numbers are close to a multiple of 100 or 1,000. Ensure children are confident in choosing which numbers to add on, i.e. rounding to the nearest 100 or 1,000 and subtracting the difference.

STRENGTHEN To support understanding, encourage children to draw their own number lines and to think about how much they should add on to get to, for example, the next ten.

DEEPEN Give children different calculations and ask them to explain which mental strategy would be most beneficial. For example, to work out $32 + 98$ it may be easier to add 100 then subtract 2 but to work out $34 + 65$ it may be easier to add the tens then the ones and combine them. Encourage children to explain why they chose each method.

ASSESSMENT CHECKPOINT Can children mentally add whole numbers using an appropriate strategy?

ANSWERS

Question **1** a): $40 + 30 = 70$
$7 + 5 = 12$
$47 + 35 = 70 + 12 = 82$

Question **1** b): $300 + 400 = 700$
$50 + 70 = 120$
$350 + 470 = 700 + 120 = 820$

Question **2** a): $74 + 69 = 143$

Question **2** b): $740 + 690 = 1,430$

Question **3** a): Andy needs to subtract 2 to find the first answer and subtract 3 to find the second answer.
$324 + 198 = 522$
$324 + 197 = 521$

Question **3** b): $672 + 99 = 771$
$426 + 397 = 823$
$296 + 3,147 = 3,443$
$7,608 + 1,998 = 9,606$
$18,790 + 39,990 = 58,780$

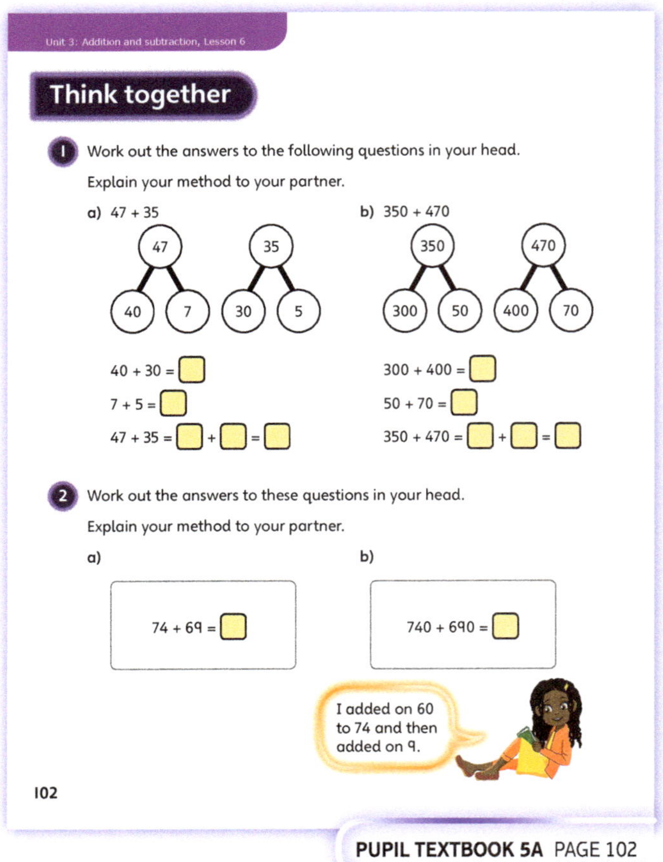

PUPIL TEXTBOOK 5A PAGE 102

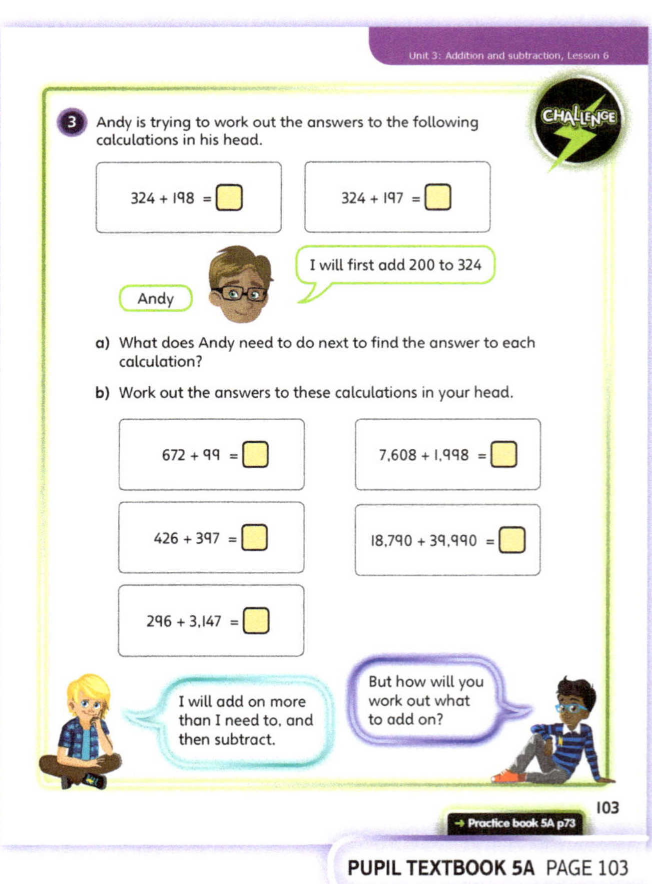

PUPIL TEXTBOOK 5A PAGE 103

Practice

WAYS OF WORKING Independent thinking

IN FOCUS Question ❶ aims to consolidate children's understanding of mentally adding two whole numbers by partitioning. In question ❷ each part builds on the previous one to encourage children to make connections between calculations. Question ❻ is more open and encourages children to write their own strategies for mental calculations. They may choose to add the 1s and add the 10s or to add on, for example, 200 and subtract.

STRENGTHEN Encourage children to use place value counters to help them partition the numbers and add the 1s and the 10s separately.

DEEPEN Question ❼ involves mentally adding more than two whole numbers. This can be explored further by giving them other missing number problems with more than two whole numbers. Say numbers aloud and ask children to add them mentally.

ASSESSMENT CHECKPOINT By the end of the practice children should be confident in choosing the most useful method for adding whole numbers mentally. Question ❻ will show how well children have grasped mental addition through the working out they show.

ANSWERS Answers for the **Practice** part of the lesson appear in the separate **Practice and Reflect answer guide**.

Reflect

WAYS OF WORKING Independent learning

IN FOCUS This **Reflect** activity checks that children can explain how to mentally add two whole numbers. Encourage children to carefully explain how they would carry out each calculation, and why, as well as actually answering it. Look out for children who need support to do this.

ASSESSMENT CHECKPOINT Make sure for each calculation children can explain how to choose the most efficient method for mentally finding the total of two whole numbers.

ANSWERS Answers for the **Reflect** part of the lesson appear in the separate **Practice and Reflect answer guide**.

After the lesson ⏸

- Can children add two whole numbers mentally?
- Do they understand that there are several mental methods to choose between?
- Are there any children who are not able to explain why you would choose one method over another?

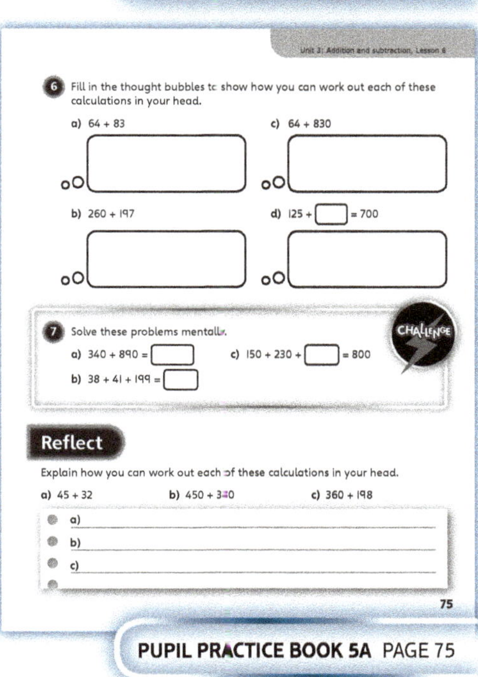

PUPIL PRACTICE BOOK 5A PAGE 73

PUPIL PRACTICE BOOK 5A PAGE 74

PUPIL PRACTICE BOOK 5A PAGE 75

Mental addition and subtraction ②

Learning focus

In this lesson, children will learn how to mentally subtract whole numbers by choosing the most efficient method from a variety of strategies.

Small steps

→ Previous step: Mental addition and subtraction (1)
→ **This step: Mental addition and subtraction (2)**
→ Next step: Using inverse operations

NATIONAL CURRICULUM LINKS

Year 5 Number – Addition and Subtraction
- Add and subtract numbers mentally with increasingly large numbers.
- Solve addition and subtraction multi-step problems in contexts, deciding which operations and methods to use and why.

ASSESSING MASTERY

Children can understand what mental methods can be used to efficiently subtract whole numbers.

COMMON MISCONCEPTIONS

Children may miscalculate when doing a subtraction in their head, for example when mentally calculating 82 – 45 children may do 80 – 40 = 40 and 5 – 2 = 3 and get an answer of 43. Ask:
- *Are you confident of your answer? Is there another method you could use?*

STRENGTHENING UNDERSTANDING

Children should use a number line to aid their understanding, and start by subtracting 1- and 2-digit numbers with no exchanges, building up to calculations that do require an exchange.

GOING DEEPER

Give children questions that involve both mental addition and subtraction. Ask questions verbally for children to work out mentally. Introduce a context to the question, for example, give children the cost of some items and ask them to mentally work out the total cost or the difference in price between two items.

KEY LANGUAGE

In lesson: mentally, subtract, partition, tens, hundreds, number line, method

Other language to be used by the teacher: addition, exchange

STRUCTURES AND REPRESENTATIONS

number lines

RESOURCES

Optional: stopwatch or egg timer, place value counters

 In the eTextbook of this lesson, you will find interactive links to a selection of teaching tools.

Before you teach 

- Do children know how to partition a number into 100s, 10s and 1s?
- Can children add and subtract multiples of 10, 100, 1,000 mentally?

Discover

WAYS OF WORKING **WAYS OF WORKING** Pair work

ASK

- Question **1** a): *What can you think about if you are doing this calculation in your heads and trying to do it quickly?*
- Question **1** b): *Could you use a similar method to Ebo to work out this calculation mentally?*

IN FOCUS Question **1** a) will help children to see why it can be useful to work out calculations mentally rather than always using a written method. Question **1** b) builds on this, with children working out a subtraction calculation mentally.

PRACTICAL TIPS Ask children to model the written method that they think Lexi might have tried to use. Time them working it out and point out how long it takes. Ask: *Do you think you can do it any quicker?* A similar context to the **Discover** picture could be carried out in class. A question is asked, and one child must use a written calculation and the other a mental method.

ANSWERS

Question **1** a): Ebo used a mental method. He counted up from 1,995 to 2,002.

Question **1** b): 700 − 200 = 500
60 − 50 = 10
760 − 250 = 510

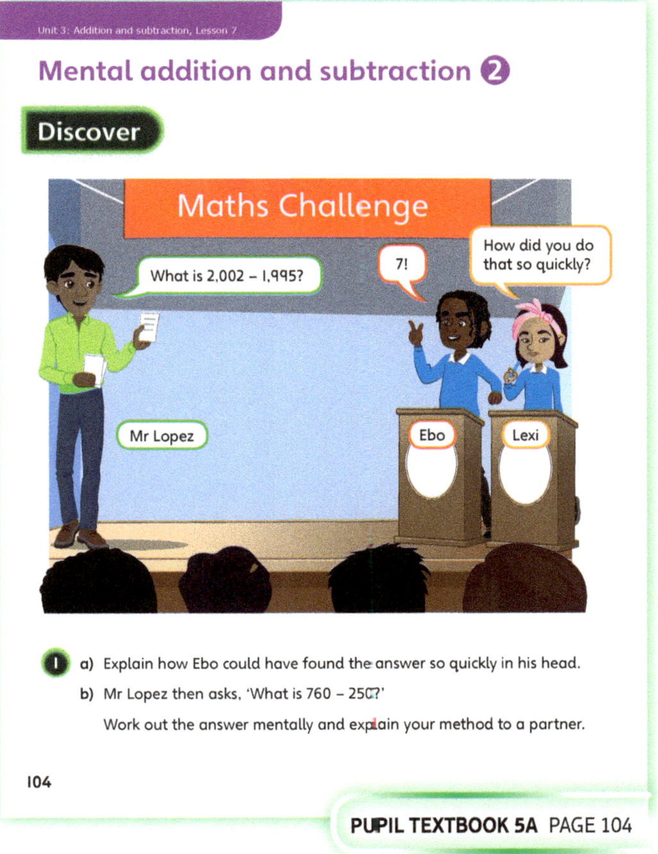

PUPIL TEXTBOOK 5A PAGE 104

Share

WAYS OF WORKING **WAYS OF WORKING** Whole class teacher led

ASK

- Question **1** a): *What is the next thousand after 1,995? How many 1s do you need to add to get to 2,000? How many do you need to add to 2,000 to get to 2,002? What should you do with these two answers?*
- Question **1** b): *What is 760 made up of? What is 250 made up of? Can you subtract the hundreds and the tens separately? What should you do with these two answers?*

IN FOCUS For question **1** a), show the number line from 1,995 to 2,000 and 2,002. Can children explain why we first add 5 ones and then 2 ones? Discuss how, overall, we have 5 + 2 = 7. Can children see why this mental method is useful and how Ebo got his answer so quickly? For question **1** b), discuss what the numbers 760 and 250 are made up of. Use correct mathematical language such as how 760 is 7 hundreds and 6 tens so 760 = 700 + 60. Encourage children to subtract the 100s and the 10s separately, and then combine their answers. Discuss an alternative strategy, for example subtracting 200 from 760 then subtracting the 50. Ask children if they can think of any other suitable mental methods.

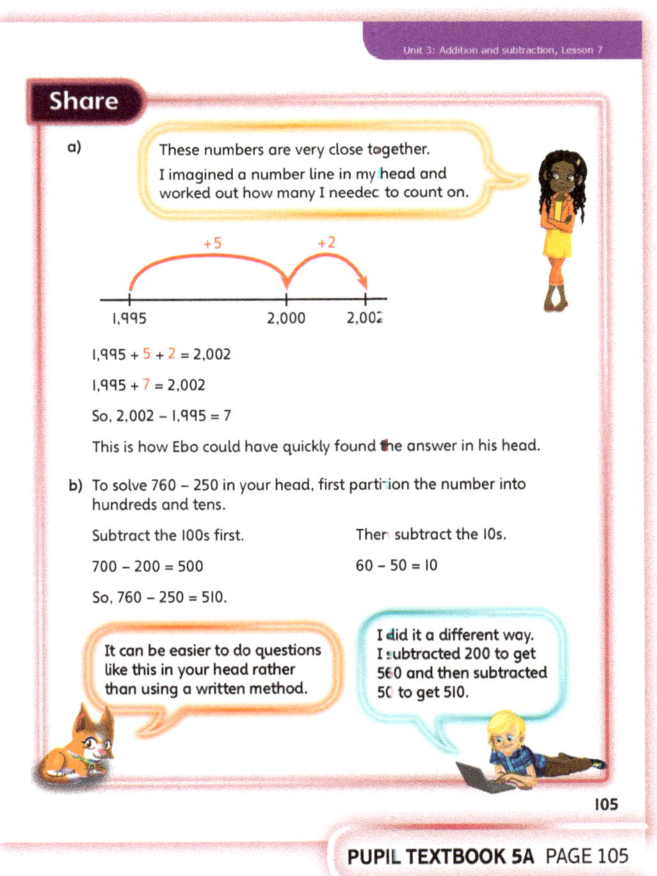

PUPIL TEXTBOOK 5A PAGE 105

Think together

WAYS OF WORKING Whole class teacher led (I do, We do, You do)

ASK

- Question ❶ : *How would you subtract these numbers mentally? Can you match this to a thought bubble?*
- Question ❷ : *What is the next hundred after 498? What do you add on to 498 to get to the next hundred?*
- Question ❸ : *How can you use adding on to mentally work this out? What about the other method?*

IN FOCUS Question ❸ allows children to explore different strategies for doing mental subtractions. Discuss the two methods and link them back to the previous lesson about what to add on to get to the nearest hundred. For the second method, ensure children are confident that this is actually a subtraction. Children need to be confident in the link between addition and subtraction, in that $360 + ? = 750$ and $750 - 360 = ?$ will both give the same answer. Encourage children to try out both methods with the calculation $640 + ? = 920$, and then decide which they prefer.

STRENGTHEN To support understanding, encourage children to imagine or draw a number line, and to think about how much they need to add on to get to the next ten, for example.

DEEPEN For question ❶, ask children if they can write a different thought bubble for any of the calculations, for example adding on to the next ten instead of subtracting. Consider setting up a class 'Maths Challenge'. Give children calculations, such as $402 - 394$, and ask them to mentally work them out as quickly as they can. Encourage them to also explain their method.

ASSESSMENT CHECKPOINT Can children choose an appropriate mental strategy for subtracting whole numbers? Answers to question ❶ will show if children have fully understood the available methods.

ANSWERS

Question ❶ : 76 – 40: $76 - 40 = 36$ (place value)

76 – 42: $76 - 40 = 36$, $36 - 2 = 34$
and $70 - 40 = 30$, $6 - 2 = 4$, $30 + 4 = 34$
(partitioning)

72 – 46: $72 - 40 = 32$, $32 - 2 = 30$, $30 - 4 = 26$
(counting back)

Question ❷ a): $506 - 498 = 8$

Question ❷ b): $710 - 697 = 13$

$4,302 - 4,299 = 3$

$10,005 - 9,987 = 18$

Question ❸ a): Both methods work because $360 + ? = 750$ and $750 - 360 = ?$ are both suitable ways to find the missing number.

$640 + \underline{280} = 920$

Question ❸ b): This will be each child's personal choice.

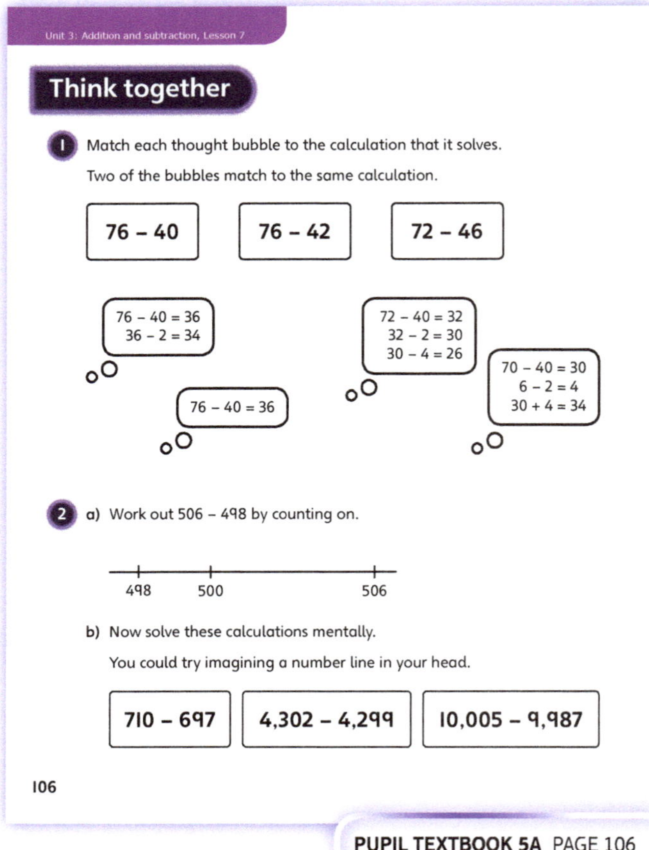

PUPIL TEXTBOOK 5A PAGE 106

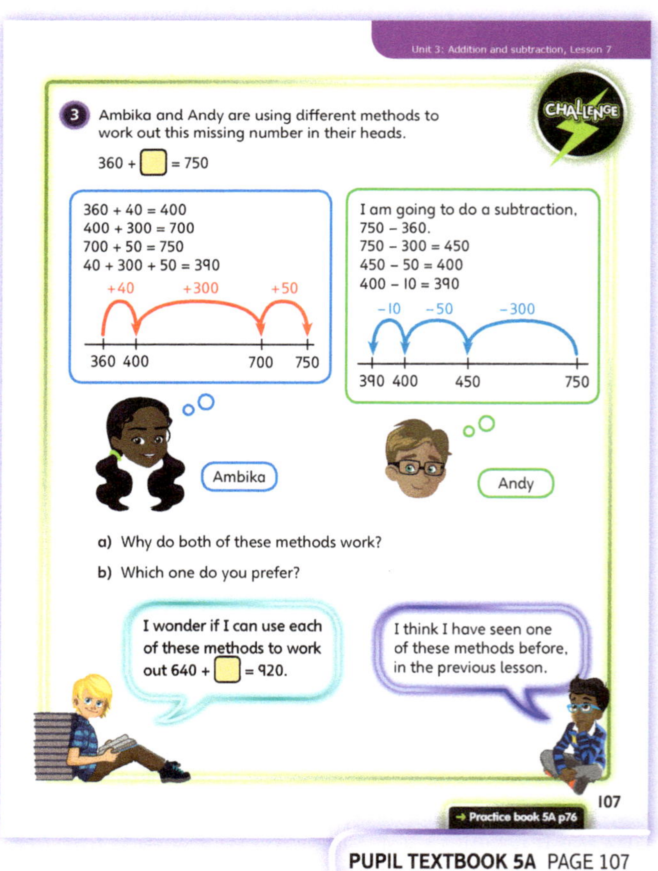

PUPIL TEXTBOOK 5A PAGE 107

Practice

WAYS OF WORKING Independent thinking

IN FOCUS Question ❶ requires children to use two different methods to work out a calculation. Encourage children to discuss both methods even if they have a preferred one. For question ❷, children can use their preferred method. Each question builds upon the previous one, to encourage children to make connections between calculations. Encourage children to compare their answer for each question and discuss how the answers have changed. In question ❹ children must think carefully about which strategy they choose for each calculation. They may realise that adding on is quicker for some questions because the numbers are quite close together.

STRENGTHEN Encourage children to use a number line to help with adding on and place value counters to help with partitioning.

DEEPEN Question ❺ can be explored further by giving children other missing number problems with more than two whole numbers, for example 740 – 260 – ? = 190. Give numbers verbally for children to add mentally. Include numbers that are not as close together or that feature a mixture of addition and subtraction.

THINK DIFFERENTLY Question ❸ asks children to fill in the missing numbers, to describe the method and to show this on a number line, making the link between subtraction and adding on.

ASSESSMENT CHECKPOINT Can children confidently subtract whole numbers mentally, choosing from a variety of strategies and identifying the most appropriate each time?

ANSWERS Answers for the **Practice** part of the lesson appear in the separate **Practice and Reflect answer guide**.

Reflect

WAYS OF WORKING Independent thinking

IN FOCUS This **Reflect** activity checks that children can explain what method to use to mentally subtract with two whole numbers. Encourage children to carefully explain how they would carry out the calculation as well as actually answering it. Encourage children to imagine or sketch a number line, if necessary.

ASSESSMENT CHECKPOINT Can children explain how to subtract with two whole numbers mentally, and why a particular mental method is appropriate?

ANSWERS Answers for the **Reflect** part of the lesson appear in the separate **Practice and Reflect answer guide**.

After the lesson ⏸

- Can children subtract with two whole numbers mentally?
- Can children identify the various mental subtraction methods that are available?
- Do children know what features to look for to choose the most sensible method?

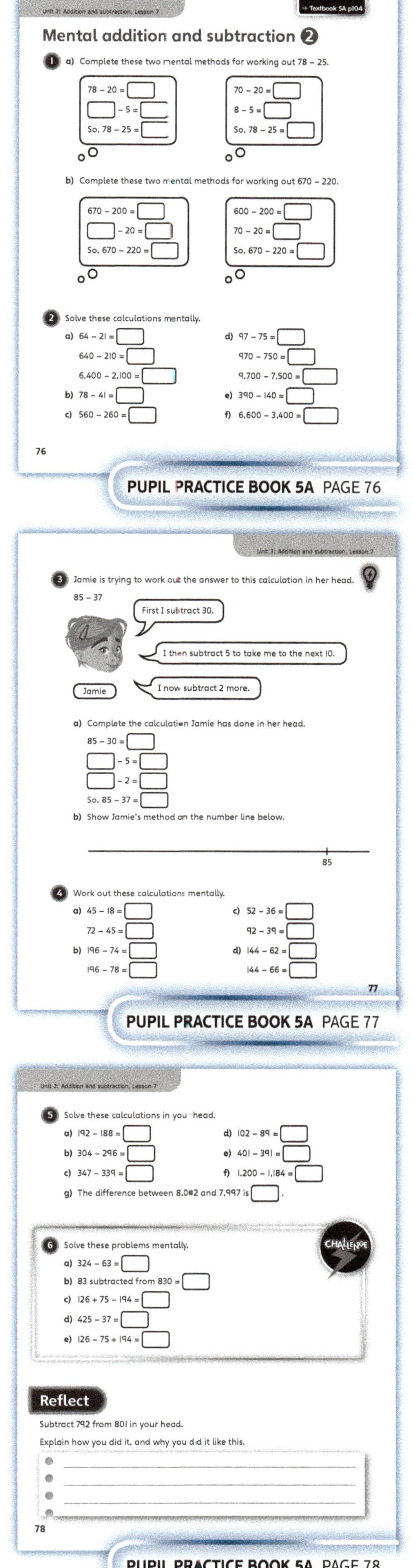

PUPIL PRACTICE BOOK 5A PAGE 76

PUPIL PRACTICE BOOK 5A PAGE 77

PUPIL PRACTICE BOOK 5A PAGE 78

139

Using inverse operations

Learning focus

In this lesson, children will learn how to use the inverse operation in order to check the answers to addition and subtraction calculations.

Small steps

→ Previous step: Mental addition and subtraction (2)
→ **This step: Using inverse operations**
→ Next step: Problem solving – addition and subtraction (1)

NATIONAL CURRICULUM LINKS

Year 5 Number – Addition and Subtraction

Estimate and use inverse operations to check answers to a calculation.

ASSESSING MASTERY

Children can use the inverse operations of addition and subtraction to check the answers to calculations.

COMMON MISCONCEPTIONS

Children may choose the wrong numbers when trying to identify the inverse calculation. For example, for $3,482 - 1,232 = 2,250$, they do $3,482 + 2,250$ instead of $1,232 + 2,250$. Ask:
• *What type of calculation is this? What are you trying to check? Does your answer make sense?*

STRENGTHENING UNDERSTANDING

Children can use a part-whole model or a bar model to help them see why addition and subtraction are inverse operations of each other.

GOING DEEPER

Give children questions that involve both mental addition and subtraction and ask them to mentally check the answer to each calculation using the inverse operation.

KEY LANGUAGE

In lesson: addition, subtraction, check, equal, exchange, inverse, operation, fact family

STRUCTURES AND REPRESENTATIONS

column method addition, column method subtraction, part-whole model

RESOURCES

Optional: place value counters

 In the eTextbook of this lesson, you will find interactive links to a selection of teaching tools.

Before you teach 

• Can children write out the four number sentences from a part-whole model?
• Can children add and subtract using the column method?

Discover

 Pair work

- Question **1** a): *Which numbers do you need to add together? What do you think the answer might be?*
- Question **1** b): *Why do you think this person has made this mistake?*

 Question **1** a) is used to demonstrate why it is useful to use the inverse operation in order to check a calculation. Question **1** b) builds on this and requires children to identify the mistake that has been made.

 Use place value counters to make the numbers in Reena and Lee's calculation. Use them to model why Lee's answer cannot be right.

Question **1** a): Reena is correct as 2,355 + 5,191 is equal to 7,546.

Question **1** b): Lee should have exchanged 1 hundred for 10 tens so that he could do the subtraction.

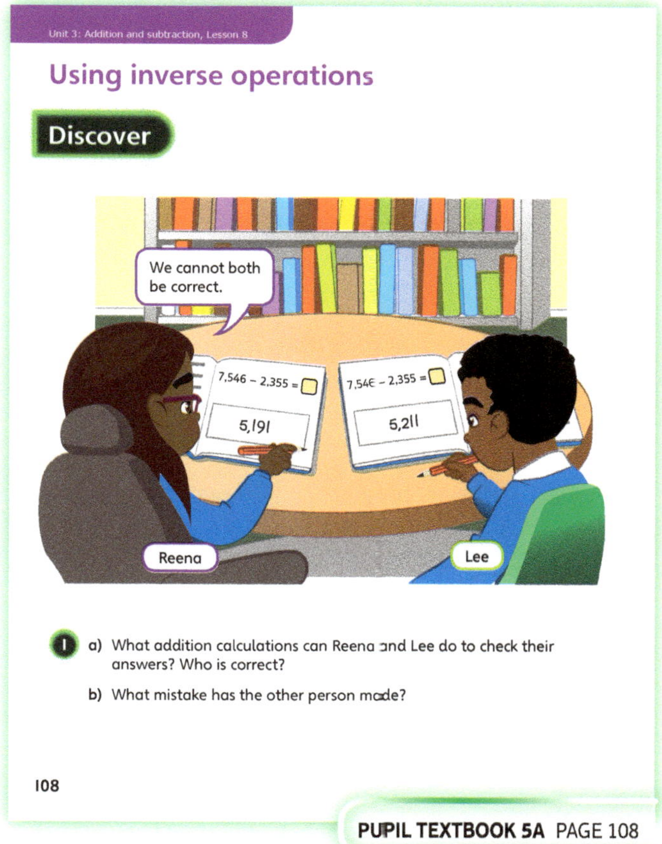

PUPIL TEXTBOOK 5A PAGE 108

Share

 Whole class teacher led

- Question **1** a): *Why should you use addition to check the calculation? What is the whole? What are the parts? What should you add together?*
- Question **1** b): *Look at the place value for each column, where has Lee made his mistake?*

 For question **1** a) show children the part-whole model and discuss why we can use addition to check the calculation. Emphasise that addition is the inverse of subtraction. Show children Reena and Lee's column additions. Look at the place value of each digit. For question **1** b) discuss with children if we can easily subtract 5 tens from 4 tens, identify the exchange that needs to take place, of 10 tens for 1 hundred, and encourage children to spot Lee's mistake here. Can children explain why the total is 7,546 and not 7,566?

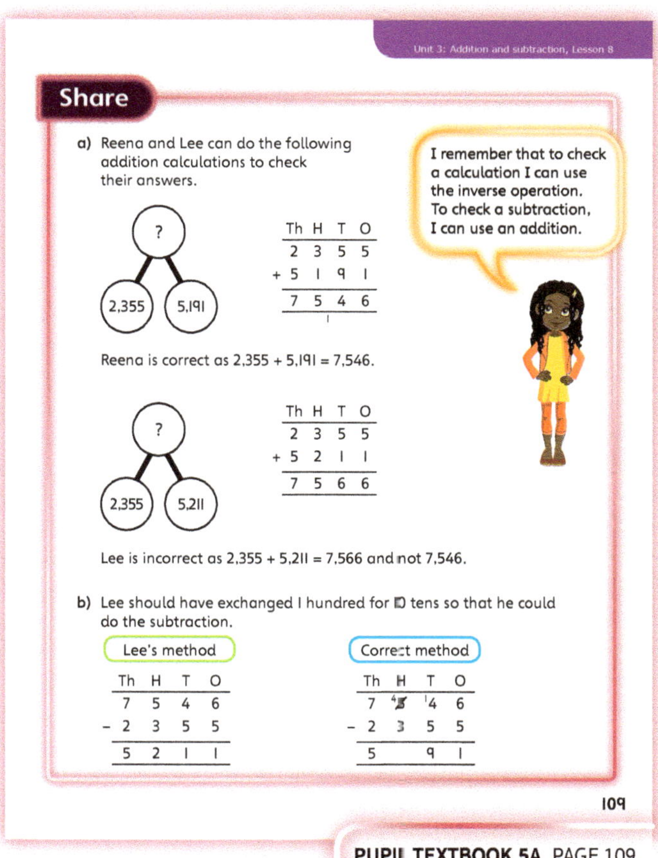

PUPIL TEXTBOOK 5A PAGE 109

Think together

WAYS OF WORKING Whole class teacher led (I do, We do, You do)

ASK

- Question **1** : *What is the inverse operation of addition? Can you use this to check Reena and Lee's answers?*
- Question **2** : *Why do you use addition to check Reena and Lee's answers? Are there any other ways that you could check the answers?*
- Question **3** : *What is the whole? What are the parts? Can you write more than one calculation?*

IN FOCUS Question **2** looks at using the inverse operation to check a calculation where two different mistakes have been made. Discuss if 11,315 is a sensible answer to 46,795 – 3,548 and link this to the lesson about using rounding to check answers. For question **3** b), discuss how it may be easier to subtract 1 from each number so that we have 9,999 rather than 10,000 and do not have to deal with an exchange. Can children explain why subtracting 1 from each number will not affect the overall answer?

STRENGTHEN To support understanding encourage children to use a part-whole model to show how addition and subtraction are inverse operations of each other and to help them identify which numbers they need to add or subtract. Provide blank boxes with the addition and subtraction signs written in, so they just need to fill in the numbers.

DEEPEN Question **3** can be explored further by giving children other part-whole models and asking them to write out all the possible number sentences then to use inverse operations to check they are correct. Give children more calculations that involve subtracting from numbers, such as 1,000 and 10,000.

ASSESSMENT CHECKPOINT Can children identify and use the inverse operation, and other useful mental methods, to check the answers to calculations and find and correct mistakes?

ANSWERS

Question **1** a): The correct answer is 23,405 + 7,892 = 31,297, so Lee is correct.

Question **1** b): Reena has not exchanged 10 hundreds for 1 thousand or 10 thousands for 1 ten thousand.

Question **2**: The correct answer is 46,795 – 3,548 = 43,247 Reena: 11,315 + 3,548 = 14,863. This is incorrect as Reena has not laid out the column subtraction correctly. The 8 ones need to be under the 5 ones, and so on.
Lee: 43,253 + 3,548 = 46,801. This is incorrect as Lee has not exchanged 1 ten for 10 ones and has just done 8 – 5 = 3 in the ones column.

Question **3** a): 770 + 230 = 1,000   230 + 770 = 1,000
1,000 – 770 = 230   1,000 – 230 = 770

Question **3** b) and c): The calculation is incorrect because 10,000 – 7,730 = 2,270 and 10,000 – 3,270 = 6,730.

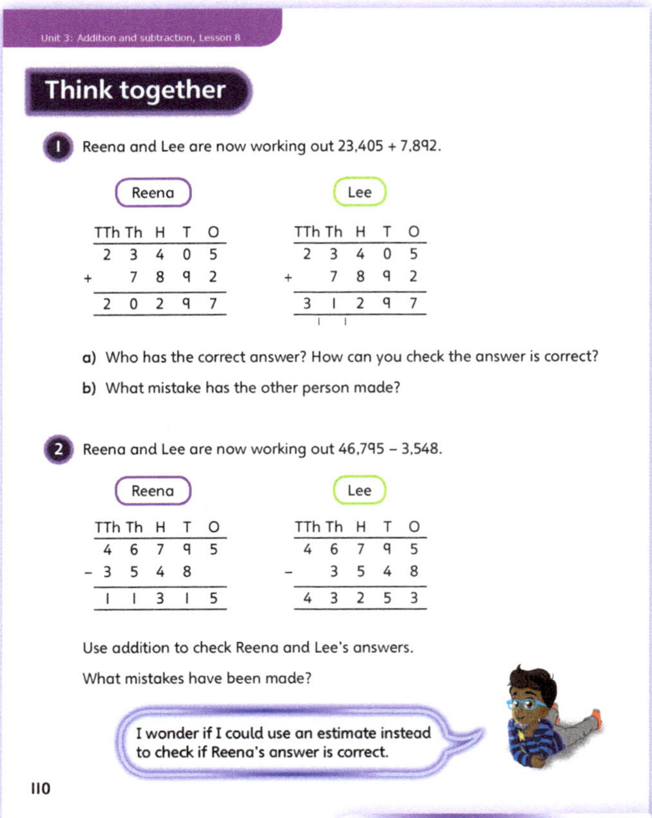

PUPIL TEXTBOOK 5A PAGE 110

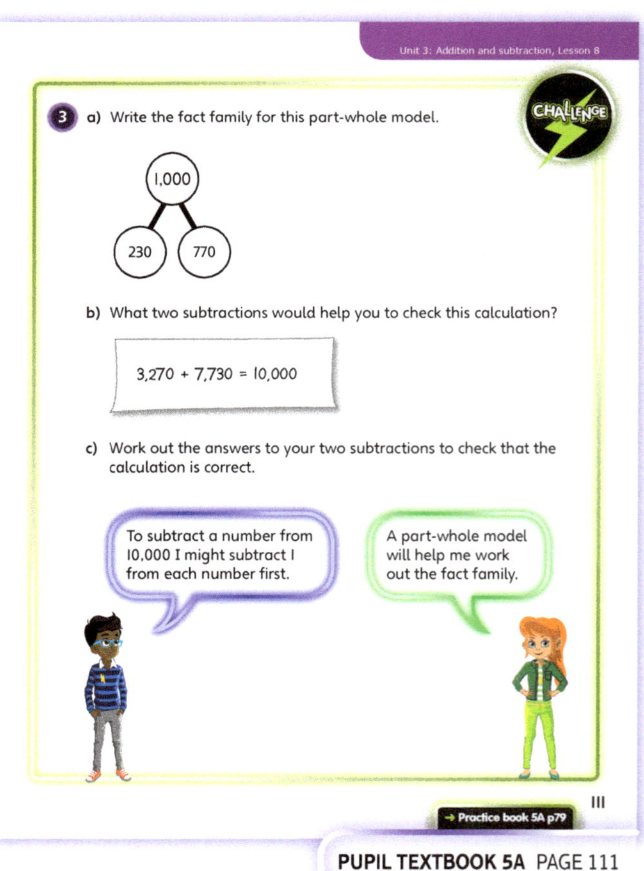

PUPIL TEXTBOOK 5A PAGE 111

Practice

WAYS OF WORKING Independent thinking

IN FOCUS Question ② consolidates children's understanding of the number sentences that can be written from a part-whole model and encourages children to use addition to check if a calculation is correct. Questions ③ and ④ require children to find the mistake that each child has made. Encourage children to use the inverse operation and not to just redo the calculation when checking it. Question ⑤ asks children to find the original calculation when given the inverse calculation. Encourage children to come up with more than one solution for this.

STRENGTHEN Give children a blank part-whole model and encourage children to fill them in with the numbers from each question, so that they can see if they need to add or subtract in order to check the inverse calculation.

DEEPEN Question ⑤ can be explored further by giving children other inverse calculations and asking them to find the original calculation. Give children missing number problems, such as _ + 346 = 742, ask them to work out the missing number and then to use the inverse operation to check they have got the correct answer.

ASSESSMENT CHECKPOINT Can children use the inverse operation to check the answers to calculations and to find and correct mistakes with confidence?

ANSWERS Answers for the **Practice** part of the lesson appear in the separate **Practice and Reflect answer guide**.

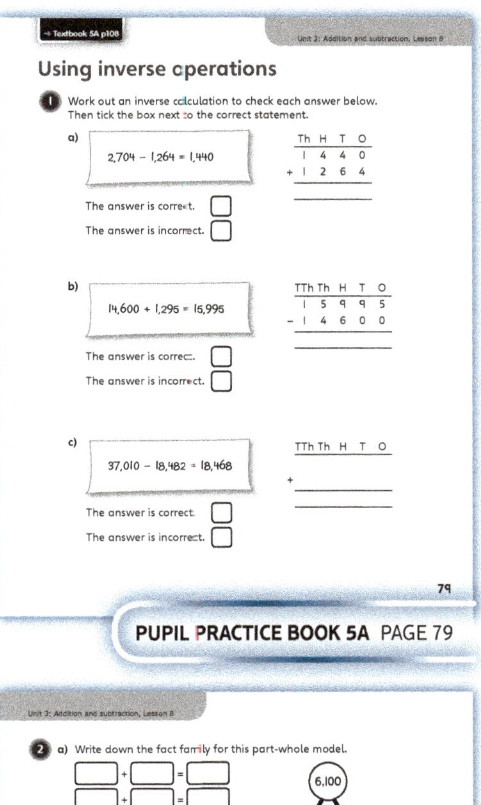

PUPIL PRACTICE BOOK 5A PAGE 79

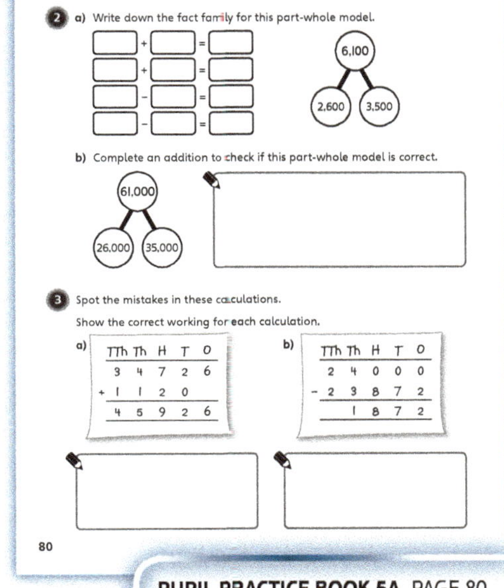

PUPIL PRACTICE BOOK 5A PAGE 80

Reflect

WAYS OF WORKING Independent thinking

IN FOCUS This **Reflect** activity checks that children are able to explain why using the inverse operation to check a calculation is more useful than just redoing the calculation, which may be more prone to mistakes.

ASSESSMENT CHECKPOINT Can children explain why using the inverse operation is an important and useful way to check and correct answers?

ANSWERS Answers for the **Reflect** part of the lesson appear in the separate **Practice and Reflect answer guide**.

After the lesson 

- Do children understand the importance of checking their answers?
- Can children identify the inverse operation for addition and subtraction?
- Which children need to use part-whole models to help work out what the inverse operation is?

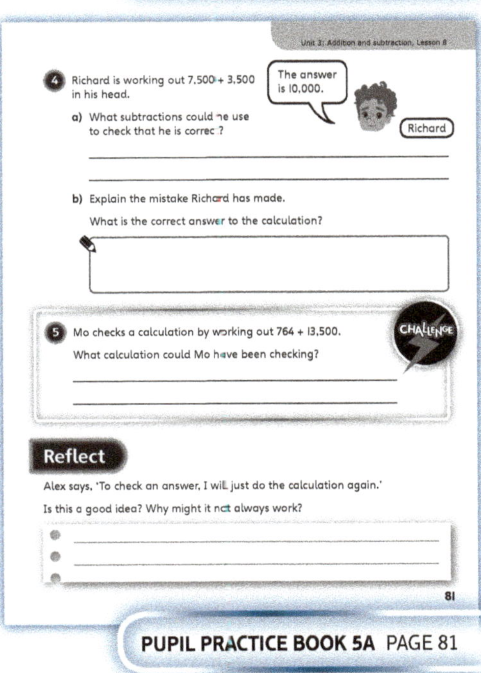

PUPIL PRACTICE BOOK 5A PAGE 81

Problem solving – addition and subtraction ①

Learning focus

In this lesson, children will learn about what strategies to use to solve problems that involve adding and subtracting whole numbers with more than 4 digits.

Small steps

→ Previous step: Using inverse operations
→ **This step: Problem solving – addition and subtraction (1)**
→ Next step: Problem solving – addition and subtraction (2)

NATIONAL CURRICULUM LINKS

Year 5 Number – Addition and Subtraction

Solve addition and subtraction multi-step problems in contexts, deciding which operations and methods to use and why.

ASSESSING MASTERY

Children can solve problems that involve a combination of adding and subtracting whole numbers with more than 4 digits and make multiple exchanges.

COMMON MISCONCEPTIONS

Children may not notice that the vocabulary used in the problem can be used to work out what calculation is needed. Ask:
• *Are there any key words that will give you a clue? What does 'total' or 'difference' mean?*

Children may not carry out both steps of a two-step problem. Ask:
• *What are you trying to find out? What kind of answer are you expecting to get? Have you used all of the information from the question?*

STRENGTHENING UNDERSTANDING

Children should first focus on solving problems with just addition or subtraction, before combining them. Encourage children to use a bar model or a number line to aid their understanding.

GOING DEEPER

Give children problems that are based on information represented in a list or table where they need to identify the correct information first before solving the problem.

KEY LANGUAGE

In lesson: more than, altogether, difference, total, method, addition, subtraction, combined, comparison

STRUCTURES AND REPRESENTATIONS

bar model, number line, column method addition, column method subtraction

 In the eTextbook of this lesson, you will find interactive links to a selection of teaching tools.

Before you teach ⏸

• Do children know how to lay out column addition and subtraction and make multiple exchanges?
• Do children understand key vocabulary, such as 'more than' and 'less than'?

Discover

WAYS OF WORKING Pair work

ASK

- Question ① a): *What is the price of the used sports car? What is the price of the new sports car? What does 'difference' mean?*
- Question ① b): *How much money does Jen have? What do you know about how much Holly has?*

IN FOCUS Question ① a) requires children to identify amounts of money and to find the difference between them, using subtraction. Question ① b) involves working out the total amount of money that Jen and Holly have between them, this is a two-step addition problem.

PRACTICAL TIPS Show an example of a real car sales website or magazine and discuss why it might be useful to compare amounts of money in this context.

ANSWERS

Question ① a): £16,725 – £7,560 = £9,165. The new sports car costs £9,165 more.

Question ① b): Jen and Holly have £6,650 altogether.

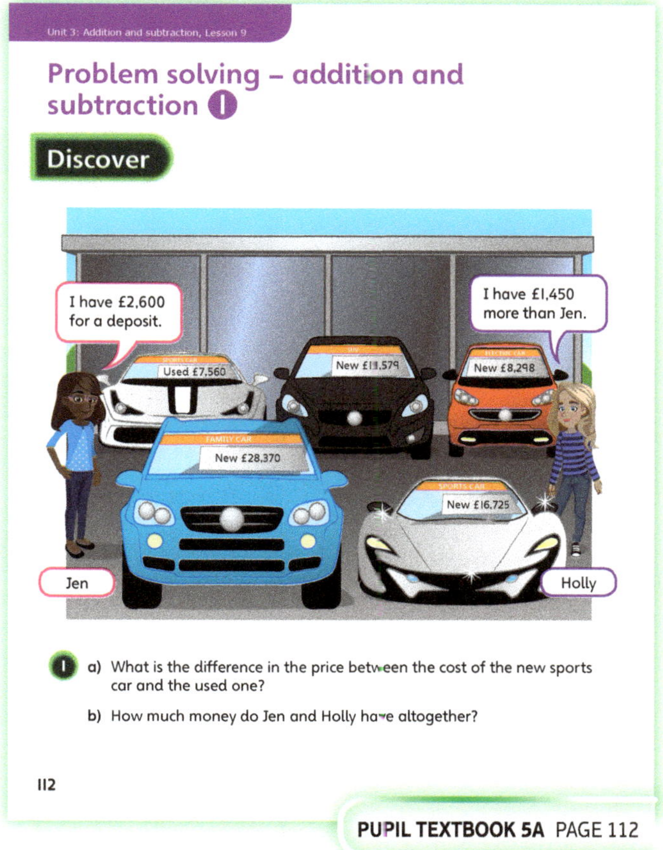

PUPIL TEXTBOOK 5A PAGE 112

Share

WAYS OF WORKING Whole class teacher led

ASK

- Question ① a): *What calculation do you need to do if you are finding the difference? What method could you use to subtract?*
- Question ① b): *If Holly has £1,450 more than Jen, how can you work out how much money Holly has? How can you work out how much money Jen and Holly have altogether?*

IN FOCUS For question ① a), show children the bar model and ask how they know that finding the difference leads us to carrying out a subtraction. Ensure children are confident in the correct layout for a 5-digit subtract 4-digit calculation. Take the opportunity to reinforce the place value of each digit when carrying out the calculation and to address the need for an exchange. For question ① b), discuss how 'more than' indicates the need for an addition, but ensure that children do not just add the two numbers together. Show the bar model, which should help children to see why this will not work. Discuss how we could work out how much money Holly has, and then ask children if this is our answer to the whole question. Discuss how we need to work out how much money Jen and Holly have altogether and so need to do another addition.

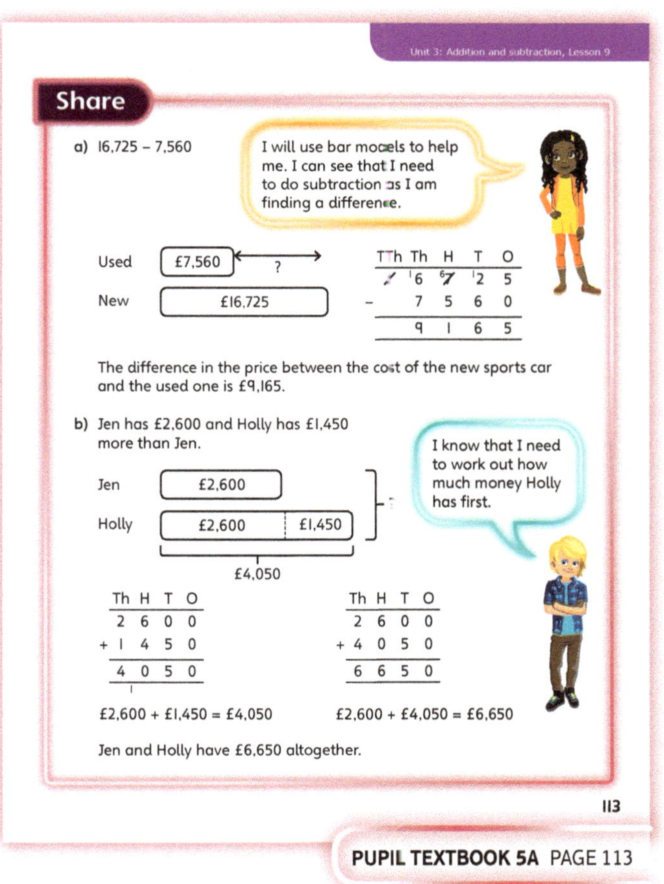

PUPIL TEXTBOOK 5A PAGE 113

Think together

WAYS OF WORKING Whole class teacher led (I do, We do, You do)

ASK

- Question **1** : *How much money do Jen and Holly have altogether? How much does the new sports car cost? What calculation do you need to do?*
- Question **2** : *How much does each car cost? What method can you use to find how much they cost in total?*
- Question **3** : *What does 'combined' mean? How can you work out the combined cost of the SUV and the electric car? How much does the family car cost? How will you find how much more this is?*

IN FOCUS In question **1**, some children may not understand that the word 'more' is now being associated with a subtraction. Use the bar model to show why we need to subtract. Encourage children to use the column subtraction method. Question **2** requires children to problem solve with more than two whole numbers. Show the bar model and discuss how we need to carry out an addition calculation. Discuss the different methods that could be used, adding all three numbers at once in a column addition or adding two numbers in a column addition then adding the other number to this answer. Question **3** encourages children to solve a problem using both addition and subtraction. Encourage children to find the combined cost of the SUV and the electric car to begin with. Discuss how we then need to do a subtraction in order to find the difference in price.

STRENGTHEN To support understanding children should represent all of the problems on a bar model, writing in any information they know onto the bar model and include a question mark to show what they are trying to find out.

DEEPEN Question **3** can be explored further by asking children to find other combined prices to compare to the price of another car by finding the difference. Give a price difference and see if children can work out which pair or group of cars you are comparing.

ASSESSMENT CHECKPOINT Can children solve problems that involve addition, subtraction and a combination of the two, with whole numbers that have 4 or more digits?

ANSWERS

Question **1** : 16,725 – 6,650 = £10,075

Jen and Holly need £10,075 more to buy the new sports car.

Question **2** : 19,579 + 28,370 + 16,725 = £64,674

The three cars cost £64,674 in total.

Question **3** : 19,579 + 8,298 = £27,877

The SUV and the electric car cost £27,877.

28,370 – 27,877 = £493

The family car costs £493 more than this.

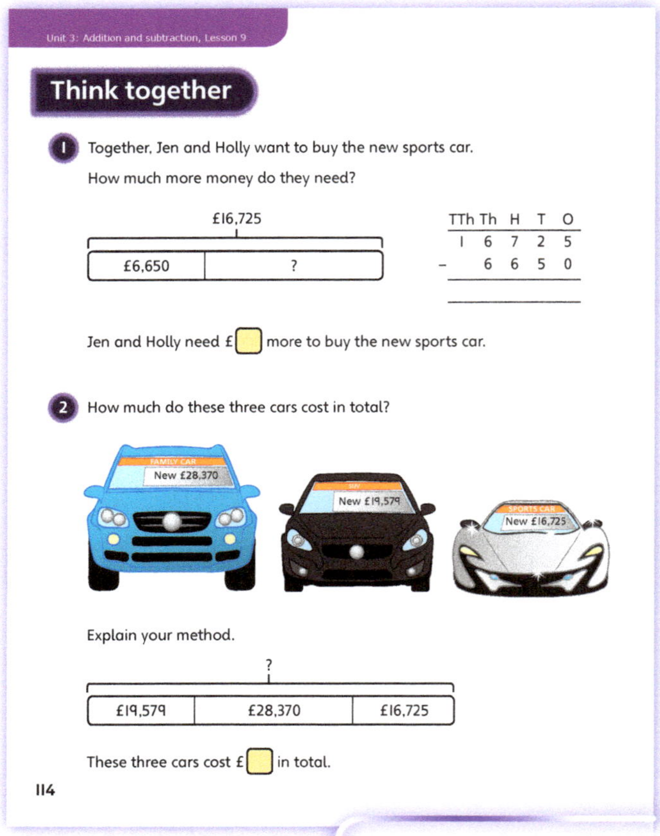

PUPIL TEXTBOOK 5A PAGE 114

PUPIL TEXTBOOK 5A PAGE 115

Practice

WAYS OF WORKING Independent thinking

IN FOCUS For question **1**, ask children to focus on the calculations that arise from each bar model and to show their method for each. Encourage children to discuss the different ways they could solve question **1** c). Question **3** aims to consolidate children's understanding of problem solving with a context. Encourage children to use the bar model and to add in any information as they complete the calculation. Question **4** involves a combination of addition and subtraction. Ensure children lay out the column subtraction correctly. Make sure they realise that 8,000 comes last in the question but needs to be at the top in the column subtraction. Question **5** focuses children on laying out column addition correctly. Discuss the place value of each digit to ensure children line each one up correctly.

STRENGTHEN For question **6**, support children in interpreting the table. Start by asking simple questions about the table, for example: *How many eggs were sold on Saturday? On Sunday? How many eggs were sold at the weekend in total?* Lead from this on to identifying what they need to do to find the answer to the actual question.

DEEPEN Question **2** can be explored further by asking children to draw a bar model from the information shown on the number line.

ASSESSMENT CHECKPOINT Can children confidently add and subtract whole numbers with 4 or more digits and solve problems that involve combinations of these calculations?

ANSWERS Answers for the **Practice** part of the lesson appear in the separate **Practice and Reflect answer guide**.

Reflect

WAYS OF WORKING Independent thinking

IN FOCUS This **Reflect** activity checks children's understanding of using addition and subtraction calculations in a problem-solving context. Encourage children to write a problem where the numbers have 4 or more digits and encourage them to use key vocabulary, such as more, total, difference. Ask children to also solve their problem once they have written it.

ASSESSMENT CHECKPOINT Can children use appropriate language to write a two-step problem that requires using both addition and subtraction to solve it?

ANSWERS Answers for the **Reflect** part of the lesson appear in the separate **Practice and Reflect answer guide**.

After the lesson ❚❚

- Can children apply what they know about addition and subtraction of whole numbers with 4 or more digits in problem-solving contexts?
- Would a class display of key vocabulary from word problems be useful?
- Which children need support to identify each step that is required to solve a two-step problem?

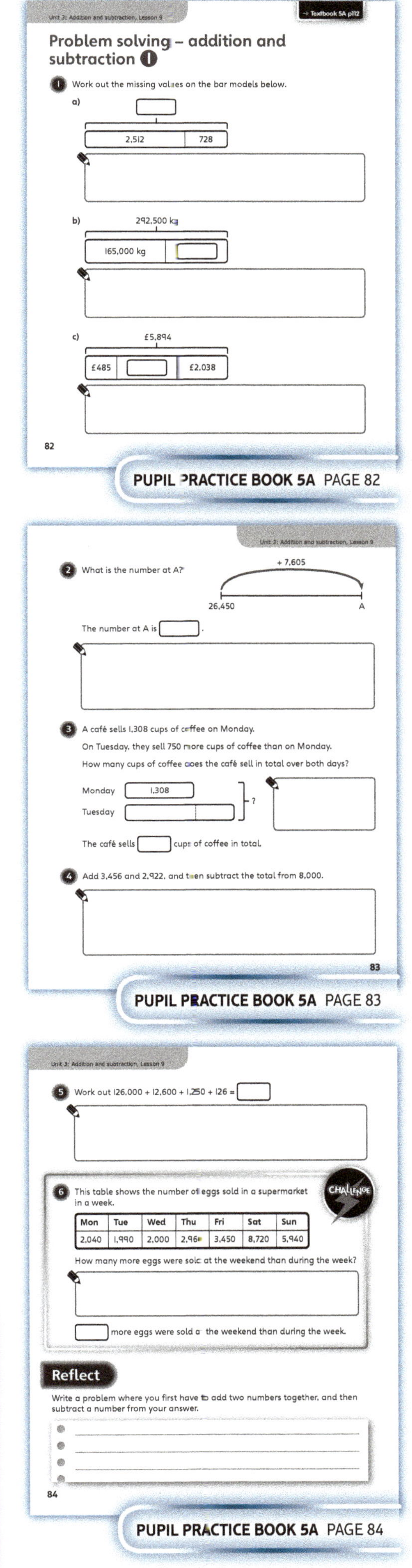

PUPIL PRACTICE BOOK 5A PAGE 82

PUPIL PRACTICE BOOK 5A PAGE 83

PUPIL PRACTICE BOOK 5A PAGE 84

Problem solving – addition and subtraction ②

Learning focus

In this lesson, children will learn how to solve more complex addition and subtraction multi-step problems that involve interpreting and identifying the information in order to solve the problem.

Small steps

→ Previous step: Problem solving – addition and subtraction (1)
→ **This step: Problem solving – addition and subtraction (2)**
→ Next step: Interpreting tables

NATIONAL CURRICULUM LINKS

Year 5 Number – Addition and Subtraction

Solve addition and subtraction multi-step problems in contexts, deciding which operations and methods to use and why.

ASSESSING MASTERY

Children can solve more complex multi-step problems that involve adding and subtracting whole numbers, where the information is represented in tables or needs to be extracted from sentences.

COMMON MISCONCEPTIONS

Children may not understand what calculations a problem is asking them to do and so will just add all of the numbers given, particularly when the information is presented in a table. Ask:
• *What does the table tell you? What information will you use from the table? What model could you draw to help you?*

STRENGTHENING UNDERSTANDING

Encourage children to draw a bar model or a number line to aid their understanding of the information provided in each problem and to draw out what they already know and what calculations they will need to do to find the answer.

GOING DEEPER

Ask children to solve problems based on other mathematical contexts that still involve addition and subtraction. For example, working out the perimeter of a field when given the length and the width.

KEY LANGUAGE

In lesson: how much, more, fewer, total, add, left, two-way table

STRUCTURES AND REPRESENTATIONS

bar model, number line, column method addition, column method subtraction

 In the eTextbook of this lesson, you will find interactive links to a selection of teaching tools.

Before you teach 

• Do children know how to layout column addition and subtraction?
• Are children confident in drawing bar models including comparison bar models?
• Do children understand key vocabulary, such as more and fewer?

Discover

 Pair work

- Question **1** a): *What model does the fuel gauge look like? Can you read it like a number line?*
- Question **1** a) *How much fuel is left? How much fuel was there to begin with?*
- Question **1** b): *How much fuel is left? How much fuel is used each hour? How much fuel is used in two hours?*

 Question **1** a) requires children to identify an amount from a scale and to use this to find the amount of fuel that has been used. In question **1** b), children will need to use more than one calculation to work out how much fuel is left after 2 hours, this is a two-step problem.

 Use water in a measuring jug to reinforce the concept of capacity measured in litres. Gradually pour the water out and demonstrate how the scale of the jug is used to work out how much has been used and how much is left. Link this to the fuel gauge in the picture.

Question **1** a): The plane has used 32,000 litres of fuel so far.

Question **1** b): There will be 37,840 litres of fuel left after two more hours of flying.

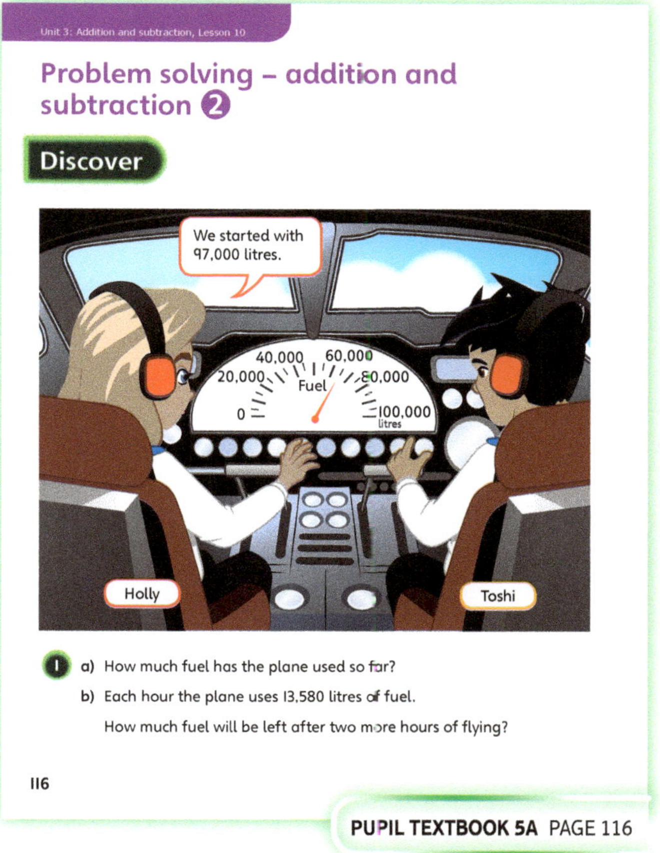

PUPIL TEXTBOOK 5A PAGE 116

Share

 Whole class teacher led

- Question **1** a): *What is the fuel gauge going up in? What number lies exactly half-way between 60,000 and 70,000?*
- Question **1** b): *How could you work out how much fuel is used in two hours? What calculation do you need to do to work out how much fuel is left after two hours?*

 For question **1** a) show the fuel gauge and the number line. Can children explain how they know they are going up in increments of 5,000 litres? Ask children how many litres of fuel there was to begin with and how we could work out how much fuel is left. Discuss the use of the number line and how we can think about the next 10,000 after 65,000 and count up from there to get our answer. Discuss the alternative method of using column subtraction and link this to the lesson about checking answers.

For question **1** b), show the bar model and ask children what calculation we need to do to work out how much fuel is used in 2 hours. Reinforce that this is not the final answer to the question. Show children the next bar model to help them understand why we now need to subtract. Some children may spot the alternative method of subtracting 13,580 twice from 65,000 which is also suitable and could be used as a check.

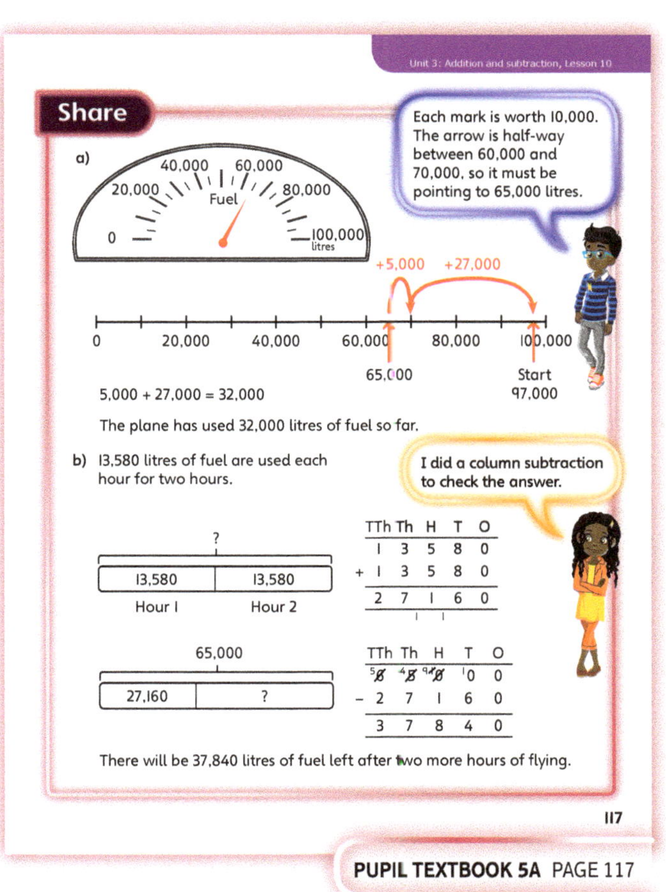

PUPIL TEXTBOOK 5A PAGE 117

Think together

 Whole class teacher led (I do, We do, You do)

ASK

- Question **1** : *Which numbers are before 2 pm? Which are after 2 pm? How will you work out how many passengers there are before 2 pm and after 2 pm?*
- Question **2** : *How will you work out how many passengers the small plane holds? What calculation could you do to work out the total number of passengers on both planes?*
- Question **3** : *Which numbers in the table are not affected by length of flight? Which numbers in the table tell you how much fuel is needed per hour? How will you calculate how much fuel is needed for 4 hours?*

IN FOCUS The information in question **3** is presented in a table, and several numbers and calculations are involved. Encourage children to think about which numbers in the table will be affected by the length of the flight and which will not. Ask children to identify the amount of flight fuel and spare fuel needed for every hour of flying and question how we could work out how much would be needed for 4 hours. Some children may try to add four times. Encourage them to think of a more efficient way. Discuss the need to add the other numbers from the table to get the total amount of fuel needed for a 4-hour flight.

STRENGTHEN Write out question **2** and ask children to highlight the key numbers (416, 280) and words (fewer, how many, total) each in a different colour. Provide a blank comparison bar model and ask children to fill in each piece of information that they know, using the same colours. Reinforce the importance of making the link between the information in the word problem and how it can be represented in a model.

DEEPEN Question **3** can be explored further by asking children to find the amount of fuel needed for other lengths of flight. Take the opportunity for children to practise other key topics, such as multiplying by 10 (for a 10-hour flight) or finding a half (for a half-hour flight).

ASSESSMENT CHECKPOINT Can children use addition and subtraction of whole numbers to solve multi-step problems that involve interpreting tables and information?

ANSWERS

Question **1** : 14,569 + 11,118 = 25,687
25,687 passengers passed through before 2 pm.
23,277 + 5,946 = 29,223
29,223 passengers passed through after 2 pm.
More passengers passed through the airport **after** 2 pm.

Question **2** : 416 − 280 = 136
416 + 136 = 552
The two planes can carry 552 passengers in total.

Question **3** : 12,500 + 2,500 = 15,000
15,000 × 4 = 60,000
60,000 + 5,600 + 5,150 = 70,750
The pilot will need 70,750 litres of fuel for a 4-hour flight.

Think together

1 The table shows the number of passengers passing through an airport on one day.

Did more passengers pass through before 2 pm or after 2 pm?

6 am – 10 am	10 am – 2 pm	2 pm – 6 pm	6 pm – 10 pm
14,569	11,118	5,946	23,277

More passengers passed through the airport _____________ 2 pm.

2 A large plane carries 416 passengers.

A small plane carries 280 fewer passengers.

How many passengers can the two planes carry in total?

The two planes can carry ☐ passengers in total.

118

PUPIL TEXTBOOK 5A PAGE 118

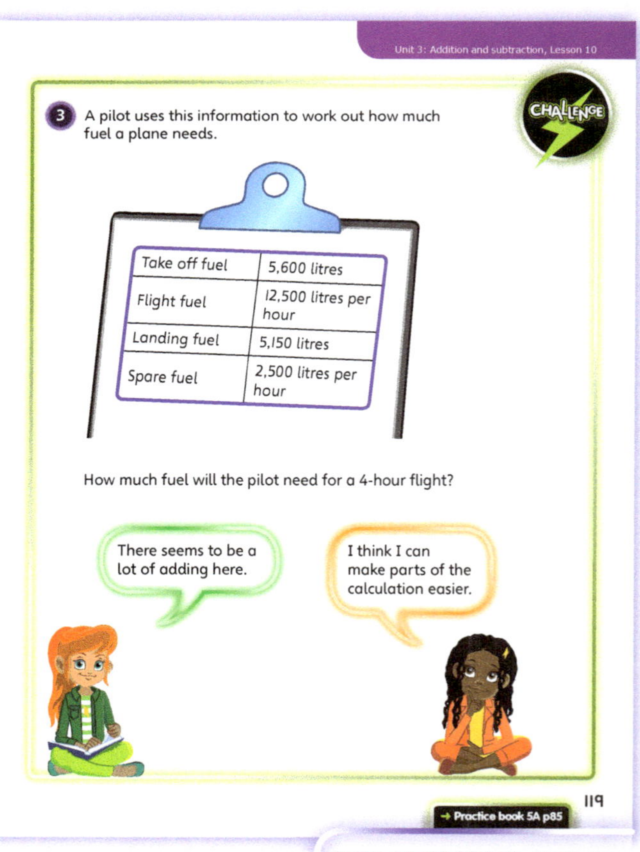

3 A pilot uses this information to work out how much fuel a plane needs.

Take off fuel	5,600 litres
Flight fuel	12,500 litres per hour
Landing fuel	5,150 litres
Spare fuel	2,500 litres per hour

How much fuel will the pilot need for a 4-hour flight?

119

→ Practice book 5A p85

PUPIL TEXTBOOK 5A PAGE 119

Practice

WAYS OF WORKING   Independent thinking

IN FOCUS   Question ① focuses on adding then subtracting to solve a problem. Encourage children to show their method for each part of their calculation. In question ② a), encourage children to think carefully about how they know Tex makes more toys (both numbers in the Tex's column are larger). For question ② b), ensure children do not just work out the total number of toys that Tex makes. Question ③ is a missing number problem with three numbers. Encourage children to draw a bar model to represent the situation and discuss the different methods that are possible to solve it, such as adding the two numbers together and subtracting from 30,000, or subtracting both numbers from 30,000. Question ④ aims to consolidate children's understanding of problem solving with a context. Encourage children to use a comparison bar model, asking questions to help them do this correctly, for example: *Which bar will be longest? How do you know? Which bar will be shortest?* Encourage children to keep re-reading the question to ensure they work out the total amount of apples.

STRENGTHEN   Encourage children to draw a bar model to help them see what calculations are needed, or provide blank bar models so they can fill the numbers in themselves.

DEEPEN   In question ⑤ the information is presented on a number line. This can be explored further by giving children a different starting number on the number line and a different jump size. This could also involve counting back on the number line and asking how many jumps are needed to get past 0.

ASSESSMENT CHECKPOINT   Can children identify the relevant information needed to confidently solve multi-step problems involving adding and subtracting whole numbers, for example in question ④?

ANSWERS   Answers for the **Practice** part of the lesson appear in the separate **Practice and Reflect answer guide**.

Reflect

WAYS OF WORKING   Independent thinking

IN FOCUS   This **Reflect** activity checks children's understanding of addition and subtraction calculations and comparison of large numbers. Encourage children to explain which method they would use to work out the answers to both calculations and how they would then compare their answers to find which is bigger.

ASSESSMENT CHECKPOINT   Can children add, subtract and compare 5- and 6-digit whole numbers?

ANSWERS   Answers for the **Reflect** part of the lesson appear in the separate **Practice and Reflect answer guide**.

After the lesson 

- Do children understand how addition and subtraction skills can be applied to solving problems?
- What mistakes are being most commonly made?
- Can you start to make any links between the tables in this lesson and the tables that children will encounter in the next unit?

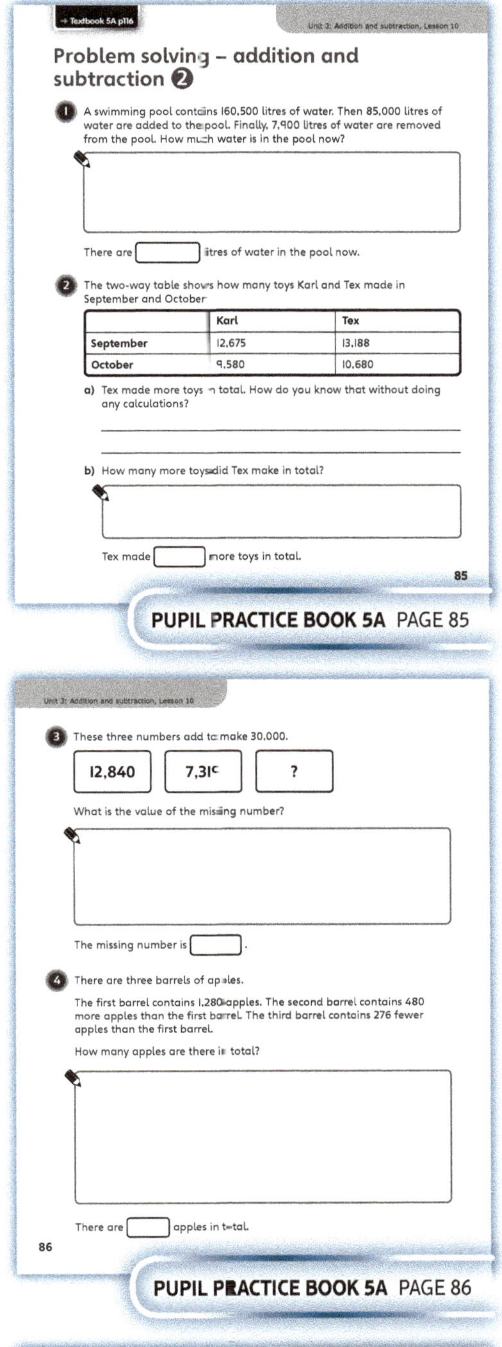

PUPIL PRACTICE BOOK 5A PAGE 85

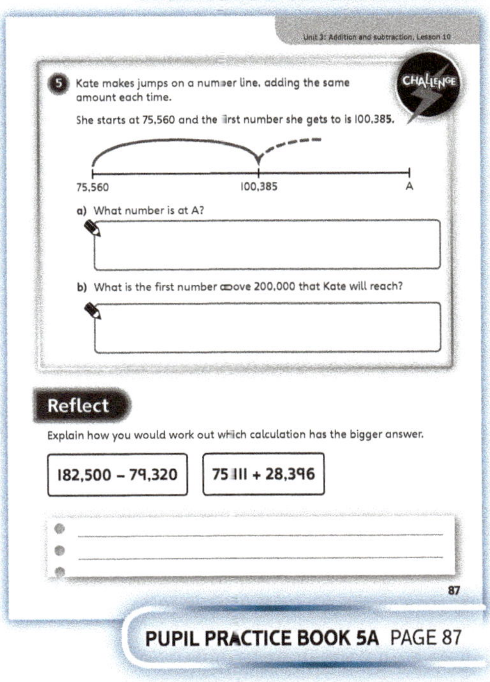

PUPIL PRACTICE BOOK 5A PAGE 86

PUPIL PRACTICE BOOK 5A PAGE 87

End of unit check

Don't forget the *Power Maths* unit assessment grid on p26.

WAYS OF WORKING Group work adult led

IN FOCUS These questions cover the whole unit and are designed to draw out misconceptions and misunderstandings.

Check that children can use the formal method for addition and can correctly align the numbers.

Children should realise that to find a missing number they will need to use an inverse operation, and be able to do this mentally using knowledge of bonds to 100.

When children are solving word problems including those with several steps, encourage them to represent each situation using a bar model to help them work out if they need to add or subtract.

ANSWERS AND COMMENTARY

Children who have mastered the concepts in this unit understand the place value of each digit in 5-digit numbers and can look for key language to see if they are requested to add or subtract in order to answer the calculation. They will have confidence in selecting efficient mental methods to solve problems and know that rounding can help them check their answer.

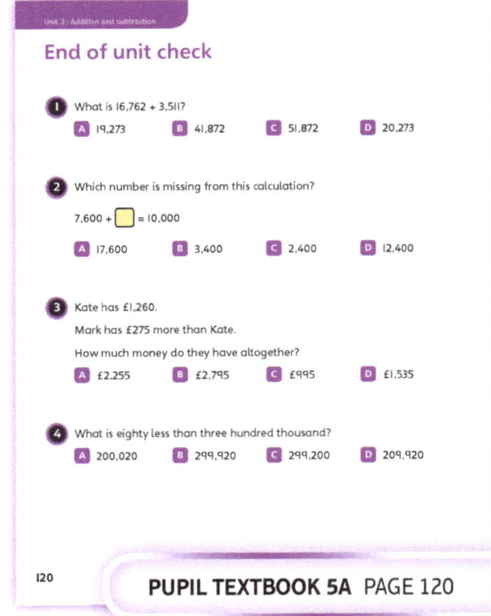

PUPIL TEXTBOOK 5A PAGE 120

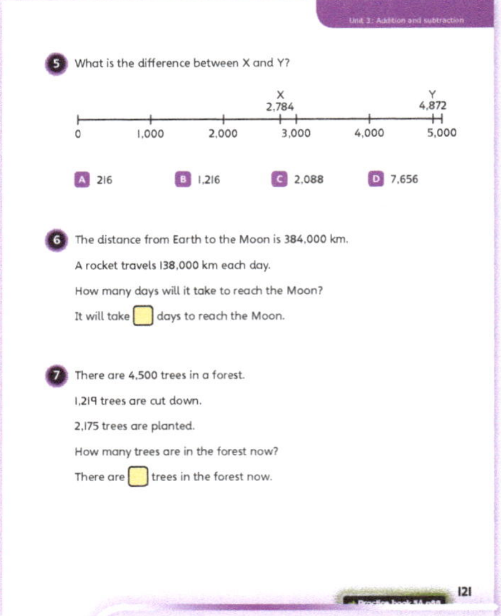

PUPIL TEXTBOOK 5A PAGE 121

Q	A	WRONG ANSWERS AND MISCONCEPTIONS	STRENGTHENING UNDERSTANDING
1	D	A suggests children have forgotten to exchange. B suggests they have lined the numbers up incorrectly and done an incorrect exchange.	Use place value grids and counters to support column methods of addition and subtraction. Work through step by step showing any counter exchanges linking the place value grid to the column method.
2	C	For A, children have added when they needed to subtract. B is an incorrect bond to 100. D is a bond to 20,000, not 10,000.	
3	B	A is an incorrect addition. For C children have subtracted instead of added. For D they have just added the two numbers.	For question 2, help children understand the part-whole method by drawing a simple bar model or part-whole model to show that they should subtract instead of add together.
4	B	For A, C or D children have made a mistake with the place value of 300,000 or 80.	
5	C	For A children have counted up to 3,000 only. B shows they have found the difference up to 4,000. D suggests they have added the numbers together.	Use a bar model to represent each situation and help children decide if they need to add or subtract.
6	3 days	Children should explore both addition and subtraction methods.	
7	5,456 trees	Do children realise they need to do both a subtraction and an addition?	

My journal

WAYS OF WORKING   Independent thinking

ANSWERS AND COMMENTARY   100,000 – 39,480 = 60,520 (Tuesday)

60,520 – 39,480 = 21,040 (difference between Monday and Tuesday)

Encourage children to work out the missing number first, by using the addition and subtraction methods that they have encountered in this unit. Look at different ways that children find the answer. Do they subtract 39,480 from 100,000 or subtract 1 from both numbers and then subtract 39,479 from 99,999?

Children's stories should use the words Monday and Tuesday and associate the number 39,480 with Monday. The number 100,000 should be associated with the sum of Monday and Tuesday. They should use language such as: *How many more on Tuesday than Monday? How many less on Monday? What is the difference between Monday and Tuesday?* Ensure that the numbers are not unrealistic (for example: 'Max cycles 39,480 km on Monday').

Power check

WAYS OF WORKING   Independent thinking

ASK

- *Are you happy to decide when and why it is more efficient to add or subtract two numbers mentally?*
- *Do you know why it is useful to work out an estimate to an addition or subtraction by rounding?*
- *Do you feel confident solving word problems by working out what information you have and what you need to find?*

Power puzzle

WAYS OF WORKING   Pair work

IN FOCUS   In this Power puzzle children have to work out the missing values using a variety of addition and subtraction methods. In the first example, children will need to start by working out numbers where there is one value missing from a row or column. Once they have worked out these missing values, they should see that they can now work out the rest.

In the second example, children are asked to make up their own answers to make the grid correct. Encourage children to double check with addition that all the row and column totals are correct.

ANSWERS AND COMMENTARY

a) Top line: 13,197   5,966,   837
   Middle line: 3,457   11,102 15,441
   Bottom line: 23,346 32,932 3,722

b) One possible answer:

   Top line: 700   150   150
   Middle line: 500   700   800
   Bottom line: 200   1,150 1,650

After the unit ❿

- How many children can confidently add and subtract using formal column methods?
- Can children choose an appropriate mental or written method to add or subtract numbers?

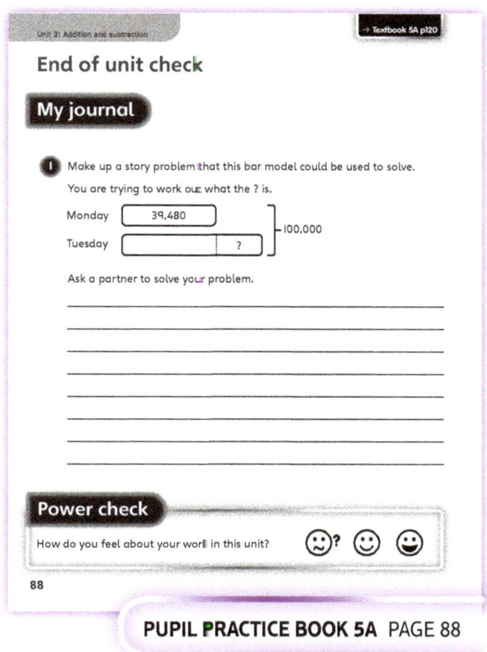

PUPIL PRACTICE BOOK 5A PAGE 88

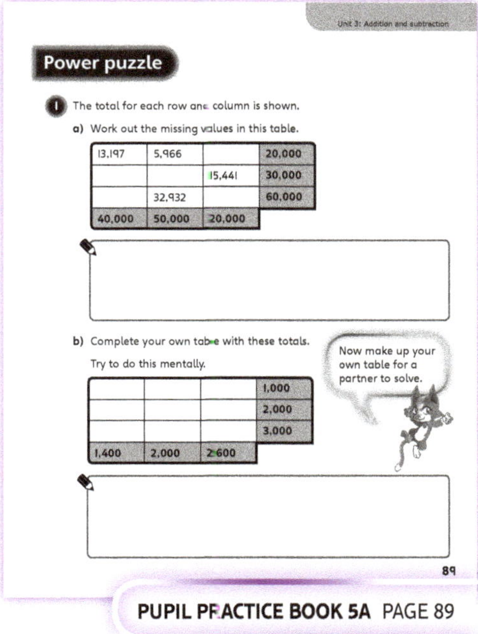

PUPIL PRACTICE BOOK 5A PAGE 89

Strengthen and **Deepen** activities for this unit can be found in the *Power Maths* online subscription.

Unit 4
Graphs and tables

WHY THIS UNIT IS IMPORTANT

This unit builds on children's statistics work from earlier units in both Year 3 and Year 4. The work covered also allows children to apply their knowledge of place value and number operations to solve simple problems based on the data presented in tables and line graphs. Children will look at examples of line graphs that display discrete and continuous data and at dual line graphs. For the first time, children will also make, complete and interpret two-way tables, which allow children to break down data into more categories.

WHERE THIS UNIT FITS

→ Unit 3: Addition and subtraction
→ **Unit 4: Graphs and tables**
→ Unit 5: Multiplication and division (1)

This unit builds on children's work in previous units on bar graphs. Also, it will bring together their understanding of tables and problem solving as they apply the four rules of calculation. Some questions will involve using their knowledge of fractions and measures.

Before they start this unit, it is expected that children:
• know how to extract information from a simple table
• know how to solve simple word problems, involving finding sum and difference
• understand how to tally information
• know how to find missing numbers on a number line and work out or estimate the location of a number.

ASSESSING MASTERY

By the end of the unit children will be able to construct, interpret and complete tables, including two-way tables, understanding the differences between a two-way table and a simple table. Children will be able to extract data from the tables to solve simple problems. Also, children will be able to read, draw and interpret line graphs, including dual line graphs, extracting the relevant data to solve sum and difference problems.

COMMON MISCONCEPTIONS	STRENGTHENING UNDERSTANDING	GOING DEEPER
Children may not accurately read the corresponding values from a line graph or table.	Encourage children to use a ruler and draw horizontal and vertical lines in order to connect and read the correct corresponding values.	Children explore and compare the same data presented in line graphs with different scales. What is the same and what is different about them?
When reading a value between two labelled intervals, children may round to the closest labelled value instead of finding or estimating a more exact value.	Encourage children to think of an axis as a number line. Give children practice in identifying numbers on a number line that lie between labelled values. Use the scales shown in the questions.	Ask children to collect their own data, to record it in a table and then present the relevant information in a line graph. Remind them to think carefully about the scale they use.

Unit 4: Graphs and tables

WAYS OF WORKING

Use these pages with the whole class to revise what children already know and remember about tables and line graphs from Year 4. Look at Ash's graph and table, and ask: *Do these both show the same information? Which representation do you find easier to understand? Do you recognise any of the key language?* Revise the words they should already know and briefly introduce and define the new vocabulary. Finally, look at the number line and give children opportunity to decide where each number should go, ascertaining first that each interval is worth 10.

STRUCTURES AND REPRESENTATIONS

Line graph: plots points using the vertical and horizontal axes to quickly and accurately read data and to easily see the relationship between each point.

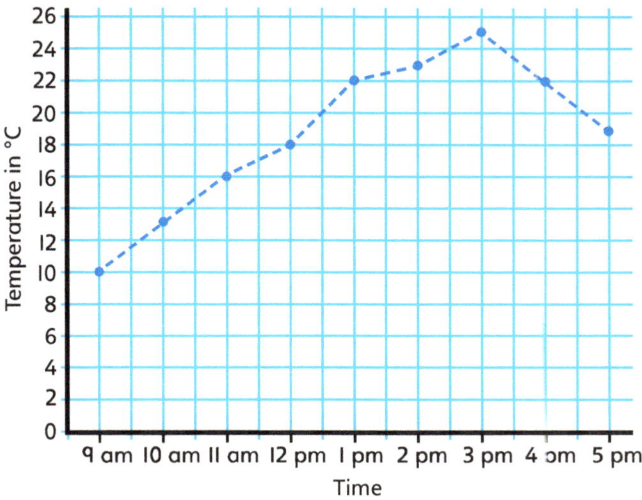

Number lines: children will recognise that reading the horizontal and vertical axes on line graphs is the same as reading number lines.

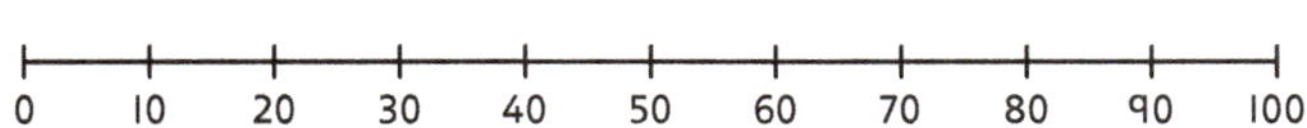

Two-way table: presents two sets of information and can include tallies or digits.

	Spots	Stripes
Socks	8	4
Hats	3	5

KEY LANGUAGE

Here is some key language that children will need to know as part of the learning in this unit:

➜ line graph, **dual line graph**

➜ horizontal axis, vertical axis, axes, scale

➜ data, information

➜ read, interpret, complete

➜ table, **two-way table**.

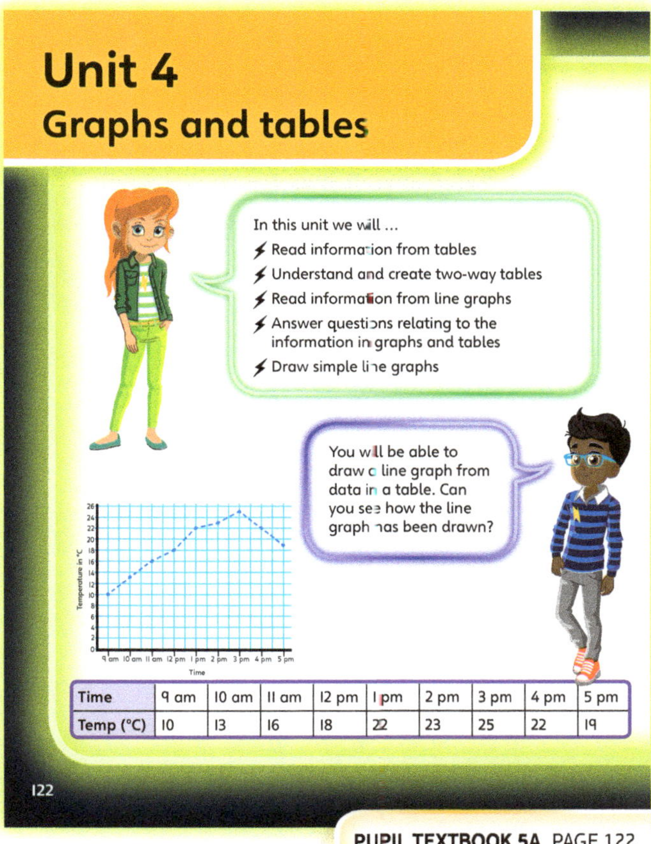

Time	9 am	10 am	11 am	12 pm	1 pm	2 pm	3 pm	4 pm	5 pm
Temp (°C)	10	13	16	18	22	23	25	22	19

PUPIL TEXTBOOK 5A PAGE 122

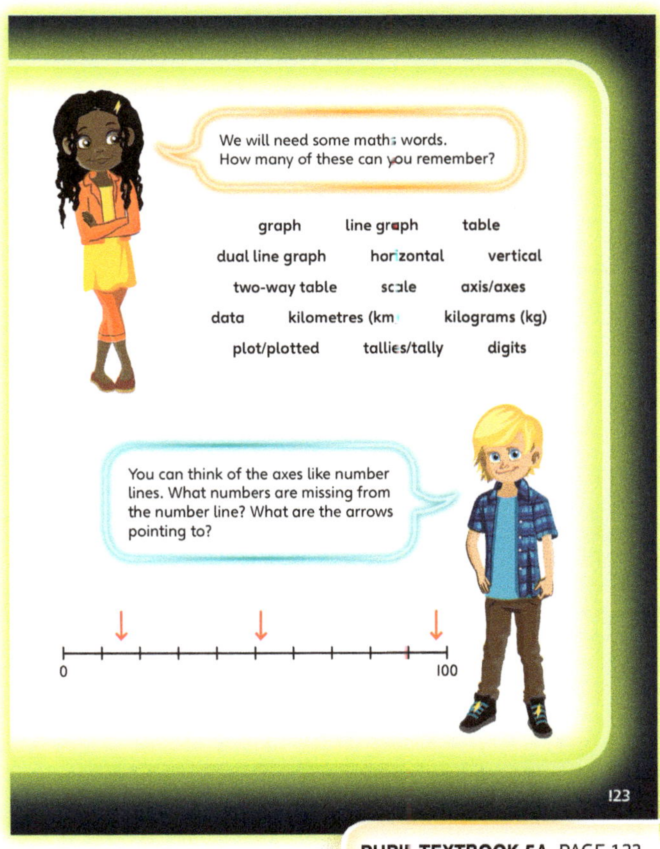

PUPIL TEXTBOOK 5A PAGE 123

Interpreting tables

Learning focus

In this lesson, children will extract information from tables to solve a range of problems involving four operations.

Small steps

→ Previous step: Problem solving – addition and subtraction (2)
→ **This step: Interpreting tables**
→ Next step: Two-way tables

NATIONAL CURRICULUM LINKS

Year 5 Statistics

Complete, read and interpret information in tables, including timetables.

ASSESSING MASTERY

Children can read and interpret information in a table and use this to solve a range of problems involving sums, differences and totals.

COMMON MISCONCEPTIONS

Children can sometimes misread the information shown in a table, retrieving the incorrect information to solve a problem. Ask:

• *What is the question asking? Which rows or columns of the table show that information? What clues are in the question to help you to decide which operation you need to use?*

STRENGTHENING UNDERSTANDING

Many of the difficulties in this lesson will be due to children not being able to work out the correct operation to use. Encourage children to draw a bar model or simple diagram to represent the problem. For tables with a large amount of data, using a ruler to keep track of the relevant row or column could help.

GOING DEEPER

Ask children to do their own research on the internet, from books or in class to create a table of results. They should write questions related to their table of data. For example: Find the cost of a product on 5 different websites; What is the eye colour or height or favourite colour of children in their class?

KEY LANGUAGE

In lesson: table, column addition, total, weight, weighs, distance, kg, km, kilometres. $\frac{1}{4}$, half, more

Other language to be used by the teacher: extract, sum

STRUCTURES AND REPRESENTATIONS

table, bar model

 In the eTextbook of this lesson, you will find interactive links to a selection of teaching tools.

Before you teach ⏸

• Can children extract information from simple tables?
• Can children add, subtract and double numbers up to 3-digits?

Discover

WAYS OF WORKING Pair work

ASK

- Question ❶ a): *What information does the table give? How does putting the information in a table help us to see the information clearly? Which rows give the information you need? What calculation do you need to do?*
- Question ❶ b): *How is this question different from the first one? What information do you need from the table? What method will you use to answer the question? How would a bar model help?*

IN FOCUS Children should be familiar with data displayed in a table so the focus is to extract the relevant information they need to solve a problem. Diagrams such as bar models can help children visualise the calculation that they need to do. Point out the clues that indicate whether the calculation is an addition (1 a) or a subtraction (1 b).

PRACTICAL TIPS You may want to copy various tables from books, the internet and newspapers for children to analyse by making up their own questions about the data shown.

ANSWERS

Question ❶ a): 127 + 152 = 279

In total, 279 loaves of bread were sold on Monday and Thursday.

Question ❶ b): 200 − 123 = 77

77 loaves were **not** sold on Friday.

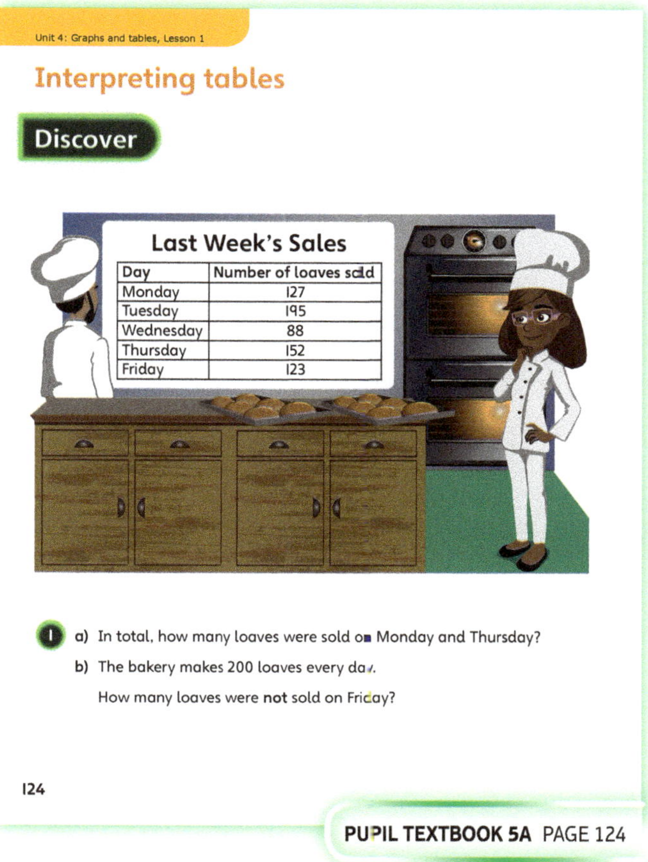

PUPIL TEXTBOOK 5A PAGE 124

Share

WAYS OF WORKING Whole class teacher led

ASK

- Question ❶ a): *What information do you use from the table? What do you do with the numbers? Why do you add? What words are in the question to help you? How does the bar model represent the situation? What method do you use to add these numbers together?*
- Question ❶ b): *What calculation do you do in this question? Why do you need to do a subtraction? Is there any information you need to use that is not in the question? What two methods are shown? Which method do you prefer?*

IN FOCUS In question ❶ a) children must decide which operation they need to use in order to answer the question. The word 'total' suggests addition, although explain to them that it is necessary to check the rest of the words in the question carefully.

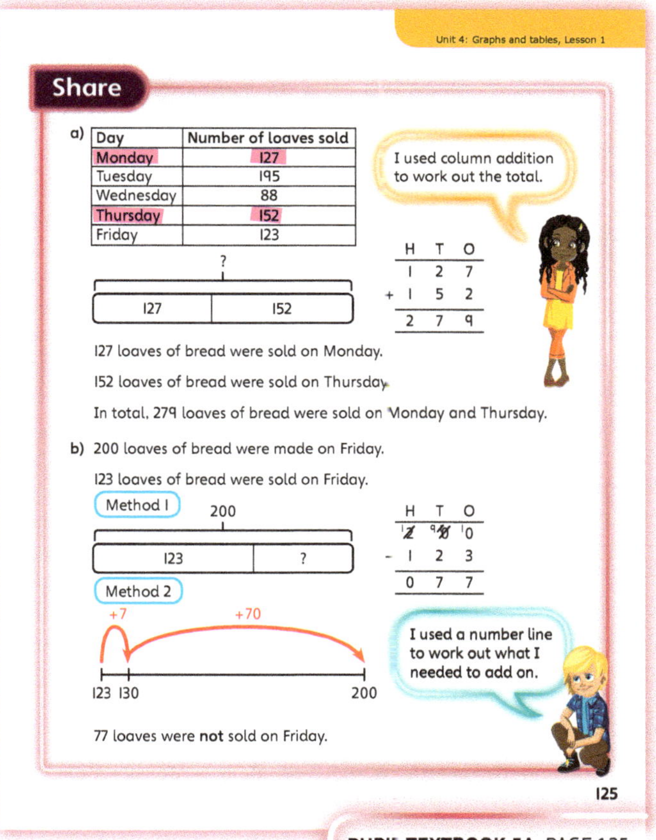

PUPIL TEXTBOOK 5A PAGE 125

Think together

WAYS OF WORKING Whole class teacher led (I do, We do, You do)

ASK

- Questions **1** and **2** : *What information is given in the table? What is the unit of measure?*
- Question **1** : *How is this table different from the table in* **Discover**? *What numbers lie between 19 and 20?*
- Question **2** : *What do the decimals mean? How can you find the missing value? What do you need to do first? What do you add up first? What do you have to do next?*
- Question **3** : *How many children took part in the survey in total? How do you know? What does $\frac{1}{4}$ mean? Can you find two ways of checking whether the statement is true?*

IN FOCUS In questions **1** and **2** the tables are presented horizontally and include a unit of measure. Children must identify what the numbers in the table represent. While question **3** involves a table presented vertically and calls upon knowledge of fractions from previous year groups.

STRENGTHEN Children are more likely to need extra help with interpreting what the question is asking rather than with extracting the data. Some may need support in question **2** with adding 3·5 and 7·5. Show this on either a number line or a bar model. All the questions could be modelled on bar models. Sometimes writing their own question about the data can strengthen children's understanding.

DEEPEN Ask children to make up their own word problems about the data in the tables or in a new table you provide. Also, you might want to ask children to bring in other areas of maths. For example, in question **2** what fraction of the total distance had Lexi run by Monday?

ASSESSMENT CHECKPOINT Are children able to extract the relevant information from a table and use it to solve addition, subtraction and multiplication problems?

ANSWERS

Question **1** a): Ernie weighs 20 kg.

Question **1** b): Both Arnie and Digga are the same weight.

Question **1** c): Charlie weighs 26 kg. Buddy weighs 19 kg.

$$26 - 19 = 7$$

Charlie weighs 7 kg more than Buddy.

Question **1** d): Charlie is overweight.

Question **1** e): Rufus could weigh [accept answer between 19 kg and 20 kg].

Question **2** : Lexi needs to walk 8 more kilometres on Friday.

Question **3** : Holly is not correct. Exactly $\frac{1}{4}$ of the children say blue is their favourite colour ($\frac{1}{4}$ of 40 = 10), so not **more than** $\frac{1}{4}$.

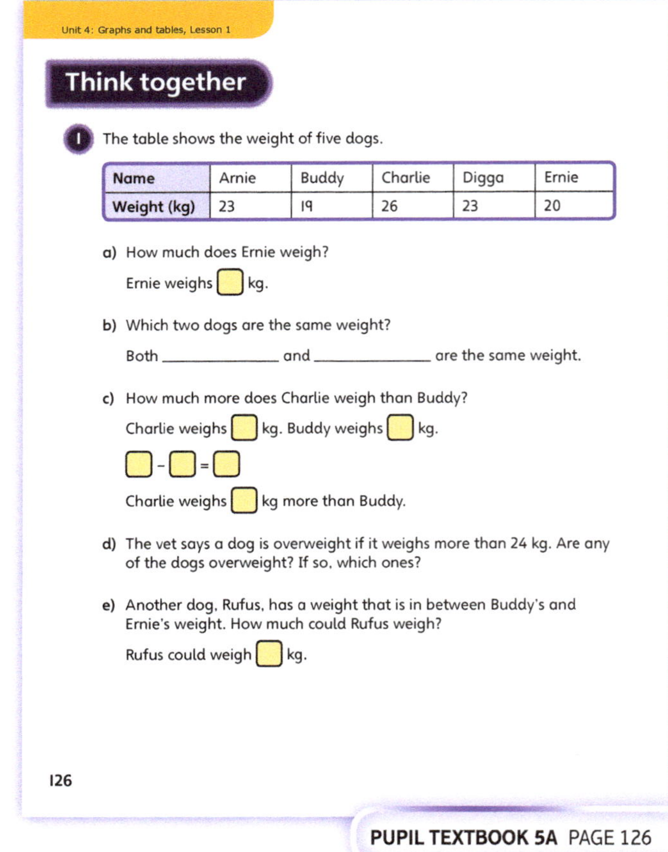

Name	Arnie	Buddy	Charlie	Digga	Ernie
Weight (kg)	23	19	26	23	20

PUPIL TEXTBOOK 5A PAGE 126

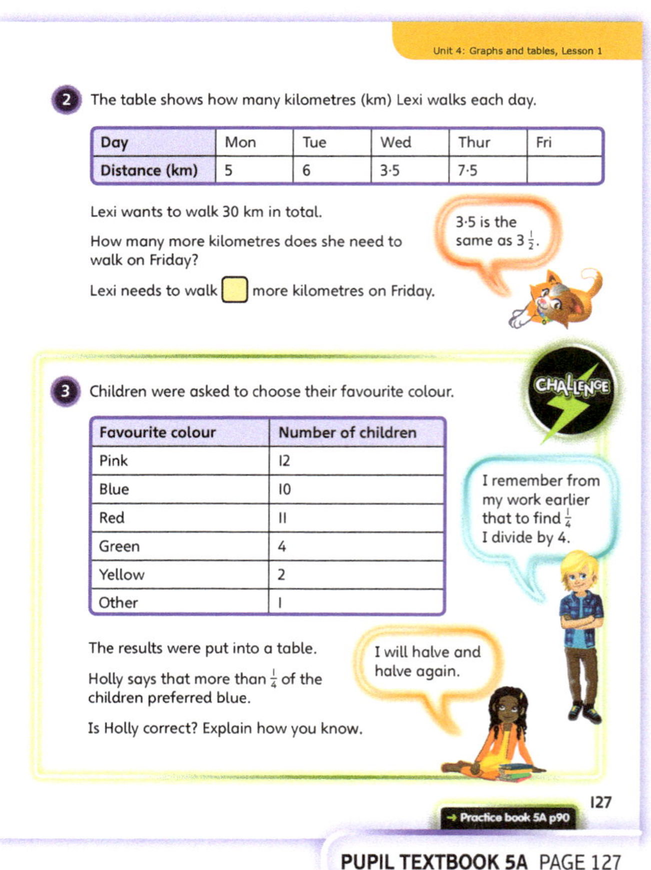

Day	Mon	Tue	Wed	Thur	Fri
Distance (km)	5	6	3·5	7·5	

Favourite colour	Number of children
Pink	12
Blue	10
Red	11
Green	4
Yellow	2
Other	1

PUPIL TEXTBOOK 5A PAGE 127

Practice

WAYS OF WORKING Independent thinking

IN FOCUS Question **1** provides practice of extracting information from a table and using it to solve a variety of problems involving addition, subtraction and multiplication.

Question **4** requires long multiplication and the challenge of working out half of £15. Encourage an efficient strategy i.e. total hours × 15, instead of multiplying each value by 15 and then adding.

STRENGTHEN The majority of difficulties with these questions will not be posed by tables, but by the calculations involved. Encourage children to explain the problem in their own words, to draw a simple diagram or bar chart to illustrate what is being asked before they then extract the relevant information from the table.

DEEPEN Ask children to make up a table with four rows of data, involving a unit of measure, similar to those in the lesson. They should write four questions for a partner to answer using information in their table. You may wish to prompt ideas: heights of four people; weights of four cats; capacity of four containers; duration of four TV programmes.

ASSESSMENT CHECKPOINT Can children extract the relevant information from a table, using it to solve a range of problems involving whole and decimal numbers?

ANSWERS Answers for the **Practice** part of the lesson appear in the separate **Practice and Reflect answer guide**.

Reflect

WAYS OF WORKING Independent thinking

IN FOCUS In this **Reflect** task children are given freedom to write their own statements about data in a table. Some prompts are given to support children who may have difficulty getting started. Encourage children to make statements of increasing complexity.

ASSESSMENT CHECKPOINT Can children extract information from the table and use it to write factual statements?

ANSWERS Answers for the **Reflect** part of the lesson appear in the separate **Practice and Reflect answer guide**.

After the lesson ⏸

- Did children know which operation they needed to use in order to answer the question?
- Can children clearly identify the type of information tables provide them with?
- Which other curriculum areas use tables that could be used to practise the skills used in this lesson?

PUPIL PRACTICE BOOK 5A PAGE 90

PUPIL PRACTICE BOOK 5A PAGE 91

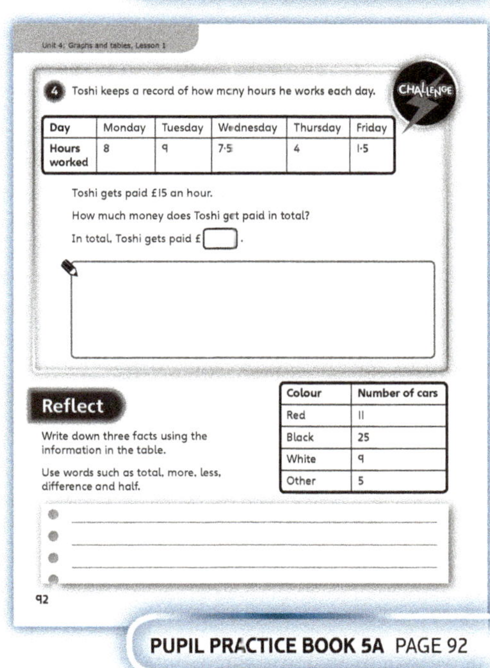

PUPIL PRACTICE BOOK 5A PAGE 92

Two-way tables

Learning focus

In this lesson, children will create and extract information from two-way tables.

Small steps

→ Previous step: Interpreting tables
→ **This step: Two-way tables**
→ Next step: Interpreting line graphs (1)

NATIONAL CURRICULUM LINKS

Year 5 Statistics

Complete, read and interpret information in tables, including timetables.

ASSESSING MASTERY

Children can create their first two-way table, understand what the numbers in each cell of the table represents and be able to find the row and column totals. Children are also able to extract information from two-way tables and use the information to solve simple sum and difference problems.

COMMON MISCONCEPTIONS

At first, two-way tables can be difficult to understand. Children may need support understanding what each of the numbers mean in the table. Discuss each cell in turn, explaining what each of the numbers refers to. Ask:
• *What are the headings for the rows? What are the headings for the columns? Which row and column meet at this number? What type of information goes in this cell?*

STRENGTHENING UNDERSTANDING

This is the first time children have seen two-way tables like this so they may initially need help to understand the numbers. Take a selection of red and blue pens and pencils, then get children to sort the pens and pencils into four groups: red pens, blue pens, red pencils, blue pencils. Record it in a two-way table, asking children to suggest the headers. Work out the row and column totals, discussing what each total shows.

GOING DEEPER

Children should generate their own two-way tables. Children have done simple surveys but they will need to think of two categories to compare. For example: girl or boy and Year 5 or Year 6; school dinner or packed lunch and left-handed or right-handed; 3-sided or 4-sided shapes and red or blue colours.

KEY LANGUAGE

In lesson: two-way table, tally, tallies, row, column, information, total, difference, fraction

Other language to be used by the teacher: extract, sum, record, cell, represent, data, compare

STRUCTURES AND REPRESENTATIONS

two-way table

RESOURCES

Optional: 2D shapes in 2 colours, counters

 In the eTextbook of this lesson, you will find interactive links to a selection of teaching tools.

Before you teach ⏸

• Can children extract information from simple tables?
• Can children sort objects into different groups based on different properties?

Discover

WAYS OF WORKING Pair work

ASK

- Question **1** a): *What are the two headings for the rows and columns? Where would a sock with stripes go? Where would a hat with spots be recorded? What item goes into this top cell?*
- Question **1** b): *How can you use the table to work out the total number of hats? What about working out the total number of socks? Does the table tell us how many items with stripes there are?*

IN FOCUS Question **1** is the first time that children have met two-way tables and so they may need help initially to understand the layout and how they work. To help children understand, print some large two-way tables (or you could use rulers to create them) and get children to physically sort groups of objects into them.

PRACTICAL TIPS Make a grid using metre sticks and write the labels: girl or boy for the columns and Jan–June birthday or July–Dec birthday for the rows. Then ask individuals to find their place in the grid. Ask questions about each cell, for example: *Where would a boy born in April go on the grid?*

ANSWERS

Question **1** a): 8, 4, 3 and 5 tallies in the respective boxes

Question **1** b): 12 – 8 = 4

There are 4 more socks than hats.

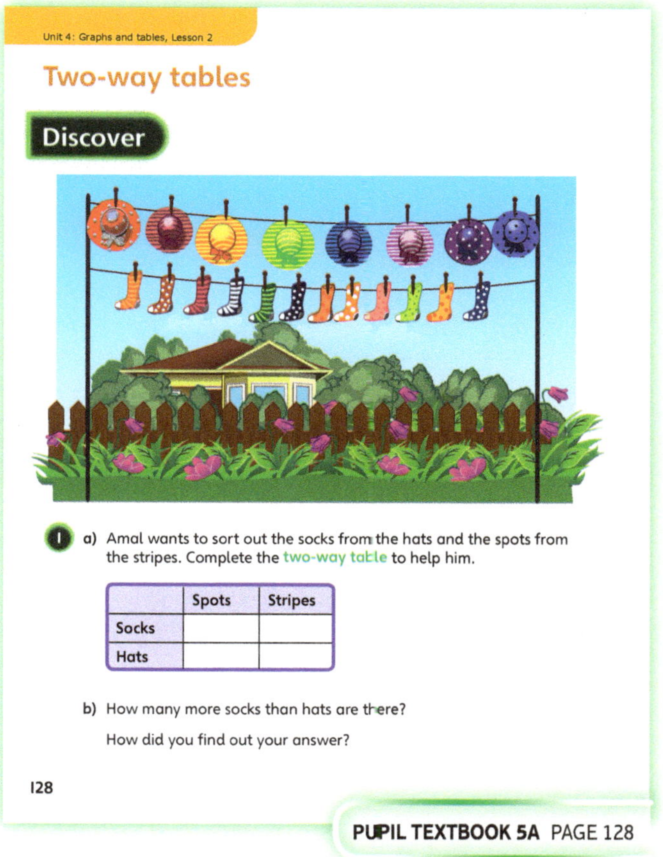

PUPIL TEXTBOOK 5A PAGE 128

Share

WAYS OF WORKING Whole class teacher led

ASK

- Question **1** a): *What is the first item on the washing line? Where should you make the tally mark for that? How do you show 5? Why do you think you show 5 that way? Which item is the tally of 8 representing in the table? How has Dexter worked out the number of each item? Which method is more reliable or quicker?*
- Question **1** b): *How can you work out the total number of hats and socks? Did anyone go back and count the answers on the number line? Where can you write the totals? How did Flo work out the totals? How can you check that you have got all of the items?*

IN FOCUS Part a) links back to previous learning about tally charts. It cements children's understanding that two-way tables show two different sets of information and so they need to be aware of both the type of clothing and the pattern.

For part b) the individual cell totals have been put in and the row and column totals have been added. Ensure children know what each of the numbers mean in the table. For example, the 12 means that there are 12 socks in total.

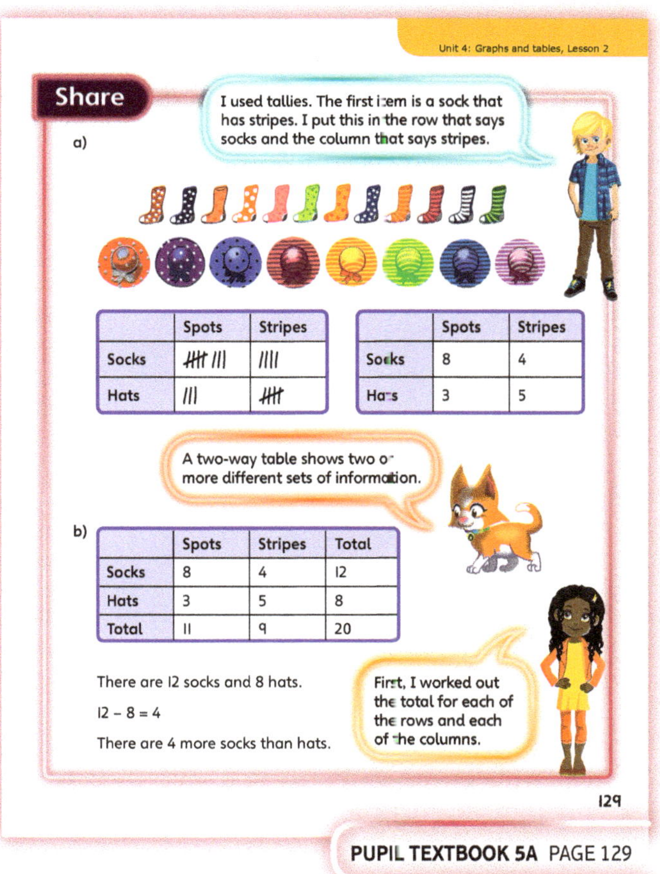

PUPIL TEXTBOOK 5A PAGE 129

Think together

ASK

- Question **2** : *What five statements can you make with the information in the table? What other information can you tell me? Can you give me a statement that involves a fraction?*
- Question **3** : *What is different about this table? Are there any cells you cannot work out straight away?*

IN FOCUS Questions **1** and **2** help children to identify which missing numbers they can work out straight away through 1-step addition or subtraction or a 2-step operation and which they need more information for. Encourage children to check their answers by checking row and column totals work in all directions. Take time to explain each table and ensure that they fully understand what the numbers mean and reassure them that the table in question **3** works in the same way.

STRENGTHEN Discuss what each number in each cell represents and what would happen to the numbers if you added more. For example, in question **1** if one more child went to the cinema on Saturday, ask: *Which numbers will change?* Also, in question **2** if two more pears are put in class 5A's fruit bowl, ask: *How will that change the information in the table?*

DEEPEN For question **1**, ask children more questions, such as what fraction of the people who went to the cinema over the weekend were children? What fraction of the children attended on Saturday? In the second question children have to realise that the denominator (total) is not 50.

ASSESSMENT CHECKPOINT Children are able to extract information from a two-way table and solve simple sum and difference problems.

ANSWERS

Question **1** a): Add 24 in the total column. Add 28 in the Sunday column.

Question **1** b): 5 children went to the cinema on Saturday.

Question **1** c): In total, 28 people went to the cinema on Sunday.

Question **1** d): 14 more children than adults went to the cinema on Sunday.

Question **1** e): The greatest number of people went to the cinema on Sunday.

Question **2** a): Add 12 in the Apples column. Add 10 in the Pears column. Add 54 in the Total column.

Question **2** b): Five statements should relate to the total number of apples or pears or pieces of fruit in each class bowl. Comparisons of the types of fruit or the amount of fruit in each class bowl can also be used.

Question **3** a): In the Small column add 8, 28 and 38. In the Medium column add 62. In the Large column add 6. In the Total column add 56 and 83.

Question **3** b): $\frac{50}{150}$ or $\frac{25}{75}$ or $\frac{5}{15}$ or $\frac{1}{3}$

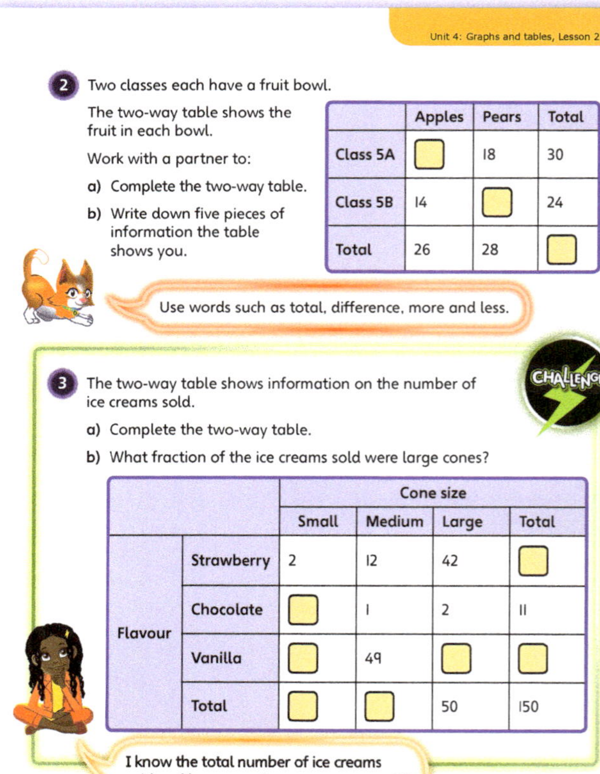

Think together

1 The two-way table shows information about 50 people who went to the cinema last weekend.

	Saturday	Sunday	Total
Children	5	21	26
Adults	17	7	
Total	22		50

a) Fill in the missing totals then answer the following questions.

b) How many children went to the cinema on Saturday?

   children went to the cinema on Saturday.

c) In total, how many people went to the cinema on Sunday?

In total,   people went to the cinema on Sunday.

d) On Sunday, how many more children than adults went to the cinema?

   more children than adults went to the cinema on Sunday.

e) On which day did the greatest number of people go to the cinema?

The greatest number of people went to the cinema on _____________ .

130

PUPIL TEXTBOOK 5A PAGE 130

2 Two classes each have a fruit bowl. The two-way table shows the fruit in each bowl.

Work with a partner to:

a) Complete the two-way table.

b) Write down five pieces of information the table shows you.

	Apples	Pears	Total
Class 5A		18	30
Class 5B	14		24
Total	26	28	

3 The two-way table shows information on the number of ice creams sold.

a) Complete the two-way table.

b) What fraction of the ice creams sold were large cones?

Flavour		Small	Medium	Large	Total
	Strawberry	2	12	42	
	Chocolate		1	2	11
	Vanilla		49		
	Total			50	150

(Column group header: **Cone size** spans Small, Medium, Large, Total)

131

→ Practice book 5A p93

PUPIL TEXTBOOK 5A PAGE 131

Practice

WAYS OF WORKING Independent thinking

IN FOCUS Question ① starts with a tally for children to complete. This reinforces the headings on both sides of the two-way table and the information held within the table.

Question ④ requires children to fill in cells from a series of statements before they can finally answer the question involving finding a fraction of a number. This requires children to recall and use past knowledge so they may need help revising.

STRENGTHEN To strengthen understanding of question ①, have children make the shapes shown to physically put each one into a two-way table that has been drawn on a piece of paper or mini whiteboard.

It is important to ensure that children understand each table before attempting to answer the questions or make statements about it, so spend time discussing it. They may also need support in filling in the missing values.

DEEPEN Ask children to draw a two-way table of their own with some values filled in and others missing. How do they ensure that enough values are filled in to find out the rest? Challenge them to fill in the minimum number of values in order for it still to be possible to complete the table. Alternatively, they could write a series of statements, similar to in question ④, and ask a partner to draw or complete a two-way table for it.

ASSESSMENT CHECKPOINT Children should be able to understand and complete a two-way table and extract relevant information from it to answer questions involving sums and differences.

ANSWERS Answers for the **Practice** part of the lesson appear in the separate **Practice and Reflect answer guide**.

Reflect

WAYS OF WORKING Independent thinking

IN FOCUS This **Reflect** task asks children to compare the tables from the previous lesson to the two-way tables covered in this lesson. Children may find it easier to explain by drawing an example of each table. Ask children what is the same and what is different about these tables.

ASSESSMENT CHECKPOINT Do children know when to record data in a simple table or when to record it in a two-way table? Are they able to show and explain the differences?

ANSWERS Answers for the **Reflect** part of the lesson appear in the separate **Practice and Reflect answer guide**.

After the lesson ⏸

- Can children create and complete a simple two-way table?
- Can they extract relevant information from a two-way table?
- Are children able to solve simple sum and difference problems relating to the data in a two-way table?

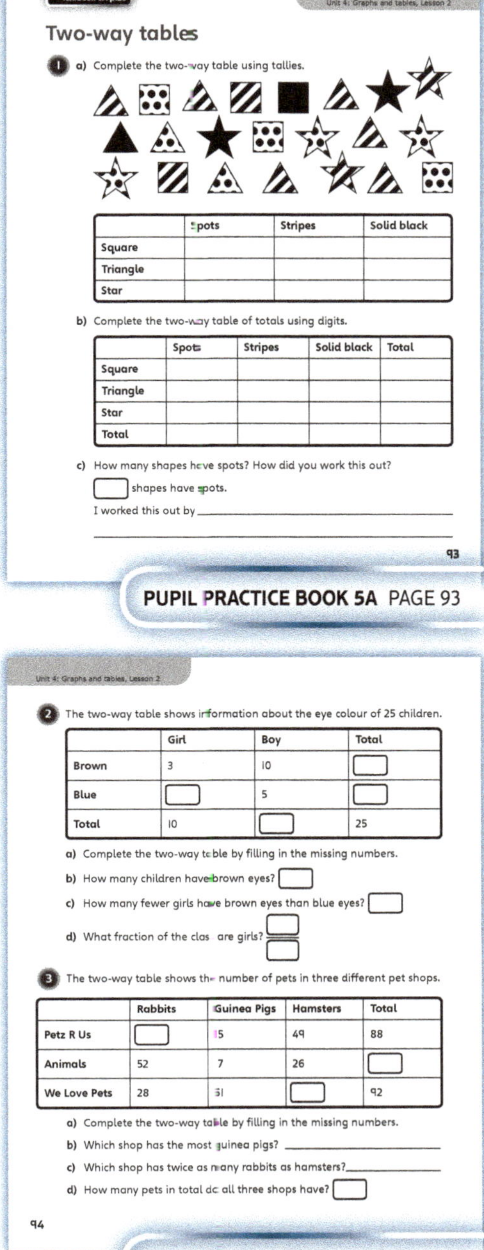

→ Textbook 5A p128

Two-way tables

1 a) Complete the two-way table using tallies.

	Spots	Stripes	Solid black
Square			
Triangle			
Star			

b) Complete the two-way table of totals using digits.

	Spots	Stripes	Solid black	Total
Square				
Triangle				
Star				
Total				

c) How many shapes have spots? How did you work this out?

[ ] shapes have spots.

I worked this out by _______________________________

____

93

PUPIL PRACTICE BOOK 5A PAGE 93

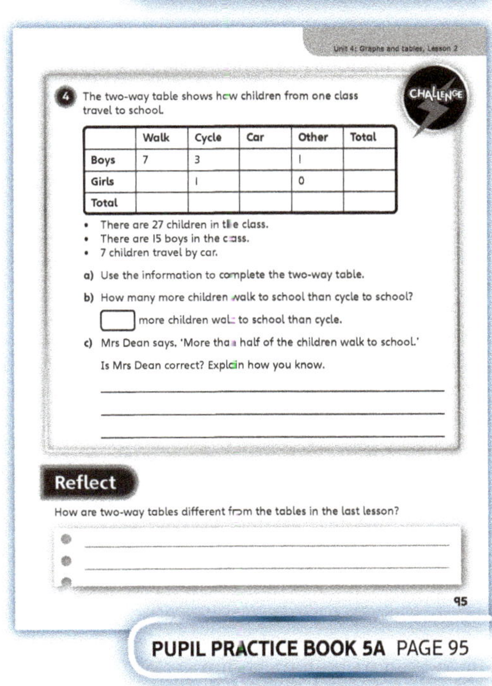

2 The two-way table shows information about the eye colour of 25 children.

	Girl	Boy	Total
Brown	3	10	[ ]
Blue	[ ]	5	[ ]
Total	10	[ ]	25

a) Complete the two-way table by filling in the missing numbers.

b) How many children have brown eyes? [ ]

c) How many fewer girls have brown eyes than blue eyes? [ ]

d) What fraction of the class are girls? $\frac{[ ]}{[ ]}$

3 The two-way table shows the number of pets in three different pet shops.

	Rabbits	Guinea Pigs	Hamsters	Total
Petz R Us	[ ]	15	49	88
Animals	52	7	26	[ ]
We Love Pets	28	31	[ ]	92

a) Complete the two-way table by filling in the missing numbers.

b) Which shop has the most guinea pigs? _______________

c) Which shop has twice as many rabbits as hamsters?_______________

d) How many pets in total do all three shops have? [ ]

94

PUPIL PRACTICE BOOK 5A PAGE 94

4 The two-way table shows how children from one class travel to school.

CHALLENGE

	Walk	Cycle	Car	Other	Total
Boys	7	3		1	
Girls		1		0	
Total					

- There are 27 children in the class.
- There are 15 boys in the class.
- 7 children travel by car.

a) Use the information to complete the two-way table.

b) How many more children walk to school than cycle to school?

[ ] more children walk to school than cycle.

c) Mrs Dean says, 'More than half of the children walk to school.'

Is Mrs Dean correct? Explain how you know.

____

____

____

Reflect

How are two-way tables different from the tables in the last lesson?

____

____

95

PUPIL PRACTICE BOOK 5A PAGE 95

Interpreting line graphs ❶

Learning focus

In this lesson, children will read line graphs with a range of scales and interpret the information to solve simple sum and difference problems.

Small steps

→ Previous step: Two-way tables
→ **This step: Interpreting line graphs (1)**
→ Next step: Interpreting line graphs (2)

NATIONAL CURRICULUM LINKS

Year 5 Statistics

Solve comparison, sum and difference problems using information presented in a line graph.

ASSESSING MASTERY

Children can read data accurately from line graphs and explain what information shows. Children can interpret intermediate values between the labelled intervals on a range of different scales and solve simple sum and difference problems using data from line graphs.

COMMON MISCONCEPTIONS

Children may not read line graphs accurately, by either not being able to work out intermediate values or by misreading the values. Ask:

- *What value is half-way between these two labelled numbers? Is this point more than half-way or less than half-way between the values you are given? Which axis do you need to look at first?*

STRENGTHENING UNDERSTANDING

To strengthen understanding of reading line graphs, discuss each graph to work out what it is about by first reading the labelling on the axes. Discuss specific points on the line. Is it the scale they have difficulty with or the wording in the question? Use two transparent rulers to show the vertical and horizontal values that connect at a particular point or use a ruler to draw a vertical line from specific points up from the *x*-axis to the graph line then a horizontal line back to the *y*-axis and vice versa. Recording the information shown in a table can help children to solve addition and subtraction problems more easily.

GOING DEEPER

Using the line graphs in the lesson or other examples you provide (with a range of scales in 5s, 10s, 50s, 100s), discuss why the different scales have been used. Can children explain why a particular scale may be chosen for a graph and why they do not always have a scale of 1? Ask children to write their own questions for the graphs in this lesson or on other line graphs provided.

KEY LANGUAGE

In lesson: line graph, read, axes, horizontal, vertical, temperature, Celsius (°C), half-way, estimate

Other language to be used by the teacher: interval, degrees

STRUCTURES AND REPRESENTATIONS

line graphs, tables

RESOURCES

Optional: number lines, transparent rulers, weather maps or forecasts, thermometers

 In the eTextbook of this lesson, you will find interactive links to a selection of teaching tools.

Before you teach ⏸

- Can children work out half-way between two numbers?
- Can children use a number line to estimate values between the labelled values?

Discover

WAYS OF WORKING Pair work

ASK

- Question **1** a): *What is this graph about? What information is shown on the horizontal and vertical axis? How can you make sure you are reading the correct value? What numbers are half-way between the labelled numbers? Where would 15 (19, 23) be on the vertical scale?*
- Question **1** b): *Which axis do you need to look at first for this question? Is that the same or different to question **1** a)? Where is 22 °C on the scale? Will the answer be a time or a temperature?*

IN FOCUS Children have met line graphs in Year 4 so these questions revise reading a value on the horizontal axis to find its corresponding value on the vertical axis (in part a) and also locating a value on the vertical axis to read its corresponding values (two corresponding times) on the horizontal axis (in part b). The scale interval is in 2s so children will need to read a value in between.

PRACTICAL TIPS With the children, watch a weather forecast recorded from TV or on the internet. Look at a weather website that shows the temperature throughout the day or look at weather maps and temperature tables from newspapers. This shows how reading these graphs can be used in a real-life context. Record the temperature in various places inside and outside school and ask children to record these in both table and line graph form.

ANSWERS

Question **1** a): The temperature at 11 am was 16 °C.

The temperature at 2 pm was 23 °C.

Question **1** b): The line graph shows that the temperature was 22 °C at 1 pm and 4 pm.

Share

WAYS OF WORKING Whole class teacher led

ASK

- Question **1** a): *What method have you used here to help us get the correct value? What happens if the line goes in between the two numbers? How do you know that the value is in the middle?*
- Question **1** b): *Where do you draw the first line this time? How can you tell the two times that the temperature is the same, at 22 °C?*

IN FOCUS For question **1** a), children understand how to accurately read the line graph by drawing horizontal and vertical lines. For question **1** b) spend time discussing the fact that the temperature was 22 °C at two specific times on the graph but that it was probably 22 °C at other times during the day, however the temperature was only recorded on the hours, so you do not know. This is why a dotted line is used. This is a good opportunity to discuss with children that line graphs are often used to show temperature over a period of time.

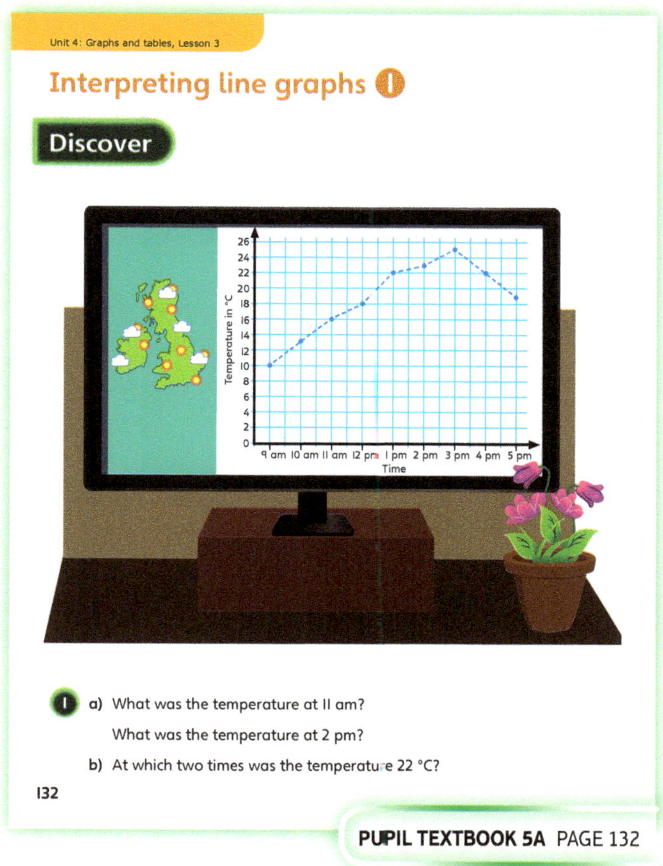

PUPIL TEXTBOOK 5A PAGE 132

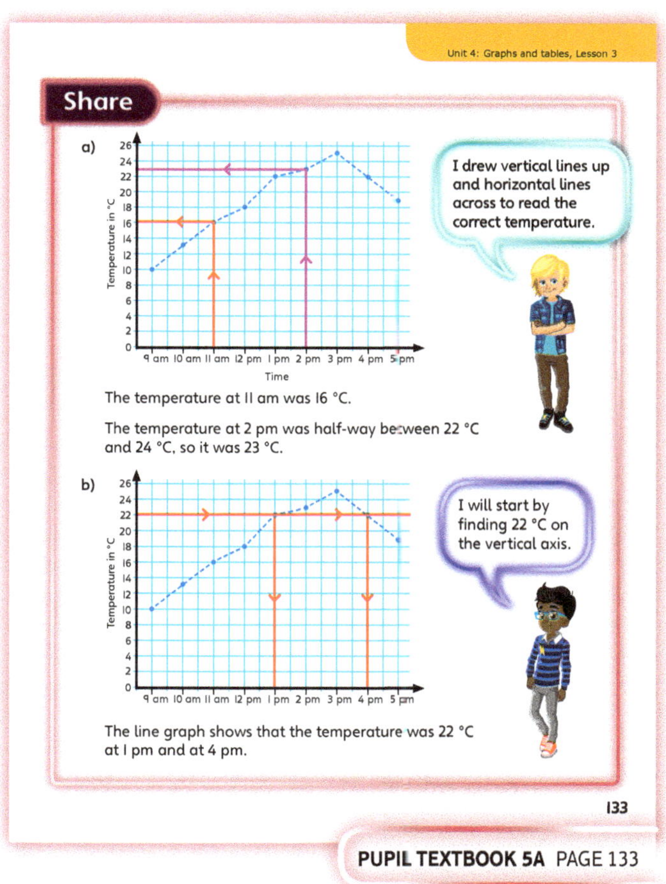

PUPIL TEXTBOOK 5A PAGE 133

Think together

WAYS OF WORKING Whole class teacher led (I do, We do, You do)

ASK

- Question **1** a): *How does the graph show the highest temperature?*
- Question **1** b): *Why is the temperature at 11:30 am only an estimate?*
- Question **1** c): *What method could you use for this question? When is the temperature first above 20 °C? How do you know? When does it stop being above 20 °C? Are these times exact? Why not?*
- Question **2** : *What information does this graph show? How is it different to the graph about temperature? Why do the numbers need to be exact? Do the parts of the line in between the points have any meaning? Could Mr Jones be correct? Could he be incorrect?*

IN FOCUS Question **2** provides a different context where the answers need to be whole numbers as you cannot have part of a child. The graph cannot show definitively whether the teacher's comment is correct or not but it is more likely that some or even many of the children were late more than one day that week.

STRENGTHEN To support understanding, ask children to tell you all they can about each graph. Ask them to tell you what each point means. Some children may find it useful to make a table for each graph so they are then just reading values from a table. Use horizontal and vertical lines on the graphs to help children read the correct corresponding values.

DEEPEN For question **1**, ask children to estimate the temperatures at all or some of the half and quarter hours. What temperature would they need to record at more or less than half-way between the given numbers? For example, at 9:30 am (11·5 °C) or 2:15 pm (23·5 °C).

ASSESSMENT CHECKPOINT Can children read the correct values from line graphs and interpret them to answer questions? Can children explain what information the line graph shows?

ANSWERS

Question **1** a): The highest recorded temperature during the day was 25 °C.

Question **1** b): The temperature at 11:30 am was about 17 °C.

Question **1** c): The temperature was above 20 °C for about [accept anything between 3·5–4] hours.

Question **2** a): 41 or 42 children were late on Thursday.

Question **2** b): 7 or 8 more children were late on Monday than on Tuesday.

Question **2** c): Mr Jones's comment is probably not correct, although the total is > 100, some children may have been absent on more than one day.

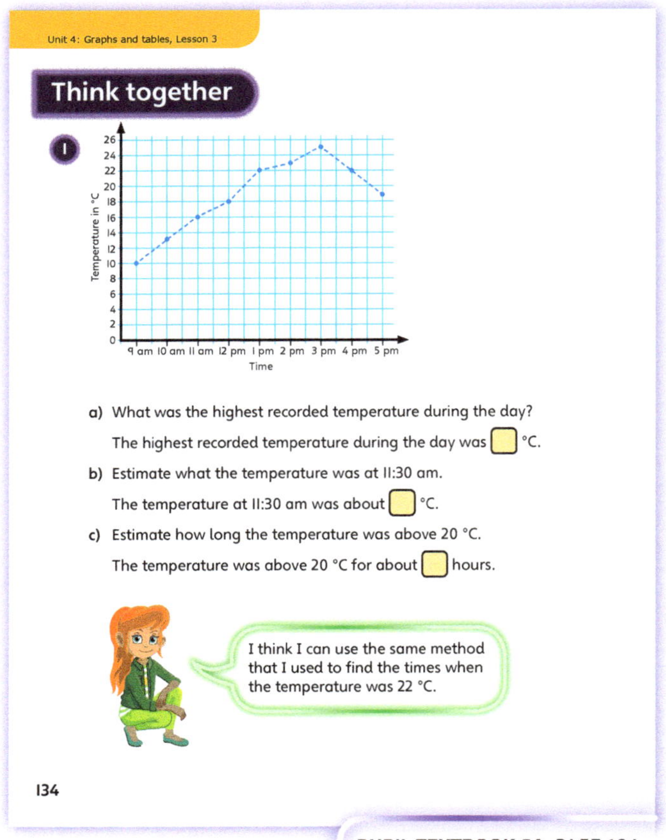

PUPIL TEXTBOOK 5A PAGE 134

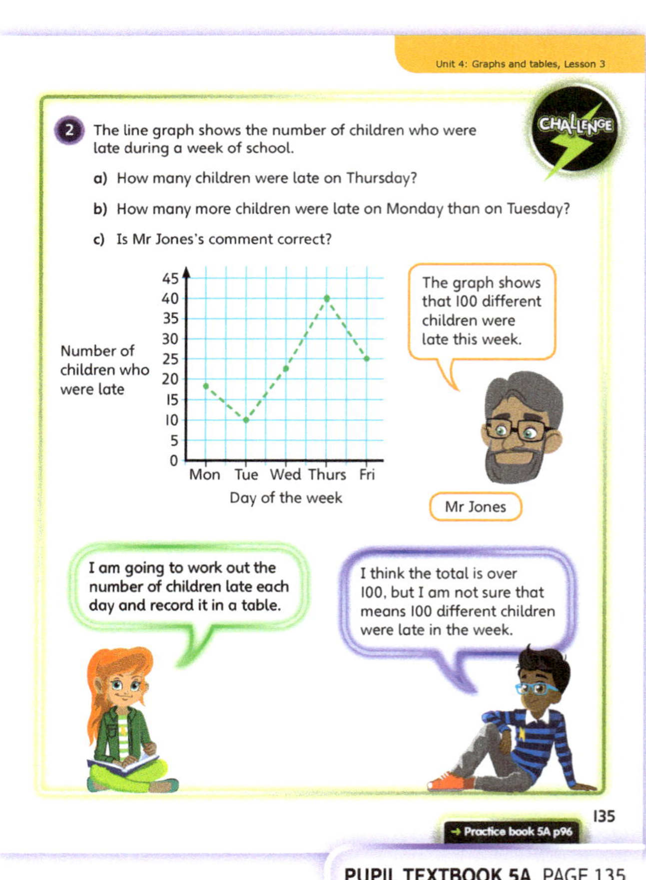

PUPIL TEXTBOOK 5A PAGE 135

Practice

WAYS OF WORKING Independent thinking

IN FOCUS Question **1** focuses on the main skills needed to read and interpret line graphs: finding the corresponding values given either an *x*-axis value or given a *y*-axis value; using the data to answer sum and difference questions and to reason about the values shown. Draw out that the in-between points shown by the dotted line have no meaning (each day has a separate discrete value, there is no trend specifically).

Question **3** is more complicated as it shows a cumulative total. You may need to discuss how this works, giving an example: after 1 minute about **12** children had completed the puzzle, after 2 minutes a total of about 25 children had completed it (25 – **12** = 13, so about 13 children completed it between 1 and 2 minutes).

STRENGTHEN Ask children to tell you all they can about a graph. What information is written on the axes? What does this axis measure? Is there a label? Use horizontal and vertical lines or two rulers on the graph to help children read the correct values and practise reading all the points from the horizontal axis to the vertical axis and back again.

DEEPEN Extend question **3** by asking children to estimate how many children completed the puzzle after $2\frac{1}{2}$ minutes, $4\frac{1}{2}$ minutes, $6\frac{1}{2}$ minutes.

THINK DIFFERENTLY Question **2** provides a different context where the scale does not start at zero and each interval is 20, making it difficult to be very accurate. Discuss the zig-zag line that shows the graph between 0 and 180 and how this allows data with larger values to be used. The lowest value is 190, so there is no need for values lower than that. Discuss the dotted line – the in-between values again have no value – each day is a separate piece of data. However, the dotted line does allow an easy comparison of each day, but it is almost impossible to know the exact km.

ASSESSMENT CHECKPOINT By this point, children should be able to read and interpret information shown on line graphs, solving associated problems with the data. Can children explain whether the points on the dotted line have a real value or not?

ANSWERS Answers for the **Practice** part of the lesson appear in the separate **Practice and Reflect answer guide**.

Reflect

WAYS OF WORKING Independent thinking

IN FOCUS Look for children's response as to how they could ensure accuracy in reading the temperature, given a specific time.

ASSESSMENT CHECKPOINT Can children explain a method to read values from the horizontal axis on a line graph? Do they mention drawing vertical and horizontal lines or using a ruler to help with accuracy?

ANSWERS Answers for the **Reflect** part of the lesson appear in the separate **Practice and Reflect answer guide**.

After the lesson

- Can children use the data on a graph to solve simple sum and difference problems?
- Can children read graphs with different scales, identifying intermediate values?

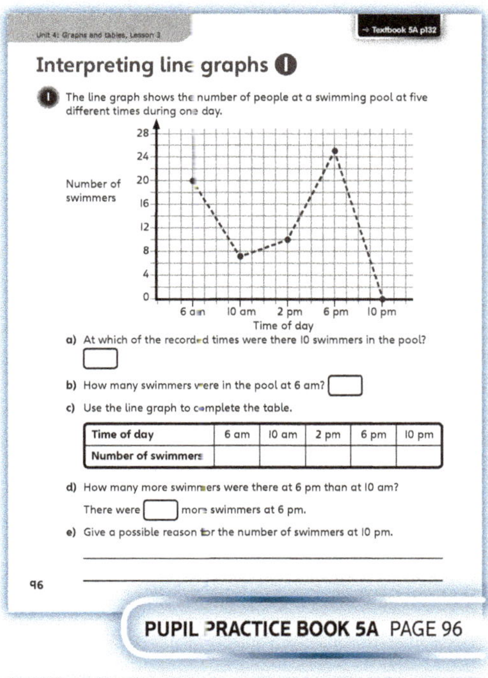

PUPIL PRACTICE BOOK 5A PAGE 96

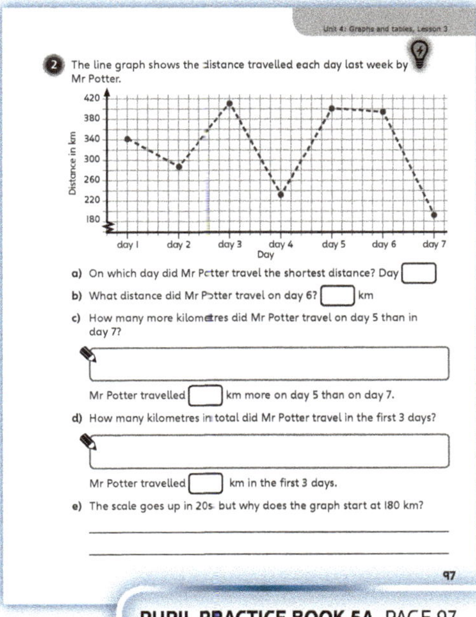

PUPIL PRACTICE BOOK 5A PAGE 97

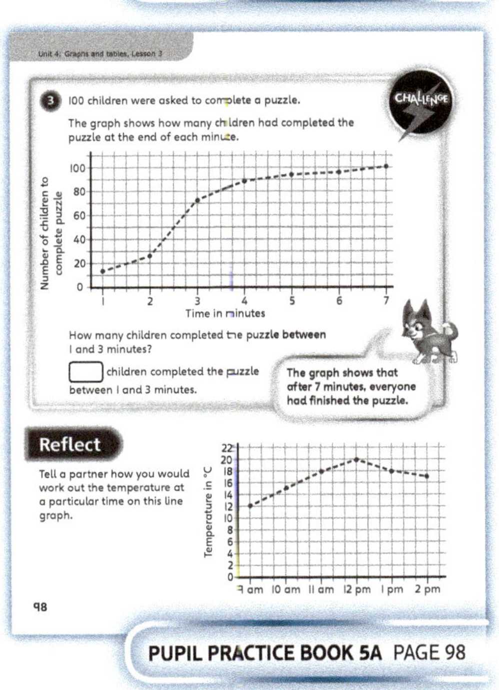

PUPIL PRACTICE BOOK 5A PAGE 98

Interpreting line graphs ❷

Learning focus

In this lesson, children will continue to develop their reading and interpretation of line graphs with more complex scales, including dual line graphs, to solve simple sum and difference problems.

Small steps

→ Previous step: Interpreting line graphs (1)
→ **This step: Interpreting line graphs (2)**
→ Next step: Drawing line graphs

NATIONAL CURRICULUM LINKS

Year 5 Statistics

Solve comparison, sum and difference problems using information presented in a line graph.

ASSESSING MASTERY

Children can accurately read information from line graphs, including dual line graphs. Children can solve simple sum and difference problems using data from the line graphs with increasingly more complex scales.

COMMON MISCONCEPTIONS

Children may not read intermediate values on a more complex scale accurately. Ask:
• *What number is half-way between these two numbers? Where would 5, 250, 500 be on this scale?*

Children may have difficulties in explaining what a line graph shows. Ask:
• *What information do the labels on the axes show? What story is this graph trying to tell you?*

STRENGTHENING UNDERSTANDING

Ask children to tell the story of the line graph. Use single line graphs until children are more secure. Discuss what the labels tell you and how the scale has been decided, identifying some of the values in between labelled intervals. Try to encourage children to draw some general conclusions about the data.

GOING DEEPER

Ask children to list the advantages and disadvantages of dual line graphs where two graphs are on the same grid. Ask: *When would this be useful? When would this be unhelpful?* Ask children to fully analyse the information shown on a dual line graph, comparing the data shown on each line. They could write more questions to compare the two lines.

KEY LANGUAGE

In lesson: line graph, **dual line graph**, axis, horizontal, average, temperature, Celsius °C, height, metre, increase, hours, minutes, warmer

Other language to be used by the teacher: colder, decrease, rise, fall, interval, scale, half-way, data, gradient, slope, constant, vertical

STRUCTURES AND REPRESENTATIONS

line graph, dual line graph, number line

 In the eTextbook of this lesson, you will find interactive links to a selection of teaching tools.

Before you teach ⏸

• Can children find values between two points on a number line?
• Can children read values from line graphs with simple scales?

Discover

WAYS OF WORKING Pair work

ASK

- Question ① a): *What does this graph show? What information is on the horizontal and vertical axis? How can you work out the height at a particular time? What is each marker worth on this axis? How does the graph show that the helicopter stays at the same height?*
- Question ① b): *When does the helicopter take off? When does it land? How do you know? How can you work out how long the trip took? How do you know it is in the air at all times between these two times?*

IN FOCUS For question ① a), the scale of this line graph is more complex than in the previous lesson. Begin with reading the height at 10 am before discussing what half-way between the intervals represents and reading the height at 11 am. Look together at the horizontal scale. Ask: *What do the markers in between the full hours mean?* Read the height at 9:45 and at 10:15. Draw attention to the horizontal line at 3,500 m, explaining that this means it stayed at the same height and to where the line finishes or meets the horizontal axis (which is not at the end of the graph).

PRACTICAL TIPS You could act out the helicopter flight with a toy to explain the graph to the class.

ANSWERS

Question ① a): The height of the helicopter at 10:30 am is 3,500 metres.

The helicopter stays at this height for 30 minutes from 10:30 am to 11:00 am.

Question ① b): The helicopter flight starts at 9:30 am and ends at 11:45 am, so it lasts 2 hours and 15 minutes ($2\frac{1}{4}$ hours).

Share

WAYS OF WORKING Whole class teacher led

ASK

- Question ① a): *Why is it helpful to draw the line up from 10:30 am? How did that help to find the height? What does Flo's horizontal line show? How can you use this to work out how long it stayed at 3,500 m?*
- Question ① b): *How can you tell what time the helicopter took off? What time does it land? How do you know? How does the number line help you work out the duration?*

IN FOCUS For question ① b) ensure children know how to find the missing times on the horizontal line, giving them opportunity to discuss it then demonstrating that, as there are 4 spaces between the hours, each one must be worth $60 \div 4 = 15$ minutes markers, so 9:15; 9:30 and 9:45 and so on. Some children may need to label the in between times. Practise reading up from a time back across to the height and across from a height down to a time. The number line shows clearly one way of finding the duration, but can children suggest a different way of getting from 9:30 to 11:45? Is the answer still $2\frac{1}{4}$ hours?

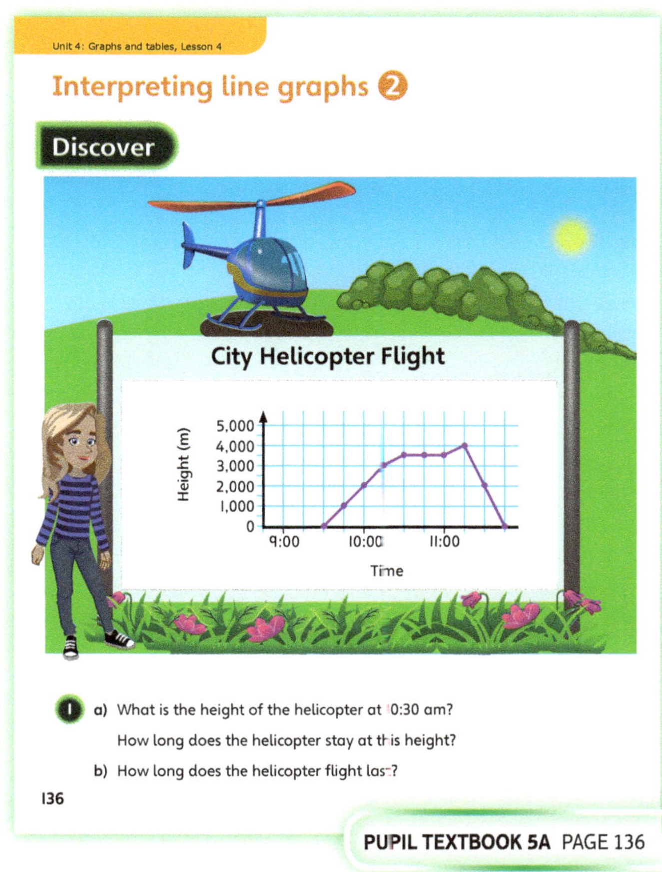

PUPIL TEXTBOOK 5A PAGE 136

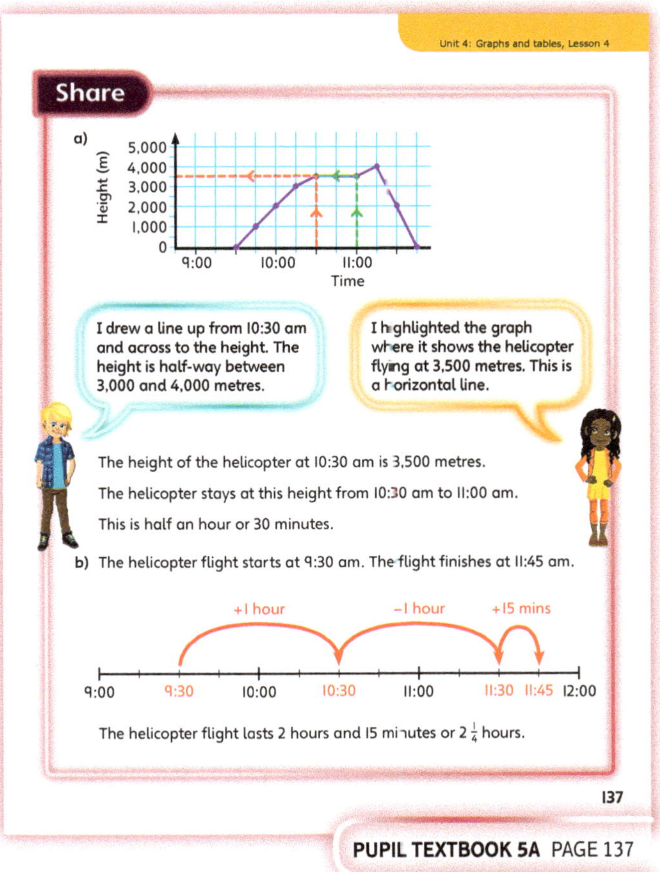

PUPIL TEXTBOOK 5A PAGE 137

Think together

WAYS OF WORKING Whole class teacher led (I do, We do, You do)

ASK

- Question ❶ : *What is different about this graph? Why are there two different lines on here? What is the graph showing? How can you tell which is City A and which is City B? How can you work out the difference in temperature? Can you read all of the temperatures accurately? How do the graph lines show which city is warmer? Why is it called a dual line graph?*
- Question ❷ : *Explain what the graph shows. How do you know the drone starts and finishes on the ground? How will Astrid work out the height and increase in height for each minute? What does Flo mean, how does the graph line show if it increases by the same amount? How does the line show that the drone stays at the same height? How do you know the drone lands before 15 minutes?*

IN FOCUS In question ❶, dual line graphs are introduced to help children read two sets of data on one graph. This encourages children to use the key to be able to read the different temperatures for City A and City B.

STRENGTHEN For question ❷, concentrate on Astrid's method to practise reading the values for each minute. Discuss each statement in turn, demonstrating how the graph would show the statement if it were true. Ask: *Does the graph show this?*

DEEPEN In question ❷ compare the two methods the characters suggest. Which method do they prefer? Ask: *Can you use the gradient (slope) of the line to answer all of the questions? Also, can children make up more statements to test?*

ASSESSMENT CHECKPOINT Can children read values from single line graphs? In question ❶, can children read values from dual line graphs? Can children solve simple difference problems from the information shown on line graphs?

ANSWERS

Question ❶ a): 15 °C

Question ❶ b): 17–18 °C

Question ❶ c): 10 °C

Question ❶ d): City A is warmer because the graph line for City A is higher every month, showing a higher temperature every month.

Question ❷ : The drone starts from the ground. – True

For the first 5 minutes, the drone's height increases by 100 metres every minute. – True (gradient of line is constant)

When it reaches 500 metres the drone flies at this height for 7 minutes. – False (it is level for 5 minutes)

The drone returns to the ground after 15 minutes in the air. – False (It lands after 12 minutes, 15 minutes is the end of the graph.)

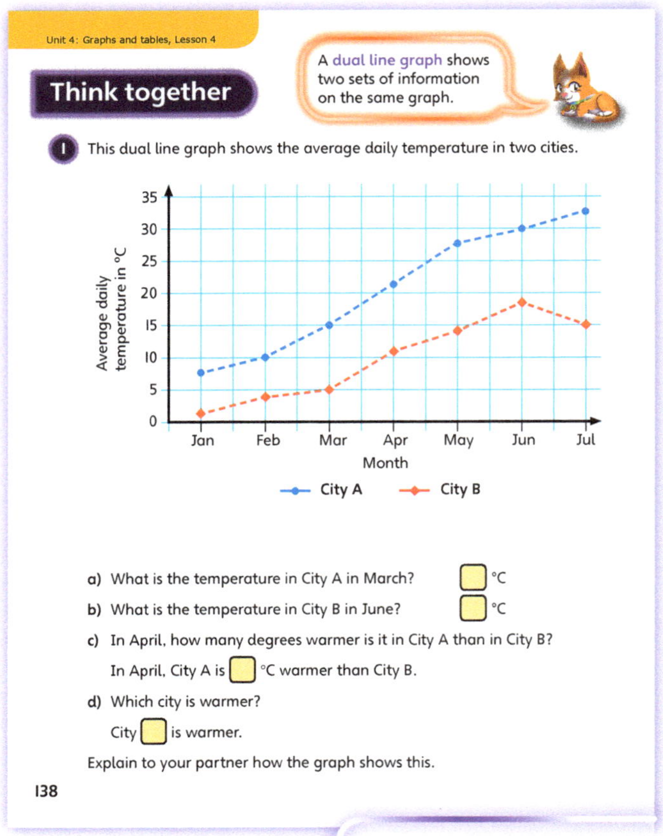

PUPIL TEXTBOOK 5A PAGE 138

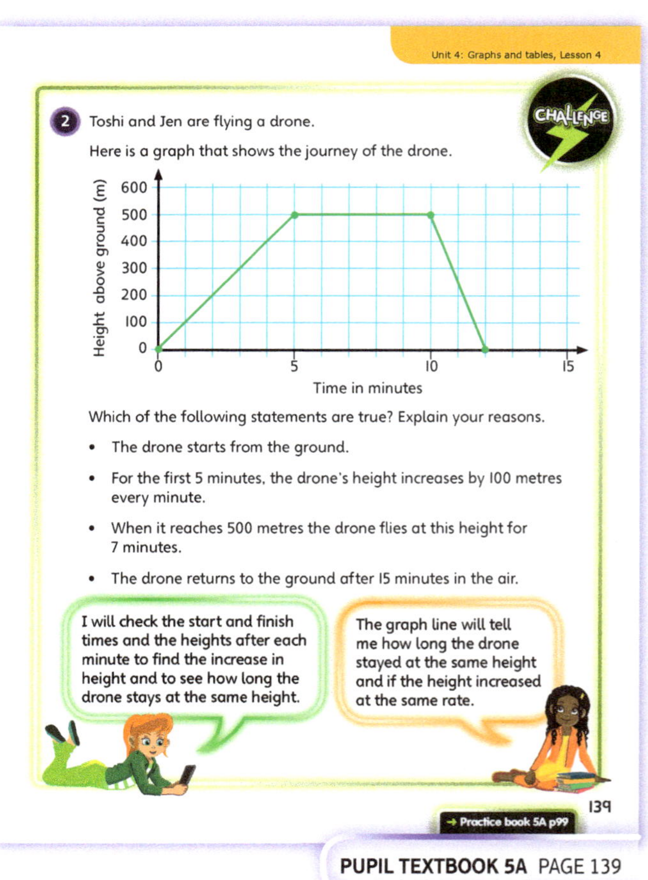

PUPIL TEXTBOOK 5A PAGE 139

Practice

WAYS OF WORKING Independent thinking

IN FOCUS In question ❸ the scale involves large numbers, so an accurate reading is impossible. Children will need to estimate, so discuss it with them. 1980 is a labelled marker but 1993 is in between 1990 and 1995, so ask: *Which year is it closer to?* The value for both the years are between markers too. Discuss how to work out the unlabelled lines (20,000 ÷ 4 = 5,000). Ask: *How accurate is it possible to be?*

STRENGTHEN Support children by reading each question together and discussing what it shows. Establish what the scale is and how to work out intermediate values, allowing children to label for these markers or reproduce them on a horizontal number line. Spend time practising the estimation of points on the line in between marked points.

DEEPEN For question ❶ ask children to describe the journey and answer or write further questions. Ask: *How long is the balloon above 8 m? How do you know the height does not increase by the same amount each minute?* For question ❸, ask children to compare their answers – are they similar or very different? Can they help each other to be even more accurate in their estimation?

ASSESSMENT CHECKPOINT Can children read and interpret line graphs with increasingly complex scales, including dual line graphs? Are children beginning to explain what the shape (gradient) of the graph line shows?

ANSWERS Answers for the **Practice** part of the lesson appear in the separate **Practice and Reflect answer guide**.

Reflect

WAYS OF WORKING Independent thinking

IN FOCUS This **Reflect** question focuses on a misconception that line graphs need to start from zero. Children will need to reflect back over the lesson to realise that they have met line graphs which do not start at zero, yet still provide accurate information and can be read easily.

ASSESSMENT CHECKPOINT Can children give examples to show that the statement is sometimes true, by giving examples where it is true and where it is not true? Can they start to make generalisations about when a line graph would start at zero and when it would not?

ANSWERS Answers for the **Reflect** part of the lesson appear in the separate **Practice and Reflect answer guide**.

After the lesson ⏸

- Can children accurately read values from line graphs with increasingly more complex scales?
- Can children find relevant information from dual line graphs?
- Can children use a line graph to solve simple sum and difference problems?

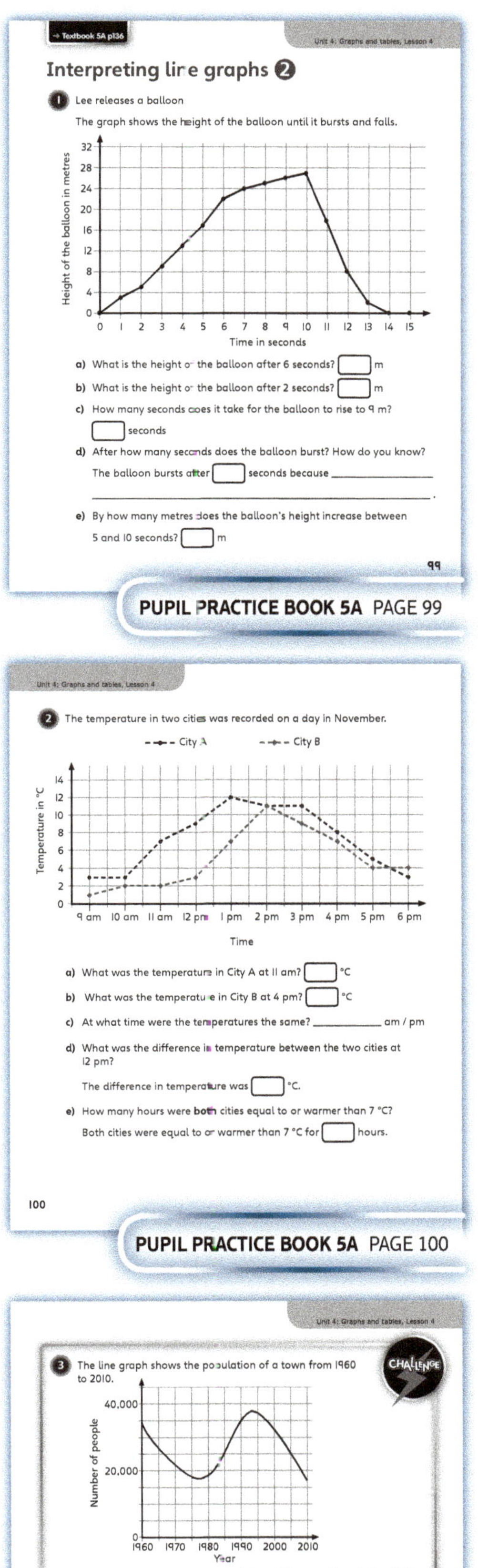

PUPIL PRACTICE BOOK 5A PAGE 99

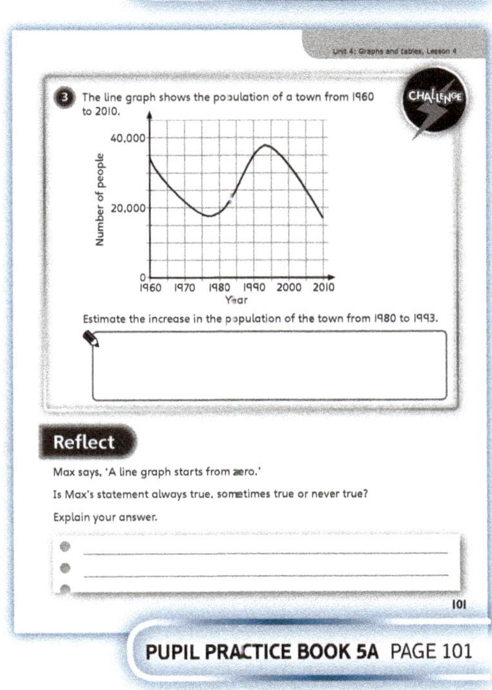

PUPIL PRACTICE BOOK 5A PAGE 100

PUPIL PRACTICE BOOK 5A PAGE 101

Drawing line graphs

Learning focus

In this lesson, children will draw simple line graphs from data that is given in a table.

Small steps

→ Previous step: Interpreting line graphs (2)
→ **This step: Drawing line graphs**
→ Next step: Multiples

NATIONAL CURRICULUM LINKS

Year 5 Statistics

Solve comparison, sum and difference problems using information presented in a line graph.

ASSESSING MASTERY

Children can draw a simple line graph from data that is given in a table. Children understand which data goes along the horizontal axis (day, time, month) and which goes on the vertical axis (the frequency, temperature, height).

COMMON MISCONCEPTIONS

Common mistakes often involve labelling the axes incorrectly. This can mean unequal spacing between values on the axes, choosing incorrect intervals or missing out values. This will then affect their ability to plot correctly. Ask:

- *Count up in 2s (5s, 10s, 50s, 100s). What number have you missed out? What is the highest number you need to plot? Is the space between these two values the same as the space between these two values? What should the spacing be?*

STRENGTHENING UNDERSTANDING

The majority of questions in this lesson provide children with axes and scale already given. Where they are not provided consider giving children the grids with the scale and axes already drawn to make plotting easier.

It is important to link each row or column of the table with the point children plot. Read the first day, week or hour aloud and plot the corresponding frequency value. Plot the next points together one at a time, joining them as you go until children are able to continue plotting the remaining points.

GOING DEEPER

Ask children to generate their own data from a survey or observation or provide some data. It is helpful if the data is over time, as in the lesson questions. Ask children to draw their own graph for this data on squared paper. This will ensure that they have practice in drawing graphs from scratch, which is an important skill that they need to develop. Ensure they label the axes accurately.

KEY LANGUAGE

In lesson: line graph, table, data, information, label, sales, scales, axes, temperature, plot, plotted

Other language to be used by the teacher: interval, row, column, frequency value, horizontal, vertical, Celsius, value

STRUCTURES AND REPRESENTATIONS

line graphs, tables

RESOURCES

Mandatory: cm squared paper

Optional: prepared grid for question ④ **Practice**; survey data in tables

 In the eTextbook of this lesson, you will find interactive links to a selection of teaching tools.

Before you teach ⏸

- Can children read a line graph?
- Can children count in 2s, 5s, 10s, 20s, 100s?

Discover

 Pair work

ASK

- Question **1** a): *What does the information in the table show? Why can you not plot the points until you have the axes? What goes on the horizontal axis? What about the vertical axis? What scale should you use?*
- Question **1** b): *What do you need to draw first? How long does each axis need to be? How and where do you mark the points? What scale have you used on your vertical axis? How have you shown this?*

IN FOCUS In question **1** a) children begin by considering what needs to be thought about when drawing a line graph from a table of data. Look back at a line graph from a previous lesson, looking at each of the elements in turn. Discuss the labelling on the axes: which information is shown on which axis; the scale on the vertical axis (where it starts, the interval, labels on the lines not in the spaces, equally spaced); and whether the line is dotted (we do not have exact data for the measurements between the points), or solid (intermediate points have a measured value). For part b) allow children to experiment with the scale they choose as this will provide useful discussion prompts.

PRACTICAL TIPS You could record the temperature at the same time each day or every hour during one day and make this into a giant class line graph.

ANSWERS

Question **1** a): Before drawing the graph the sales people need to think about:
  - what they show on each of the axes
  - the scale on each of the axes
  - whether to use dotted or solid lines to connect the points.

Question **1** b): Accept children's graphs that are suitable scales, correctly labelled on both axes and points plotted accurately.

Share

 Whole class teacher led

ASK

- Question **1** a): *What do you need to work out before you plot any points?*
- Question **1** b): *Which information is on the horizontal axis and the vertical axis? What is the scale on the vertical axis? Could you use a different scale? What do you need to draw first to start your own line graph? How did Dexter know where to plot the number for Monday? Which point do you think he will plot next?*

IN FOCUS Question **1** b) is important as it is the first time children will draw their own line graph. Ensure that children consider what the maximum value is to be able to decide a suitable scale; that labels are equally spaced and ensure the labels are on the lines, not the spaces.

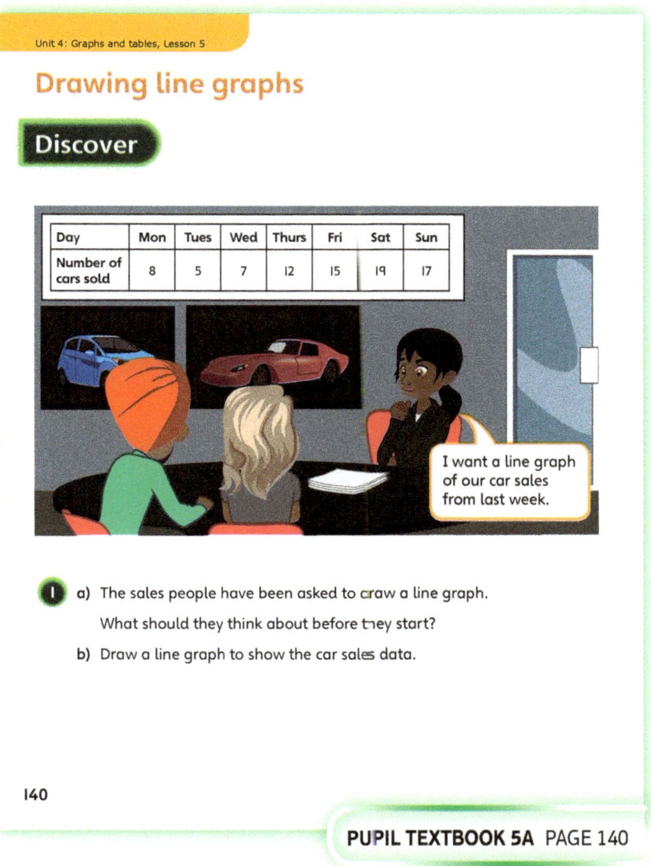

Day	Mon	Tues	Wed	Thurs	Fri	Sat	Sun
Number of cars sold	8	5	7	12	15	19	17

PUPIL TEXTBOOK 5A PAGE 140

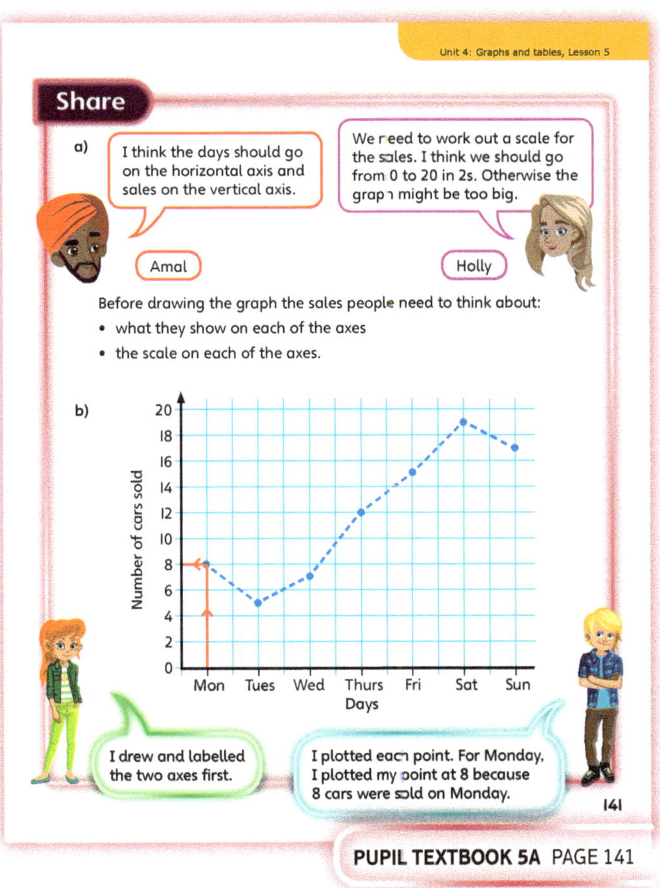

PUPIL TEXTBOOK 5A PAGE 141

Think together

WAYS OF WORKING Whole class teacher led (I do, We do, You do)

ASK

- Question **1** : *What scale should you use?*
- Question **2** : *How is this data different from the others? Why is the horizontal axis not at the bottom of the graph? Where do you plot the points for Monday and Friday?*
- Question **3** : *What should you remember when drawing a line graph? Can you spot where information is missing? Are all of the numbers on the vertical axis equally spaced? Has Danny written all of the numbers? Are all of the points plotted accurately? Is the line the correct type for this data?*

IN FOCUS Question **2** involves negative temperatures. Point out that the horizontal line is still drawn through 0 and not along the bottom because you need to plot a temperature below zero.

STRENGTHEN Draw one of the line graphs as a class on the whiteboard, with children coming up to draw different elements on. Discuss together the accurate way of presenting each element.

DEEPEN For question **1** you could increase the number of sales significantly so that they are all 3-digit multiples of 5 or 10, between 210 and 450, requiring a scale that does not start at 0 and intervals of 20.

Ask children to use all of the mistakes they have spotted to check their own graphs for questions **1** and **2**. Have they made any of the same mistakes?

ASSESSMENT CHECKPOINT Can children draw a line graph from information supplied in a table? Can children construct a line graph from scratch or do they need the axes and scale supplied?

ANSWERS

Question **1** : Check children have a graph that is plotted correctly.

Question **2** a): Check children have plotted Mon at ⁻3 and Fri at 11.

Question **2** b): The warmer the temperature, the more ice creams were sold.

Question **3** : • *x*-axis labelled 'months' (instead of years)
- *y*-axis mistakes: the interval between 100 and 200 is half the size of the gap between all of the other years
- 300 is missed out completely
- no label (title) for the axis
- 2015 has been plotted incorrectly – this shows 495 on the *y*-axis (should be 395).
- 2013 has been plotted incorrectly – does not line up with 2013 on the *x*-axis.
- graph line is a solid line, but it should be dotted.

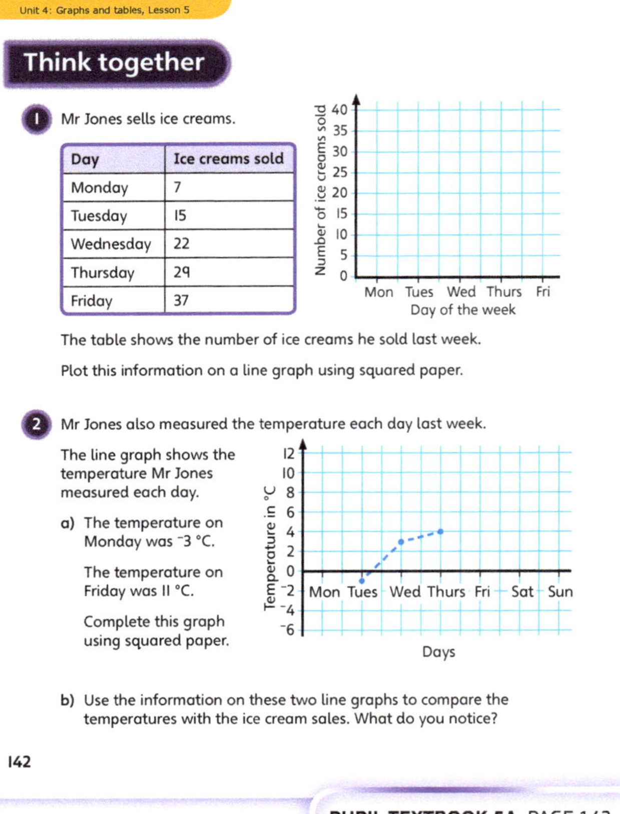

PUPIL TEXTBOOK 5A PAGE 142

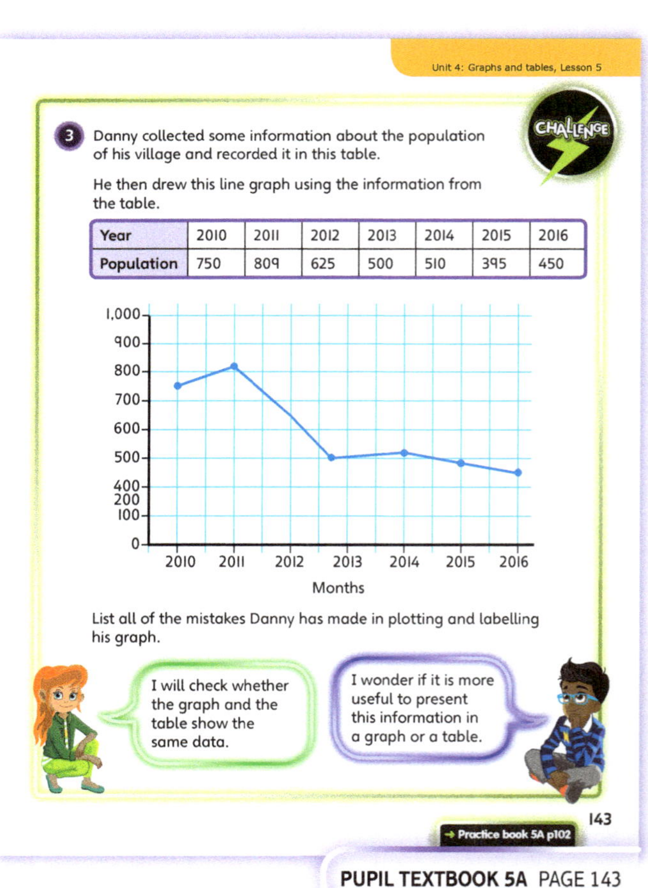

PUPIL TEXTBOOK 5A PAGE 143

Practice

WAYS OF WORKING   Independent thinking

IN FOCUS   Questions **1** to **4** give children the opportunity to practise constructing line graphs from a table of data. The questions progress in difficulty with less and less of the line graph already completed and with increasingly more difficult scales. Encourage children to check each other's work or to use Danny's mistakes from the textbook lesson to check their own work. Look for children plotting points accurately on the graphs as this is often the biggest reason for children losing marks in assessments.

STRENGTHEN   Help children to plot points accurately, making a very clear link between the two values in the table that connect to make that point on the graph. Help children when plotting points that are not on a grid line and lie between two values on the vertical scale. Support children in completing the table in question **4** so that they can draw the line graph.

DEEPEN   Ask children to adapt one of the tables to make it more challenging (such as, with larger numbers, decimals, more negative numbers) and then to construct their own line graph from scratch. They can check with the textbook what they need to remember.

ASSESSMENT CHECKPOINT   By the end of the practice, children should be able to plot line graphs accurately.

ANSWERS   Answers for the **Practice** part of the lesson appear in the separate **Practice and Reflect answer guide**.

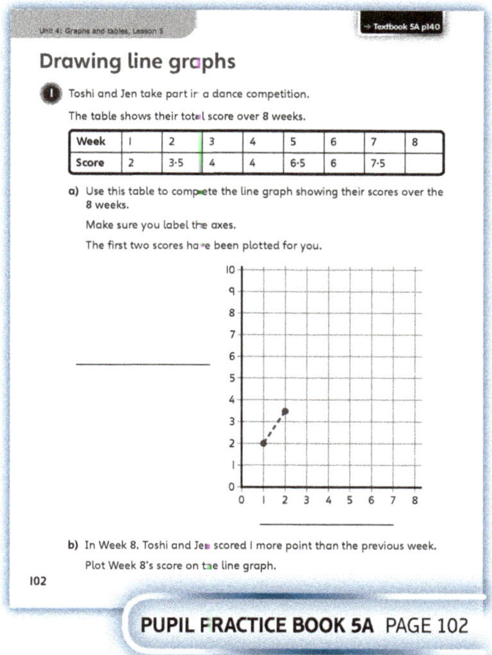

PUPIL PRACTICE BOOK 5A PAGE 102

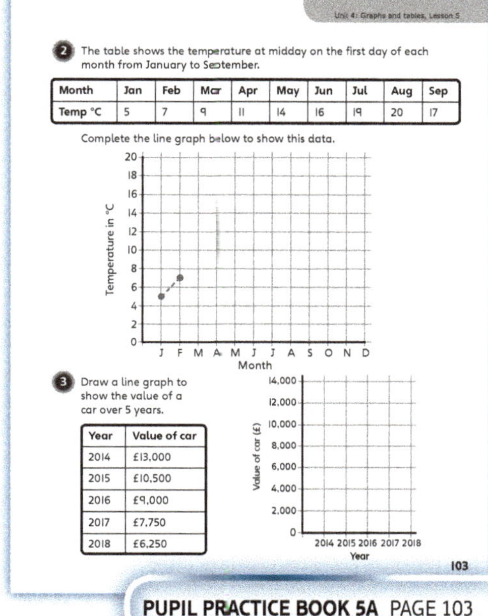

PUPIL PRACTICE BOOK 5A PAGE 103

Reflect

WAYS OF WORKING   Independent thinking

IN FOCUS   For this task children are asked to reflect on what they need to consider when drawing a line graph from a table of data. Discussion throughout the lesson will help cement these ideas, so that they have a full understanding of why these elements are important when drawing a line graph.

ASSESSMENT CHECKPOINT   Do children remember to mention the scale, the correct labelling of the axes, equal spacing, accurate plotting and whether the line should be dotted or solid?

ANSWERS   Answers for the **Reflect** part of the lesson appear in the separate **Practice and Reflect answer guide**.

After the lesson ⏸

- Can children work out an appropriate scale for given data?
- Are they able to plot points accurately to construct or complete a line graph for data given in a table?
- Where else in the curriculum could children construct line graphs to present data?

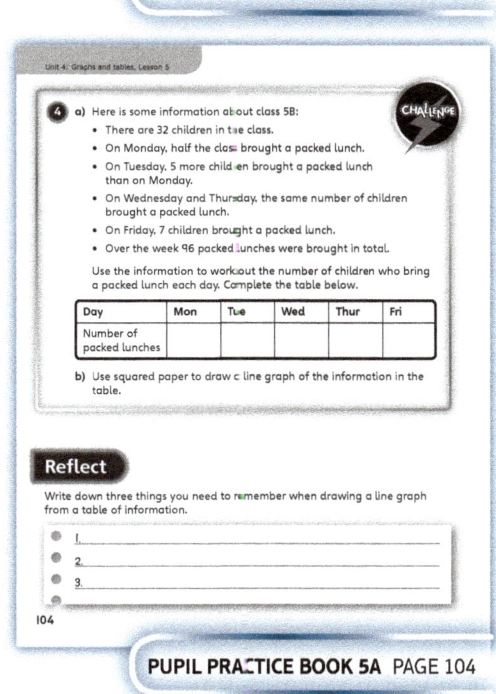

PUPIL PRACTICE BOOK 5A PAGE 104

End of unit check

Don't forget the *Power Maths* unit assessment grid on p26.

WAYS OF WORKING Group work adult led

IN FOCUS Questions **1** and **2** use the same line graph and involve interpreting the scale correctly. Question **1** checks children can read the correct value. Question **2** involves finding a difference between values that are both between labelled intervals, so involves an element of estimating to identify the correct answer.

Questions **3** and **4** involve the same table of data. Children will need to ensure they read all of the information given, especially the opening statement which tells them that the table shows the results for 100 children. They will need this information to answer question **3**.

The SATs-style question involves a two-way table. Children will have to work out the totals to help them work out whether the statement is true of false.

ANSWERS AND COMMENTARY

Children who have mastered the concepts in this unit will be able to construct, interpret and complete tables, including two-way tables, understanding the differences between a two-way table and a simple table. Children will be able to extract data from the tables to solve simple problems. Children will be able to read, draw and interpret line graphs, including dual line graphs, extracting the relevant data to solve sum and difference problems.

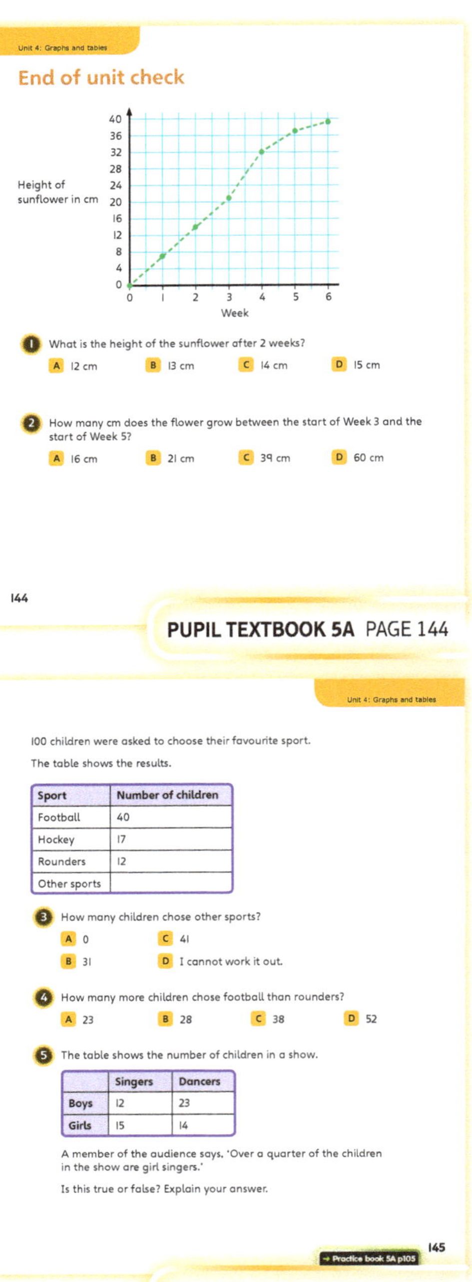

PUPIL TEXTBOOK 5A PAGE 144

PUPIL TEXTBOOK 5A PAGE 145

Q	A	WRONG ANSWERS AND MISCONCEPTIONS	STRENGTHENING UNDERSTANDING
1	C	The other answers show children's misconception of reading a scale.	For questions 1 and 2 work with children to work out the missing values. Show the scale on the axis as a number line and discuss how they would find the missing values. Get them to check by counting out loud to see if the scale fits in with the numbers already marked on the axis. Focus on the language of the questions. Which clues are there to show whether to add or subtract? Practise number bonds to 100.
2	A	B shows children have just read off the Week 3 example. C is just the Week 6 value. D added the two values, instead of subtracting.	
3	B	A shows they think the answer is 0 because there is a missing value. C added the values incorrectly or not subtracted them from 100 correctly. D shows they have not seen the total for the table.	
4	B	A shows they have found the difference between football and hockey. C is a mistake with the subtraction. D shows they added instead of subtracting.	
5	F	This is the SATS-style question. Children need to identify the correct cell number (15) and to work out the total (64) $\frac{1}{4}$ of 64 is 16, so this is false. Computational errors will be common or using the wrong piece of data or not being able to find $\frac{1}{4}$ of 64.	

My journal

WAYS OF WORKING Independent thinking

ANSWERS AND COMMENTARY

Question ① involves a line graph relating to the number of people in a shop at any given time. Children should be asked to explain why they have given the answers they have to the first two parts, as both involve an element of reasoning. For the third part encourage children to also include one statement that involves a possible range of answers. Question ② involves reasoning about a simple table. Children need to justify their answers. Try to encourage children to think beyond just simple facts involving simply reading of the data.

Question ① a): Look for answers between 12 and 16, referring to the numbers at 3 pm and 4 pm or any number with a relevant explanation – you cannot know definitively.

Question ① b): Look for answers between 6 pm and 7 pm with explanations.

Question ① c): Accept children's answers that correctly interpret the data.

Question ② a): Children need to give an explanation that it doesn't show data over time. Some children may suggest drawing a bar graph.

Question ② b): Accept various possible answers that compare the table data.

As the data is from the same shop, you may want to ask children whether there is a connection between them, because there was with the ice cream sales in Lesson 5 of the textbook. (There isn't a connection as you cannot tell if more sales were made when there were more people in the shop.)

Power check

WAYS OF WORKING Independent thinking

ASK

- *Are you happy to read information from a simple table or line graph?*
- *How confident are you drawing your own line graph for given data?*
- *Do you know how to draw a two-way table and answer related questions?*

Power puzzle

WAYS OF WORKING Pair work

IN FOCUS This Power puzzle asks children to construct a two-way table using information given. Children may find it useful to add total rows and columns, as these are deliberately not given. Children will have to use their knowledge of number operations and fractions to work out the missing values. Encourage children to read the information one line at a time and not to try to do too much at once. This question can appear very confusing unless children break it down into small parts.

ANSWERS AND COMMENTARY

First column: 34, 16, 50
Second column: 26, 74, 100
Total column: 60, 90, 150

After the unit ⏸

- Can children draw a simple line graph and read information from a line graph, using this to solve simple sum and difference problems?
- Can children understand and construct two-way tables?

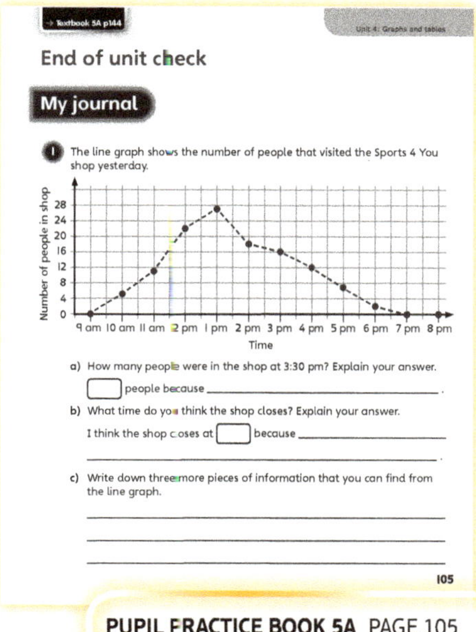

PUPIL PRACTICE BOOK 5A PAGE 105

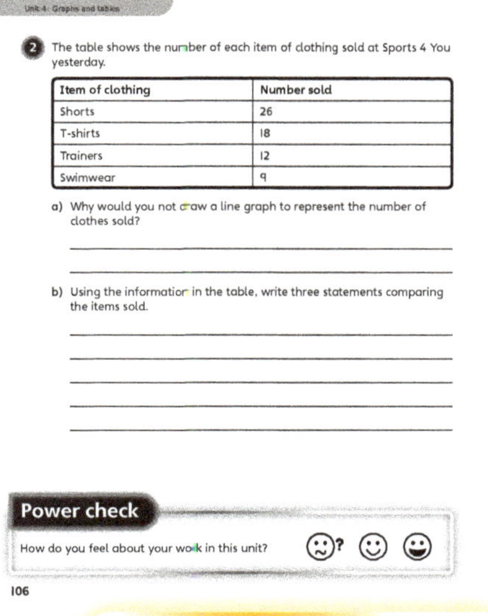

PUPIL PRACTICE BOOK 5A PAGE 106

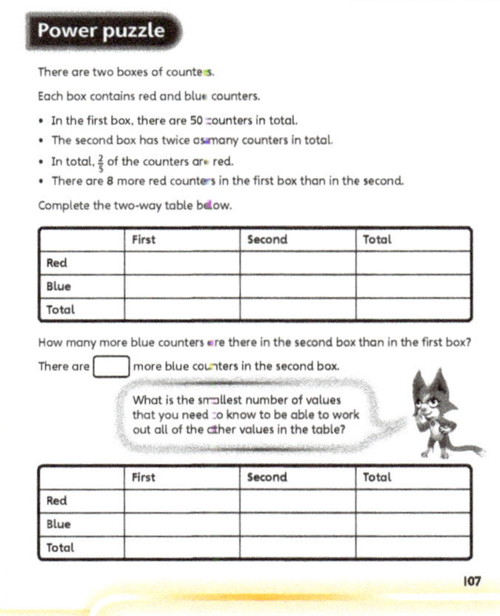

PUPIL PRACTICE BOOK 5A PAGE 107

Strengthen and **Deepen** activities for this unit can be found in the *Power Maths* online subscription.

Unit 5
Multiplication and division ❶

Don't forget to watch the Unit 5 video!

WHY THIS UNIT IS IMPORTANT

This unit will develop children's multiplicative reasoning. Children will begin by developing their understanding of multiples and factors, recognising what they are and how they are found. These concepts will be closely linked to familiar and new concrete and pictorial representations to secure their understanding. Following this, children will learn about prime numbers and how they are different to composite numbers. Using this they will then investigate how to use factors, including prime factors, to investigate and manipulate numbers. Children will then investigate square and cube numbers, linked to their concrete understanding of the shape namesakes.

Having learnt about properties of numbers, children will learn how identifying and using inverse operations can help solve and check calculations when calculating the multiplication and division of whole numbers by 10, 100 and 1,000.

WHERE THIS UNIT FITS

→ Unit 4: Graphs and tables

→ **Unit 5: Multiplication and division (1)**

→ Unit 6: Measure – area and perimeter

In this unit, children develop their understanding of the multiplicative properties of numbers. This unit follows their learning about data handling and precedes their work on measure and perimeter.

Before they start this unit, it is expected that children:
- know their times-tables to 12 fluently and can count in 10s, 100s and 1,000s
- understand and use the operations of multiplication and division confidently
- recognise the place value of 4-digit numbers.

ASSESSING MASTERY

Children will be able to reliably find multiples and factors of given numbers. They will be able to explain the unique properties of prime, square and cube numbers and represent them in concrete, pictorial and abstract ways including using the notations (^{2}) and (^{3}). They will be able to reliably multiply and divide whole numbers by 10, 100 and 1,000, recognising patterns and explaining how these affect a number's place value. Finally, they will be able to confidently use their understanding of these concepts to multiply and divide whole numbers by multiples of 10, 100 and 1,000. They will show their ability to solve problems using their new learning.

COMMON MISCONCEPTIONS	STRENGTHENING UNDERSTANDING	GOING DEEPER
Children may confuse factors and multiples.	To secure understanding of factors and multiples, use factor trees and arrays to demonstrate the different factors a given number has.	Encourage children to investigate which types of numbers have the most factors. Can they make generalities about the numbers with the most or least factors?
Children may recognise multiplying and dividing by 10, 100 and 1,000 as just 'adding or taking off a zero' from a number.	Encourage children to make or draw the calculations they are trying to solve. Discuss what their representation shows. Does the system of adding a zero always work? Why not?	Encourage children to create their own video tutorial or poster, explaining why the 'just add a zero' procedural method does not work consistently and what happens to a number's place value.

Unit 5: Multiplication and division 

WAYS OF WORKING

Use these pages to introduce the focus to children. You can use the characters to explore different ways of working too!

STRUCTURES AND REPRESENTATIONS

Array: Arrays are a visual representation of multiplication and division. They are an excellent tool for showing equal groups within a number.

Bar model: The bar model enables children to more easily represent a problem. In the context of this unit, it is used to show different types of calculations.

70	70	70	70	70	70	70

Factor tree: Factor trees are used to show the factors a given number has.

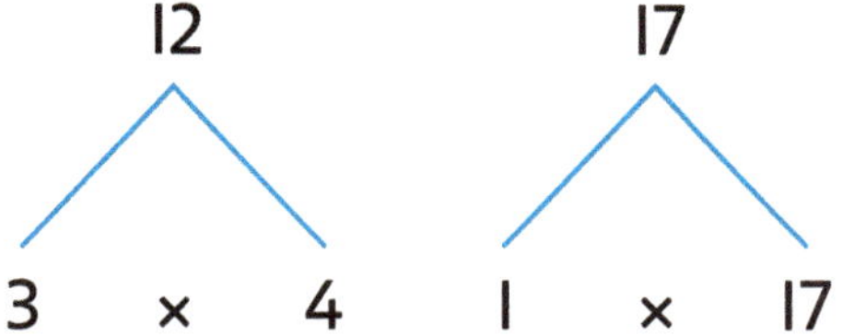

Multiplication square: Multiplication squares are used in this unit to demonstrate and investigate the patterns found in different types of numbers.

×	1	2	3	4	5	6	7	8	9	10	11	12
1	1	2	3	4	5	6	7	8	9	10	11	12
2	2	4	6	8	10	12	14	16	18	20	22	24
3	3	6	9	12	15	18	21	24	27	30	33	36

KEY LANGUAGE

Here is some key language that children will need to know as part of the learning in this unit:

→ multiple
→ factor
→ prime number
→ composite number
→ square (x^2)
→ cube (x^3)
→ multiply, multiplication, times
→ divide, division
→ inverse operation
→ place value
→ ones, tens, hundreds, thousands, tens of thousands

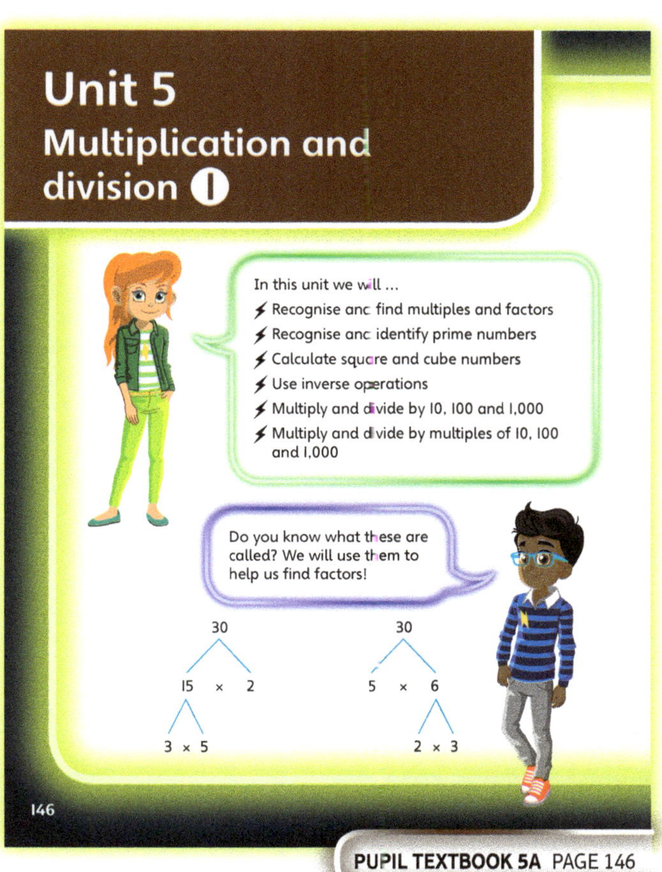

PUPIL TEXTBOOK 5A PAGE 146

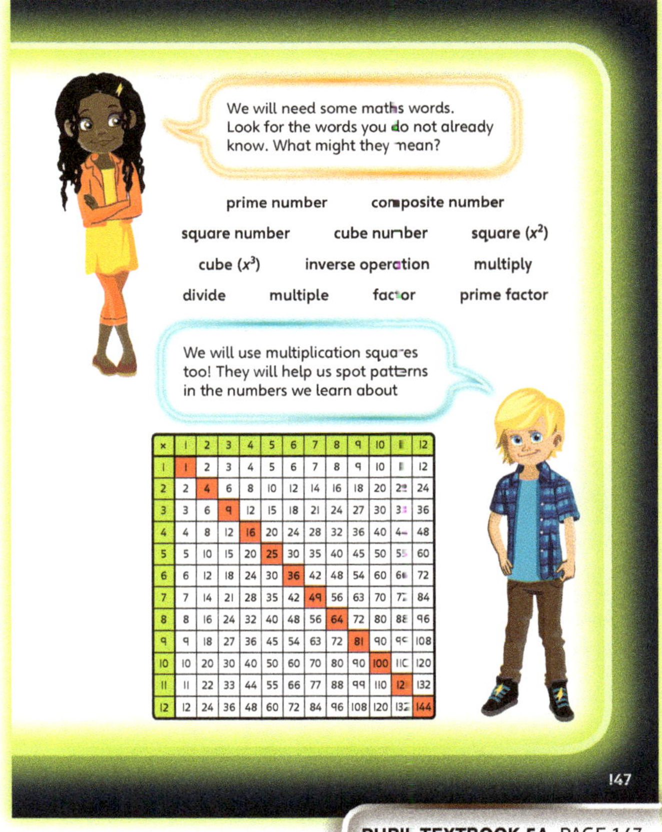

PUPIL TEXTBOOK 5A PAGE 147

Multiples

Learning focus

In this lesson, children will learn the meaning of the mathematical term 'multiple'. They will spot patterns in multiples of numbers and use these to make generalisations and predictions.

Small steps

→ Previous step: Drawing line graphs
→ **This step: Multiples**
→ Next step: Factors

NATIONAL CURRICULUM LINKS

Year 5 Number – Multiplication and Division
- Identify multiples and factors, including finding all factor pairs of a number, and common factors of two numbers.
- Solve problems involving multiplication and division including using their knowledge of factors and multiples, squares and cubes.

ASSESSING MASTERY

Children can understand and explain the term 'multiple' and can reliably find multiples of numbers, explaining where and why there are patterns. They can link this with their understanding of times-tables.

COMMON MISCONCEPTIONS

Children may assume that any number ending in X is a multiple of X (for example, 6, 16 and 26 are all multiples of 6). While this may be the case with some multiples (multiples of 2 and 5), it is not always the case. Ask:
- *Can you count in X on this 100 square? Colour in all of the numbers that are a multiple of X.*
- *Are all numbers ending X coloured? Why or why not?*

STRENGTHENING UNDERSTANDING

For children whose fluency with multiplication tables, or understanding of multiplication as an operation may be weaker, use arrays to help make concrete their understanding. Ask: *Can you show me a group of 4? What would 2 groups of 4 look like as an array? How about 3? What do you notice about the array? Can you record the multiplication as a calculation? Can you record the linked division?*

GOING DEEPER

To deepen understanding of multiples, children can be given clues about a number that they need to identify. For example: *My number is a multiple of 5, 6 and 2 but not of 7. What is my number?* Also, children could be encouraged to create their own 'guess my number' challenge for their partner.

KEY LANGUAGE

In lesson: multiple, remainder

Other language to be used by the teacher: multiply, times, multiplication

STRUCTURES AND REPRESENTATIONS

100 square, arrays, number lines

RESOURCES

Mandatory: 100 square

Optional: blank counters, sorting circles

 In the eTextbook of this lesson, you will find interactive links to a selection of teaching tools.

Before you teach 

- How confident are children with their multiplication tables?
- Could this limit their understanding or progress in this lesson?
- How will you mitigate this risk?

Discover

 Pair work

ASK

- Question **1** a): *What are the multiples of 4?*
- Question **1** a): *Could you use the multiples you have found so far to predict higher multiples?*
- Question **1** b): *Is Luis correct? How can you prove your ideas?*

IN FOCUS Question **1** b) will help to approach the potential misconception that all numbers ending in X are a multiple of X. Children could be encouraged to find other similar examples. Ask: *Why is 14 not a multiple of 4 even though it has a 4 in it? How can you prove 14 is not a multiple of 4?*

PRACTICAL TIPS Children should have access to a 100 square and follow the process shown in the picture. If possible, this could be done outside in the playground as in the picture, or, if not, then with counters and printed 100 squares.

ANSWERS

Question **1** a):

1	2	3	4	5	6	7	8	9	10
11	12	13	14	15	16	17	18	19	20
21	22	23	24	25	26	27	28	29	30
31	32	33	34	35	36	37	38	39	40
41	42	43	44	45	46	47	48	49	50
51	52	53	54	55	56	57	58	59	60
61	62	63	64	65	66	67	68	69	70
71	72	73	74	75	76	77	78	79	80
81	82	83	84	85	86	87	88	89	90
91	92	93	94	95	96	97	98	99	100

Question **1** b): Luis is incorrect. 74 is not a multiple of 4.

Share

 Whole class teacher led

ASK

- Question **1** a): *How can you find the multiples of 4?*
- Question **1** a): *How do the counters show the multiples clearly?*
- Question **1** a): *Did you find all of the multiples shown?*
- Question **1** a): *Did you notice any patterns in the multiples? How did they help you to continue finding more multiples?*

IN FOCUS It will be important to focus on the pictorial representations of the multiples of 4 shown in question **1** a). This will help children link their understanding of multiplication and arrays with the new vocabulary they are learning. It may help to give children the opportunity to build the arrays shown.

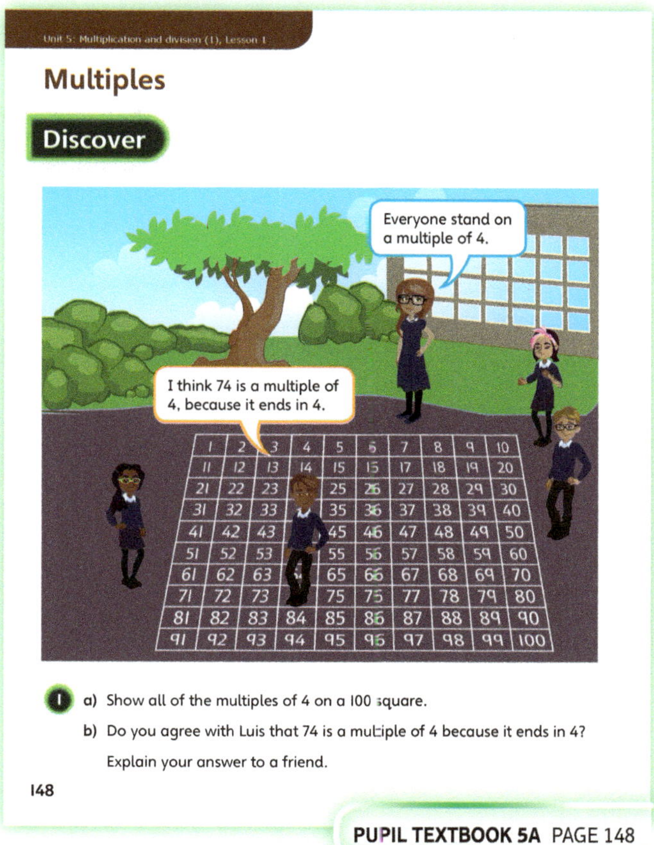

PUPIL TEXTBOOK 5A PAGE 148

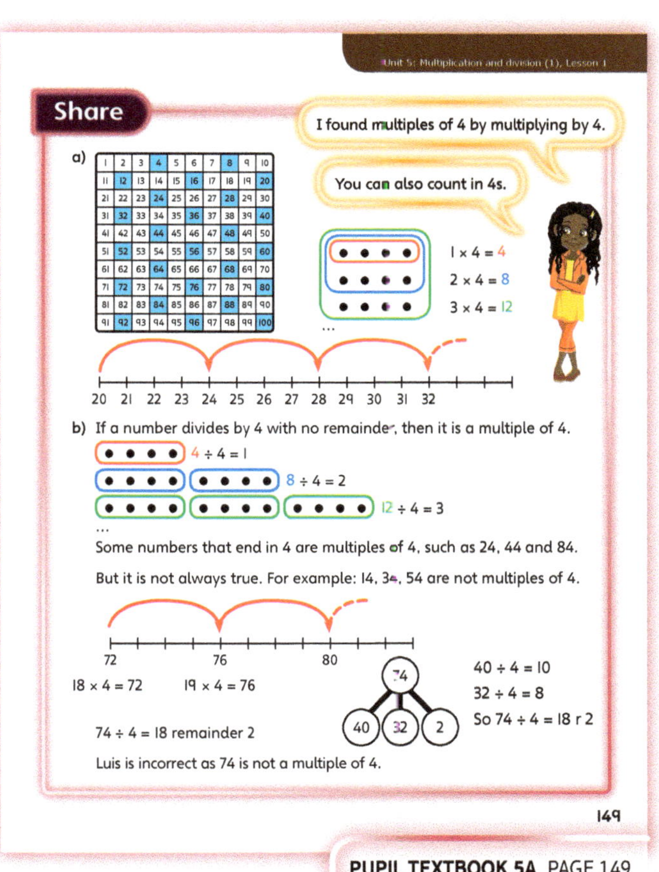

PUPIL TEXTBOOK 5A PAGE 149

Think together

WAYS OF WORKING Whole class teacher led (I do, We do, You do)

ASK

- Questions ❶ and ❷: *Can you see a pattern in the multiples already? How will this continue? What can you say is the same about all of the multiples of X?*
- Question ❷: *How can you find multiples after 100?*

IN FOCUS Question ❶ scaffolds children's ability to find multiples by allowing them to continue the already started pattern. It also scaffolds their ability to generalise about multiples of a number through the sentence starters.

STRENGTHEN Strengthen understanding of question ❸ a) by providing children with a 100 square. Ask: *How does the 100 square help? Can you use it to predict a 3-digit multiple of 6? From the numbers you have found, can you see whether the statement in ❸ b) is always, sometimes or never true?*

DEEPEN While solving question ❸ c), develop children's reasoning by asking: *What numbers have you found? What is special about the numbers in the middle segment of the sorting circles? How can you use this to help you find multiples of 4 more easily in the future? Why are there no numbers in the multiple of 4 only segment?*

ASSESSMENT CHECKPOINT Questions ❷ and ❸ will assess children's ability to find multiples of numbers and record them in different ways. Question ❸ will also be useful to help assess children's recognition that not all multiples ending in X are a multiple of X.

ANSWERS

Question ❶: Multiples of 2 have 0, 2, 4, 6 or 8 in the ones digit.
Numbers that are multiples of 2 are all even.
Numbers that are **not** multiples of 2 are all odd.

Question ❷: 5, 10, 15, 20, 25, 30, 35, 40, 45, 50, 55, 60, 65, 70, 75, 80, 85, 90, 95, 100

Multiples of 5 have 0 or 5 in the ones digit.

Even multiples of 5 all end in 0.

Odd multiples of 5 all end in 5.

Question ❸ a): For example:

	Multiple of 6	Not a multiple of 6
Ends in a 6	6, 36, 66, 96	16, 26, 46, 56, 76
Does not end in a 6	12, 18, 24, 30, 42	1, 2, 3, 4, 5, 7, 8, 9, 10, 11, 13, 14, 15

Question ❸ b): The statement is sometimes true, as evidenced in the chart in question ❸ a).

Question ❸ c): For example:

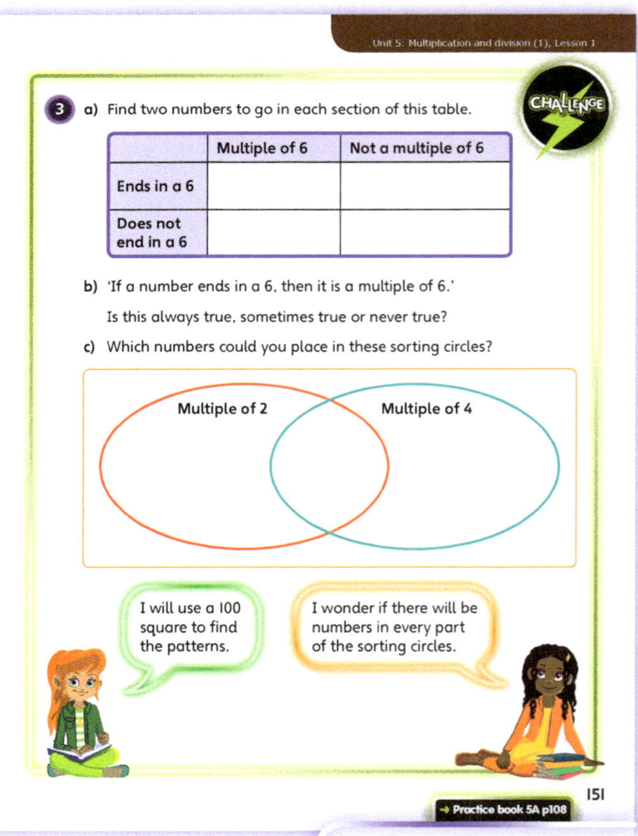

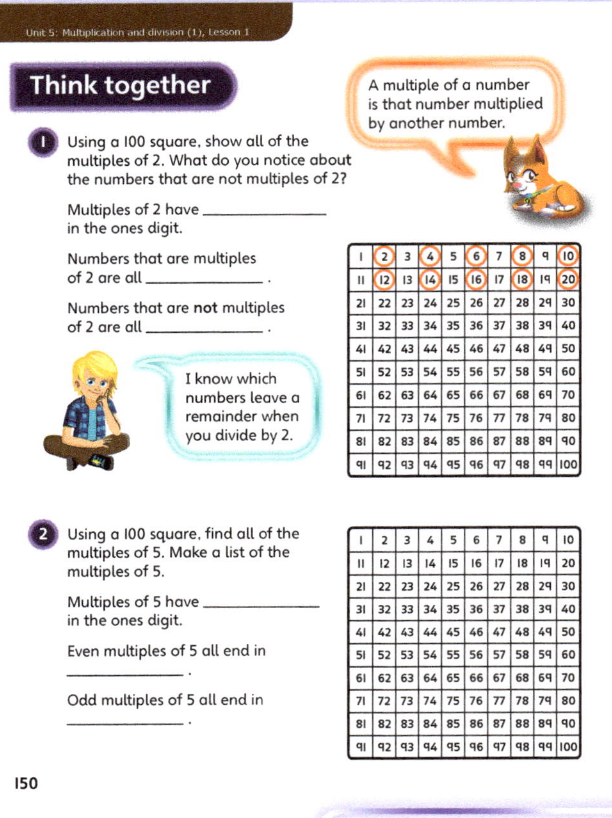

PUPIL TEXTBOOK 5A PAGE 150

PUPIL TEXTBOOK 5A PAGE 151

Practice

WAYS OF WORKING Independent thinking

IN FOCUS Question ① helps scaffold children's independent understanding of multiples by offering them different pictorial representations of them. It may help children to build their own versions using cubes or counters and then draw what they have made.

Question ④ helps to demonstrate to children the link between multiples and division. This is important for their multiplicative reasoning, as children may otherwise only make the link between multiples and multiplication.

STRENGTHEN If children need help to begin investigating question ⑤, ask: *What will you need to find first? Why? What resource could you use to help you find the multiples you need? How will you know if the number you have created is a multiple of 9?*

DEEPEN Children could deepen their understanding of the generalisations that can be made around multiples by investigating question ⑤ in more detail. Ask: *What happens if you change the question from multiples of 2 and 6 to multiples of 3 and 5? What is different and what stays the same? Is it the same section that remains empty? Why or why not? Is it possible to find combinations where all boxes are filled? Explain your ideas.*

ASSESSMENT CHECKPOINT Question ⑦ will assess children's understanding of how multiples work. If they agree that, because the pattern is increasing by 7 therefore they are multiples of 7, it will be clear that they have not quite grasped the concept.

ANSWERS Answers for the **Practice** part of the lesson appear in the separate **Practice and Reflect answer guide**.

Reflect

WAYS OF WORKING Independent thinking

IN FOCUS This question will demonstrate whether children have a secure enough understanding of multiples and how to find them to be able to diagnose another child's error.

ASSESSMENT CHECKPOINT It will be important to look for children's ability to discriminate between 70 as the multiple of 7 and 10 as the amount of sevens needed to have 70.

ANSWERS Answers for the **Reflect** part of the lesson appear in the separate **Practice and Reflect answer guide**.

After the lesson ❙❙

- Having concluded the lesson, are there any multiplication tables that children will benefit from more practice in than others?
- How will you adapt the next lesson on factors to account for this?
- Do children understand the patterns between the digit in the ones position and the multiple?

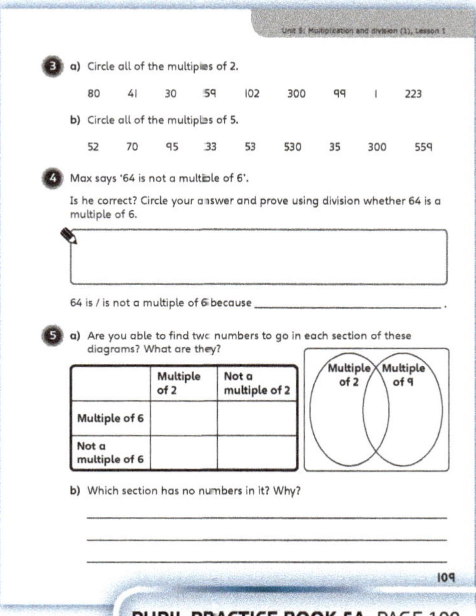

PUPIL PRACTICE BOOK 5A PAGE 108

PUPIL PRACTICE BOOK 5A PAGE 109

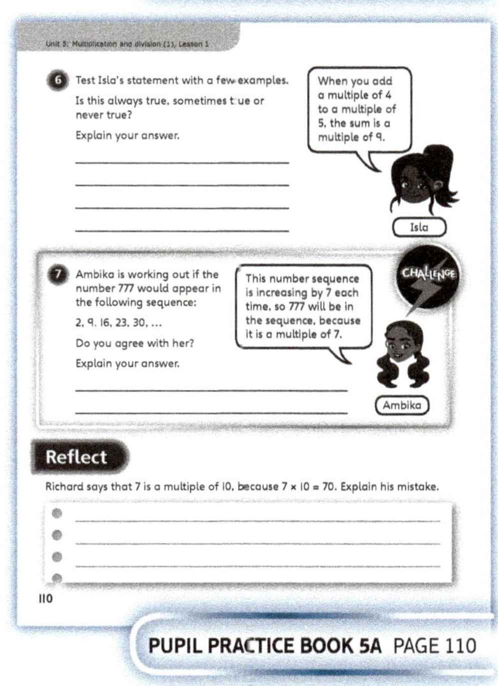

PUPIL PRACTICE BOOK 5A PAGE 110

Factors

Learning focus

In this lesson, children will learn the meaning of the mathematical term 'factor' and use multiplication and division to find factors. They will spot patterns in factors of numbers and use these to make generalisations and predictions.

Small steps

→ Previous step: Multiples
→ **This step: Factors**
→ Next step: Prime numbers

NATIONAL CURRICULUM LINKS

Year 5 Number – Multiplication and Division

Identify multiples and factors, including finding all factor pairs of a number, and common factors of two numbers.

ASSESSING MASTERY

Children can understand and explain the term 'factor', reliably find factors of numbers and explain where and why there are patterns in factors. They can link their understanding of times-tables with their new understanding of factors and the concept of dividing numbers into equal groups.

COMMON MISCONCEPTIONS

Children may confuse the term 'factor' with the term 'multiple'. Ask:
• *What did you do last lesson to find a multiple? How is that different to what you are doing today to find a factor? How would you explain the difference to someone who was confused?*

STRENGTHENING UNDERSTANDING

To help children more fluently find factors of numbers, it would be useful to identify times-tables they are less confident with and practise these with them. This could be achieved through a variety of methods including using times-tables rhymes, songs and chants, pairs games matching calculations with numbers and creating arrays, and other concrete and pictorial representations of a multiplication table.

GOING DEEPER

Children could be encouraged to investigate whether a number can be both a factor and a multiple of itself. Ask:
• *Can it be true that 3 is both a factor and a multiple of 3? Explain your ideas.*

KEY LANGUAGE

In lesson: factors, remainder, division, multiplication

Other language to be used by the teacher: multiple, divide, multiply

STRUCTURES AND REPRESENTATIONS

number lines, arrays

RESOURCES

Mandatory: squared paper, counters

Optional: multilink cubes

 In the eTextbook of this lesson, you will find interactive links to a selection of teaching tools.

Before you teach ⏸

• What methods did you use to find multiples in the last lesson?
• How can you develop the connection between multiples and factors in this lesson?
• What real-life contexts can you draw on to help children secure their understanding in this lesson?

Discover

WAYS OF WORKING Pair work

ASK

- Question ① a): *How can you show what equal groups can be made with 24 chairs?*
- Question ① a): *What mathematical representation could show your thinking best?*
- Question ① b): *How could you investigate what equal groups are in 25?*

IN FOCUS Question ① a) will allow children to experiment with factors. This will offer them a good opportunity to begin linking factors with their times-table knowledge.

Question ① b) is important as it enables children to begin making conjectures about factors, looking for patterns and making generalisations about how factors change depending on the given number.

PRACTICAL TIPS This activity could be linked to setting the school hall out for an assembly or for lunch. Children could be encouraged to rearrange the chairs in their classroom to accommodate a number of people with all rows of chairs being equal.

ANSWERS

Question ① a): 1 × 24 = 24; 1 and 24 are factors of 24
2 × 12 = 24; 2 and 12 are factors of 24
3 × 8 = 24; 3 and 8 are factors of 24.
4 × 6 = 24; 4 and 6 are factors of 24.

Question ① b): There are fewer arrangements as the factors of 25 are 1, 5 and 25.

Share

WAYS OF WORKING Whole class teacher led

ASK

- Question ① a): *How do the arrays show the factors clearly?*
- Question ① a): *Is there another calculation for each set of factors? Explain your ideas.*
- Question ① b): *How many factors of 25 did you find? How did you show them?*
- Question ① b): *Why does 25 have fewer factors than 24?*

IN FOCUS For questions ① a) and ① b), it is important to link the factors of the numbers given to children's understanding of arrays. This will ensure children make the link between their concrete understanding and the concept of dividing numbers into equal groups to give factors. Children should be encouraged to build or draw the appropriate arrays for any given numbers they are investigating. Children need to understand the link between factors and commutativity. Discuss how knowing 3 × 8 = 24 also implies 8 × 3 = 24, so we know both 3 and 8 are factors of 24.

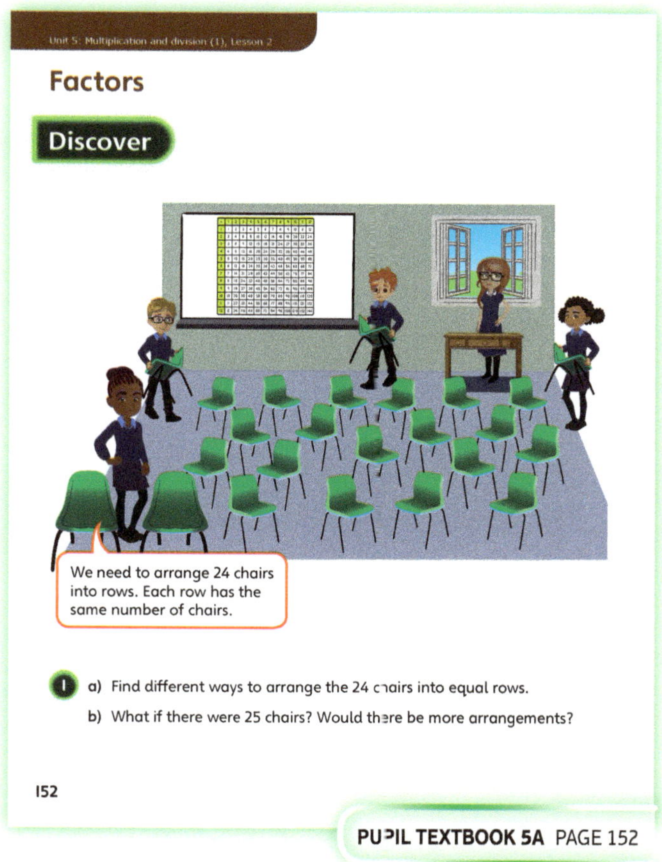

PUPIL TEXTBOOK 5A PAGE 152

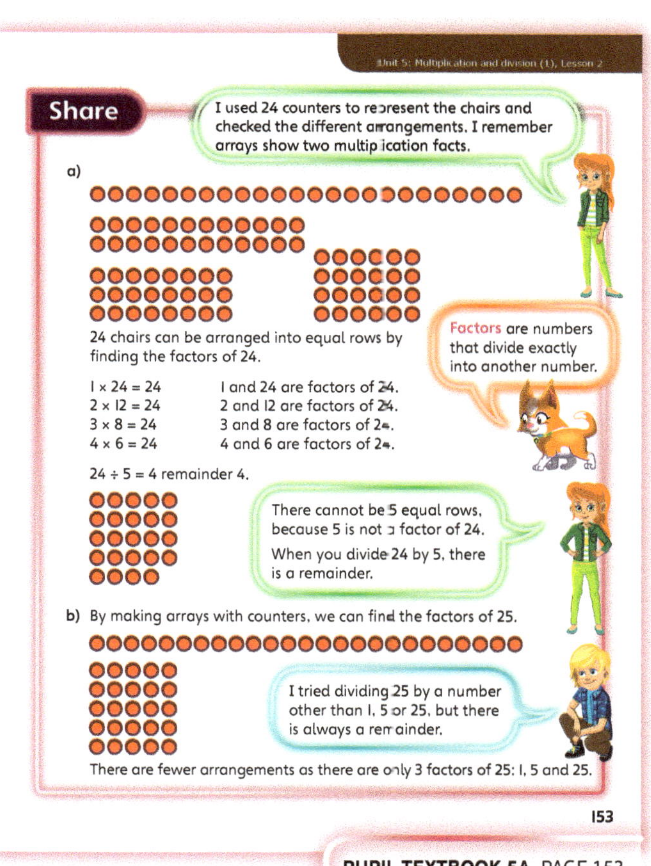

PUPIL TEXTBOOK 5A PAGE 153

Think together

WAYS OF WORKING Whole class teacher led (I do, We do, You do)

ASK

- Question **1**: *What factors does the picture show? Explain how you know. How will you prove if there are any more factors of 16?*
- Question **2** a): *How will the number line help us find if 5 is a factor of 16? How does using multiplication help us find factors?*
- Question **2** b): *How will the number line help us find if 4 is a factor of 22? How does using division help us find factors?*
- Question **3**: *How will the lists be similar for 30 and 40? How will they be different for 30 and 40?*

IN FOCUS Question **1** is important as it again links children's understanding of arrays to their new understanding of factors. It also demonstrates multiple ways of recording factors of a number.

Question **2** a) helps children recognise how factors can be found using increasingly formal methods of recording and calculating. It will also demonstrate clearly the link between multiples and factors and how, when counting in 5s, 16 is not one of the multiples, thereby proving that 5 is not a factor of 16.

STRENGTHEN To strengthen understanding and to help draw the correct arrays on squared paper for question **1**, offer children counters or multilink cubes to allow them to build the arrays first. Ask: *Can you use the squares on the paper to draw the same array accurately?*

DEEPEN When solving question **3** b), deepen children's reasoning by encouraging them to generalise about the factors they have been investigating. Ask: *Is there a way to always reliably know when you have found all of the factors of a number? Is there any way of predicting how many factors a number will have?*

ASSESSMENT CHECKPOINT Question **3** will assess children's ability to reliably find the factors of a given number using both multiplication and division. Look for children's recorded evidence of their thinking as this will demonstrate their recognition of the link between factors and multiplicative reasoning.

ANSWERS

Question **1**: The factors of 16 are 1, 2, 4, 8, 16.

Question **2** a): 5 is not a factor of 16 because 16 is not in the 5 times-table. The nearest multiple of 5 is 15 and the following multiple is 20.

Question **2** b): 4 is not a factor of 22 because 22 is not divisible by 4 without a remainder.

Question **3**: They will be able to stop their lists when they find their calculations contain the same numbers as before. In this case it will be at 5×6 or $30 \div 5$.

Question **3** a): The factors of 30 are 1, 2, 3, 5, 6, 10, 15 and 30.

Question **3** b): The factors of 40 are 1, 2, 4, 5, 8, 10, 20 and 40.

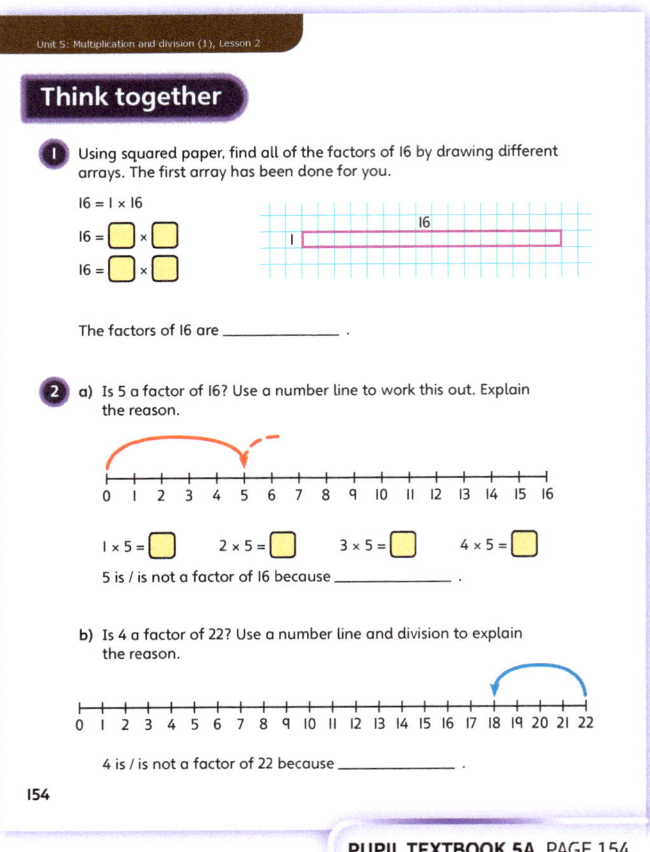

PUPIL TEXTBOOK 5A PAGE 154

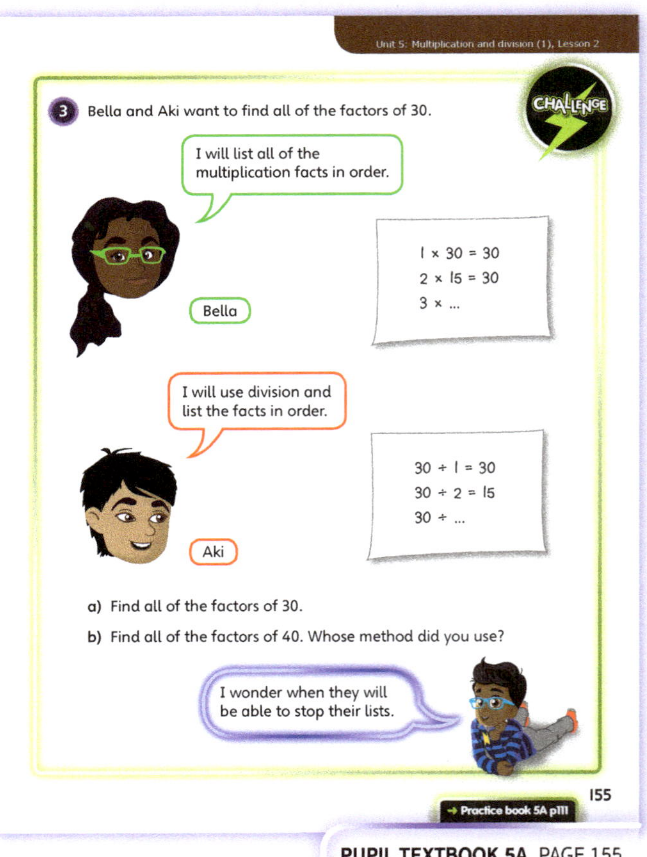

PUPIL TEXTBOOK 5A PAGE 155

Practice

WAYS OF WORKING Independent thinking

IN FOCUS Questions ❶ and ❷ link children's independent understanding of the concrete and pictorial representations of multiplication with their understanding of factors. Question ❷ develops this by requiring children to draw the arrays as well as listing the factors shown by them.

Question ❸ is important as it encourages children to provide their reasoning as to why a number either is or is not a factor of another number. It will be important to observe carefully children's responses to ensure their full understanding.

Question ❼ is important as it makes the link between factors and multiples clear and overt.

STRENGTHEN If children are not sure whether they have found all of the factors of a number in questions ❹, ❺ and ❻, ask: *How could you organise your thinking so that you can prove you have found all of the factors? How will you know when you have tested all of the possible factors?*

DEEPEN While children solve question ❻, deepen their understanding of using factors and multiples in context by asking: *Which type of number is more useful, factors or multiples? Why? Is there a time where finding multiples would be more useful than factors and vice versa?*

ASSESSMENT CHECKPOINT Question ❹ will assess children's ability to work systematically to find all of the factors of a number using both multiplication and division. They should be able to demonstrate their ability to record their findings clearly and logically and then use their written recording to justify how they know they have found all possible factors.

ANSWERS Answers for the **Practice** part of the lesson appear in the separate **Practice and Reflect answer guide**.

Reflect

WAYS OF WORKING Independent thinking

IN FOCUS This question will give children the opportunity to show their understanding of what a factor is, that it is the number multiplied together and does not have a bearing on whether a number is odd or even.

ASSESSMENT CHECKPOINT Children can reliably explain what a factor is and how to work out what the factors of 70 are.

ANSWERS Answers for the **Reflect** part of the lesson appear in the separate **Practice and Reflect answer guide**.

After the lesson ⏸

- How confident were children with the concept of factors?
- Did the lesson make the link between factors and multiples explicit? If not, how will you ensure this link is clear in future lessons?
- How could you include the concepts in this lesson in other areas of the curriculum?

PUPIL PRACTICE BOOK 5A PAGE 111

PUPIL PRACTICE BOOK 5A PAGE 112

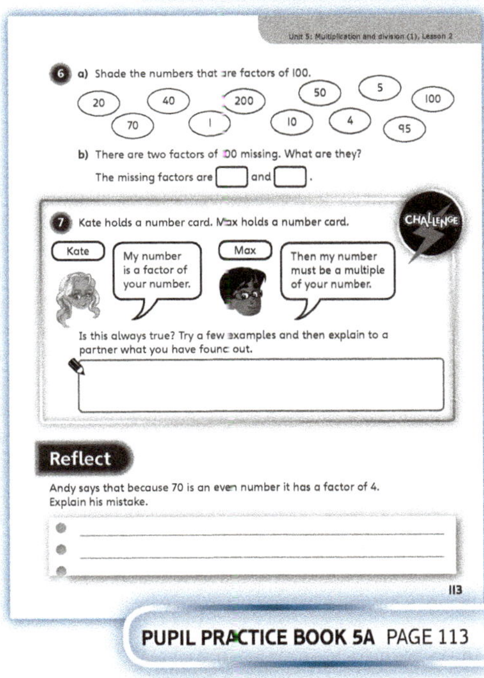

PUPIL PRACTICE BOOK 5A PAGE 113

Prime numbers

Learning focus

In this lesson, children will learn about prime numbers and how they are different to other numbers. They will learn the correct vocabulary of prime and composite numbers and how to differentiate between them.

Small steps

→ Previous step: Factors
→ **This step: Prime numbers**
→ Next step: Using factors

NATIONAL CURRICULUM LINKS

Year 5 Number – Multiplication and Division
- Know and use the vocabulary of prime numbers, prime factors and composite (non-prime) numbers.
- Establish whether a number up to 100 is prime and recall prime numbers up to 19.

ASSESSING MASTERY

Children can reliably identify prime and composite numbers, recognising and explaining that prime numbers have only two factors, 1 and itself. They can identify all prime numbers up to 100 and can use this understanding to solve problems.

COMMON MISCONCEPTIONS

Children may recognise that almost every even number is composite. Having recognised this pattern they may mistakenly assume that 2 is composite because it is even. Ask:
- *Show me how many arrays you can make with 2 counters. How many factors does 2 have?*

Children may think that 1 is a prime number as its factor is 1 and is also itself. Ask:
- *How many different factors does 1 have? How many different factors does a prime number have?*

STRENGTHENING UNDERSTANDING

To strengthen understanding of the concept of prime numbers, give children as many practical opportunities as possible to investigate and play with prime numbers. For example, setting up teams, sharing toys or food, setting out table places. When sharing in these different contexts, children should be encouraged to use the vocabulary of 'prime' if they find a number that has only 2 factors: 1 and itself.

GOING DEEPER

Children can be encouraged to investigate whether the generalisations and patterns they have made and found in the lesson continue in numbers over 100. Ask: *What patterns or rules did you find when finding prime numbers? Do you think these patterns will carry on after 100? Why? Try to prove it.*

KEY LANGUAGE

In lesson: prime numbers, composite numbers, factors, remainder, division

Other language to be used by the teacher: divide, even, multiples, arrays

STRUCTURES AND REPRESENTATIONS

arrays, 100 square

RESOURCES

Mandatory: counters

 In the eTextbook of this lesson, you will find interactive links to a selection of teaching tools.

Before you teach ⏸

- Are children confident with finding factors?

Discover

WAYS OF WORKING Pair work

ASK

- Question ❶ a): *How many rugby players are there? How many ways can you share 13 equally?*
- Question ❶ a) and b): *How can you prove that you have found all of the factors?*
- Question ❶ b): *How is 9 different from 7? How are the factors different?*

IN FOCUS Questions ❶ a) and b) will provide children with opportunities to discuss and experiment with the factors in prime numbers. It will be important for them to begin noticing patterns in prime numbers, so they can begin generalising about prime number properties.

PRACTICAL TIPS This concept could be introduced in PE while setting up team games. Alternatively, for a more class-based approach, children could be asked to share stickers out equally between them. Give them a challenge that allows them to keep the stickers only if they can share them out equally without splitting any into smaller pieces.

ANSWERS

Question ❶ a): The players can only be in 1 group of 13 or 13 groups of 1.

Question ❶ b): The team of tennis players can split into equal groups in more ways than the team of basketball players.

PUPIL TEXTBOOK 5A PAGE 156

Share

WAYS OF WORKING Whole class teacher led

ASK

- Question ❶ a): *What factors have you found for 13?*
- Question ❶ a): *How can you prove that you have found all of the factors?*
- Question ❶ b): *How do you know that 7 is a prime number?*

IN FOCUS Question ❶ a) introduces children to the vocabulary of 'prime numbers'. Also, Sparks provides a definition saying 'Numbers with only two factors are called prime numbers. These are special numbers that can only be divided by themselves and 1.' It will be important for children to practise using this information in their explanations and that they are aware of how other numbers are 'composite numbers'.

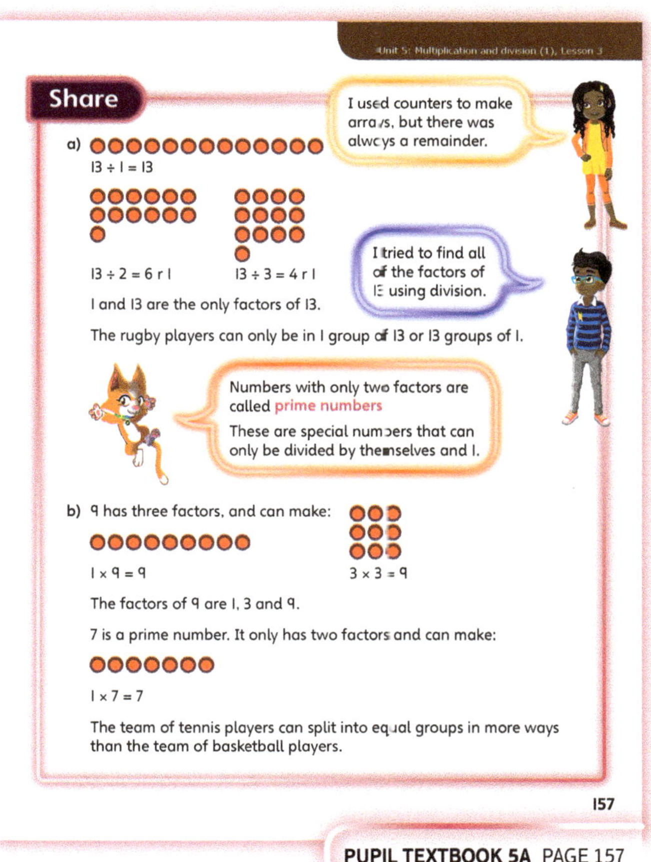

PUPIL TEXTBOOK 5A PAGE 157

Think together

WAYS OF WORKING Whole class teacher led (I do, We do, You do)

ASK

- Question **1** a): *Can you use counters to help you? How can you prove that you have found all of the factors?*
- Question **1** b): *What was the most efficient method of finding out if the number was prime?*
- Question **2**: *Which numbers do you predict will be prime?*
- Question **3**: *Can you see a pattern emerging?*

IN FOCUS Question **1** will help to tackle the assumption that all odd numbers are prime. It is also important as it gives children a larger number, closer to 100, to investigate. Discuss with children how they went about finding out if the number was prime and which was the most efficient method.

STRENGTHEN To strengthen understanding in question **3**, encourage children to use arrays to help find the factors of each number. Encourage them to look for patterns that may make identifying primes and composites easier. Ask: *Can you show me the arrays that X can make? How will finding patterns help you to find the next prime number?*

DEEPEN In question **3**, use children's investigation findings to draw out generalisations. Ask: *What do you notice about the numbers you have circled? Can you see any patterns that might help you identify prime numbers in the future?*

ASSESSMENT CHECKPOINT Question **3** will assess children's ability to find prime numbers up to 100. When explaining their findings they should show recognition that all the circled numbers have only two factors.

ANSWERS

Question **1** a): Factors of 63 are 1, 3, 7, 9, 21, 63.

Question **1** b): 63 is a composite number.

Question **2**:

Number of players	What different arrays can they make?	How many factors?	Is it a prime or composite number?
12	12 × 1, 6 × 2, 4 × 3, 3 × 4, 2 × 6, 1 × 12	6	Composite
11	11 × 1, 1 × 11	2	Prime
10	10 × 1, 5 × 2, 2 × 5, 1 × 10	4	Composite
9	3 × 3, 1 × 9	3	Composite
8	8 × 1, 4 × 2, 2 × 4, 1 × 8	4	Composite
7	1 × 7	2	Prime
6	6 × 1, 3 × 2, 2 × 3, 1 × 6	4	Composite
5	5 × 1, 1 × 5	2	Prime
4	4 × 1, 2 × 2, 1 × 4	3	Composite
3	3 × 1, 1 × 3	2	Prime
2	2 × 1, 1 × 2	2	Prime

Question **3**: The prime numbers between 0 and 100 are 2, 3, 5, 7, 11, 13, 17, 19, 23, 29, 31, 37, 41, 43, 47, 53, 59, 61, 67, 71, 73, 79, 83, 89, 97. There are 25 prime numbers between 0 and 100.

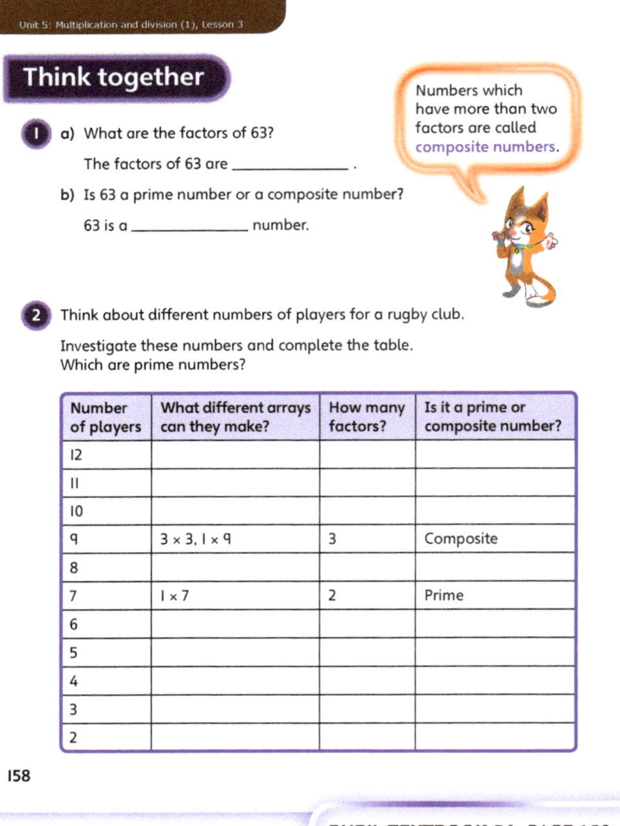

Number of players	What different arrays can they make?	How many factors?	Is it a prime or composite number?
12			
11			
10			
9	3 × 3, 1 × 9	3	Composite
8			
7	1 × 7	2	Prime
6			
5			
4			
3			
2			

PUPIL TEXTBOOK 5A PAGE 158

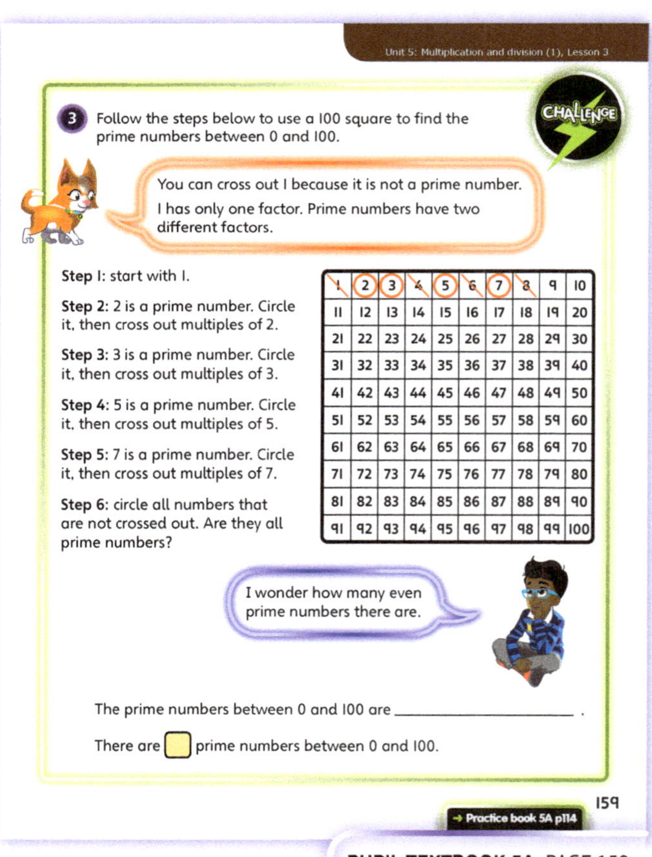

PUPIL TEXTBOOK 5A PAGE 159

Practice

WAYS OF WORKING Independent thinking

IN FOCUS Questions **1** and **2** let children independently show pictorially how a prime number cannot be shared into equal groups without having a remainder. These questions are important as they continue to strengthen the link between the pictorial representation of the array and the abstract understanding of prime numbers.

Question **3** is important as it helps to tackle the misconception that all even numbers are composite numbers. It also strengthens the understanding that, after 2, all even numbers are composite.

Questions **4** and **5** offer a valuable opportunity to continue developing generalisations about prime numbers.

STRENGTHEN When solving question **6**, children may be put off by the new and unfamiliar arrangement of the number grid. Ask: *What do you notice about the number grid? How is it similar and different to number grids you have used before? What might change in the patterns of prime numbers because of the new grid?*

DEEPEN When considering their ideas for question **5**, deepen children's reasoning by asking: *What conjecture could you make about these statements? How could you prove your ideas? Can you prove your thinking using resources or a picture? Can you make your explanation clear enough so that someone new to prime numbers would understand?*

ASSESSMENT CHECKPOINT Questions **4** and **5** give a good opportunity to assess children's fluency with prime numbers. They should be able to demonstrate understanding of the generalisations that can be made about prime numbers and prove them using the resources in the lesson.

ANSWERS Answers for the **Practice** part of the lesson appear in the separate **Practice and Reflect answer guide**.

Reflect

WAYS OF WORKING Pair work

IN FOCUS This question is important as it will give a quick assessment of children's ability to find prime numbers and how confident they are with the link between prime numbers and their knowledge of factors and multiplication. Children could be encouraged to share their diagram with their partner and discuss how and why the pictures are the same and different.

ASSESSMENT CHECKPOINT Children should be able to use arrays to demonstrate whether a number is prime or composite and explain what the diagram proves clearly and concisely.

ANSWERS Answers for the **Reflect** part of the lesson appear in the separate **Practice and Reflect answer guide**.

After the lesson

- Were children able to confidently find prime numbers all the way to 100?
- How was children's independence and resilience developed in this lesson?
- Were children able to sufficiently demonstrate their problem-solving abilities? How would you develop this area when you teach this lesson again?

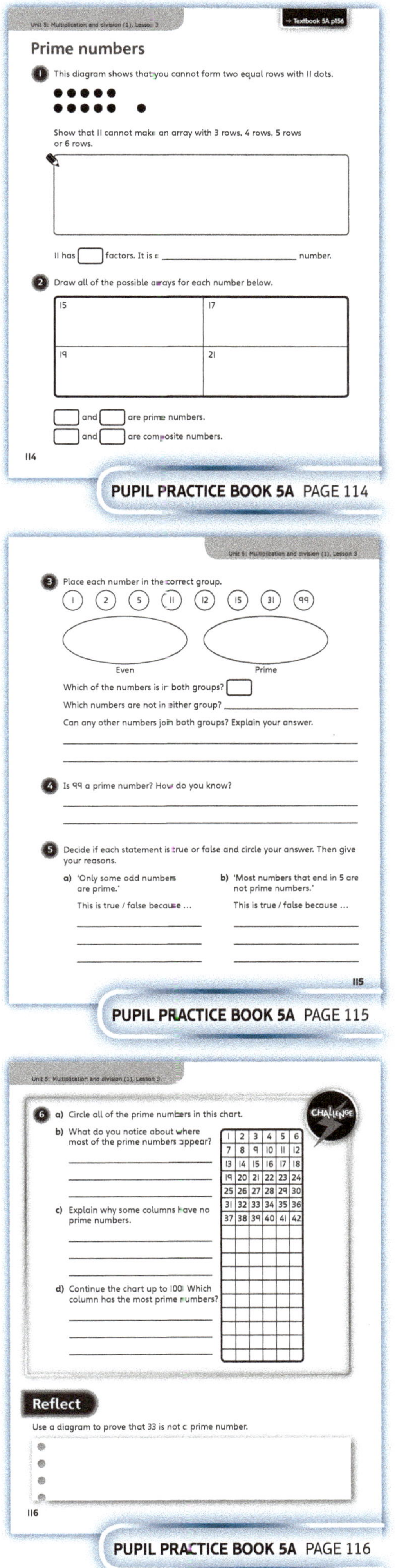

PUPIL PRACTICE BOOK 5A PAGE 114

PUPIL PRACTICE BOOK 5A PAGE 115

PUPIL PRACTICE BOOK 5A PAGE 116

Using factors

Learning focus

In this lesson, children will use their learning about prime numbers to help them solve mathematical problems and puzzles involving breaking down numbers into factors. They will use the knowledge and understanding to explain their reasoning.

Small steps

→ Previous step: Prime numbers
→ **This step: Using factors**
→ Next step: Squares

NATIONAL CURRICULUM LINKS

Year 5 Number – Multiplication and Division

Solve problems involving multiplication and division including using their knowledge of factors and multiples, squares and cubes.

ASSESSING MASTERY

Children can reliably find factors and confidently solve mathematical puzzles and problems. They understand that the order of multiplication does not matter (the associative property of multiplication) and that each number can be broken down into factors and eventually prime numbers. They can use their understanding to explain their reasoning clearly, demonstrating their thinking with concrete, pictorial and abstract evidence.

COMMON MISCONCEPTIONS

Children may confuse the term 'factor' with the term 'multiple' and find multiples instead of factors. Ask:
• *What does the term 'factor' mean? How is it different to a 'multiple'? How do you find a factor or a multiple?*

STRENGTHENING UNDERSTANDING

Children could be encouraged to find factors of numbers by investigating how objects, such as sweets, could be shared equally into groups. Ask: *Into how many groups can these sweets be shared equally? Prove it.*

GOING DEEPER

Children could be given a context similar to the following to deepen their understanding and their problem-solving skills: *I have 40 sweets and 20 yoyos. I want to share these equally in party bags so that each bag holds the same number of sweets and yoyos. How many bags could I fill? How many solutions can you find?*

KEY LANGUAGE

In lesson: factor, prime factor, factor tree

Other language to be used by the teacher: multiply, multiplication

STRUCTURES AND REPRESENTATIONS

arrays, factor trees

RESOURCES

Mandatory: counters

 In the eTextbook of this lesson, you will find interactive links to a selection of teaching tools.

Before you teach ⏸

• Was children's understanding of factors sufficient to be able to problem-solve in context?
• Do children fully understand the concept of primes and that 1 is not a prime number?
• How will you support those children who are still developing their understanding of factors and primes?

Discover

 Pair work

ASK

- Question **1** a): *How many doughnuts are in 1 bag? How many bags will Aki buy?*
- Question **1** a): *What calculation will you need to use to solve this? Explain how you know.*
- Question **1** b): *Can you draw a picture to explain this problem? Explain your picture.*
- Question **1** b): *What methods will you use to solve this? Why? Which do you prefer?*

IN FOCUS Question **1** a) will introduce children to using their understanding of factors and multiples in contextual problems. Using hands-on resources to help them solve the problem will support this understanding.

Question **1** a) and b) will help to develop children's fluency and flexibility when problem-solving by encouraging them to consider different methods of solving the same problem.

PRACTICAL TIPS This scenario could be carried out in the classroom with practical resources. Beginning the lesson by role-playing the scenario pictured would be an engaging start for children.

ANSWERS

Question **1** a): Aki will buy 120 doughnuts.

Question **1** b): 4 × 5 × 6 solved in different steps. For example, 6 lots of 4 × 5 or 4 lots of 5 × 6, due to the context of the problem. (Do not expect children to show these calculations in this way – instead observe their explanations to link to one of the three.)

Share

 Whole class teacher led

ASK

- Question **1** a) and b): *How are the methods similar and different?*
- Question **1** a) and b): *Which method do you think is most efficient? Why?*
- Question **1** a) and b): *What mathematical knowledge did you need to use to solve these problems?*

IN FOCUS It will be important, when looking at both questions **1** a) and **1** b), to discuss the different methods shown. Children should be encouraged to discuss the benefits and downsides of both methods to help them develop a deeper understanding of how the problems were solved. This will help them to think more critically about their approaches in the future. When discussing the efficacy of the methods, children should be encouraged to use the multiplication facts they are most comfortable with, such as using multiples of 10. Ensure children understand that the order in which they multiplied the numbers did not matter.

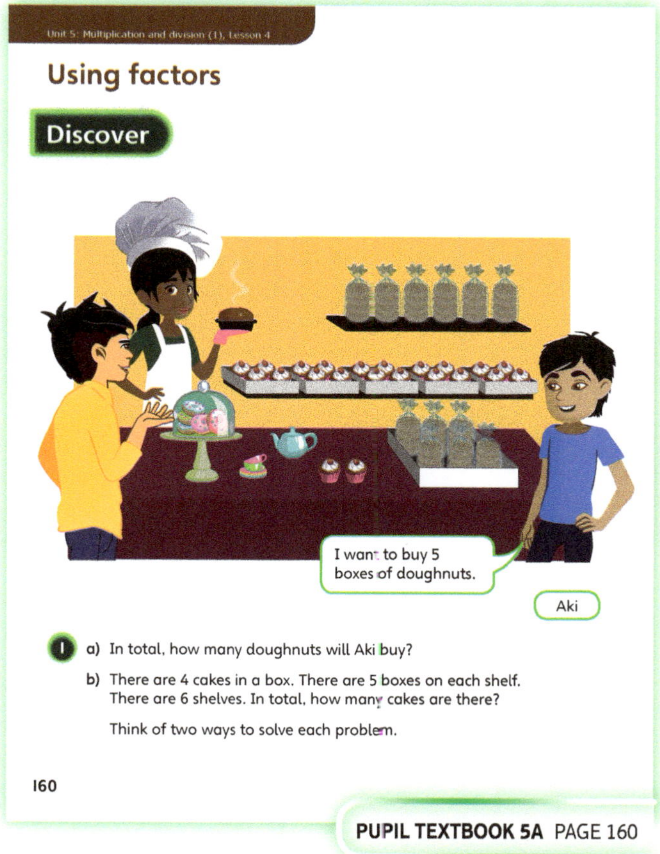

PUPIL TEXTBOOK 5A PAGE 160

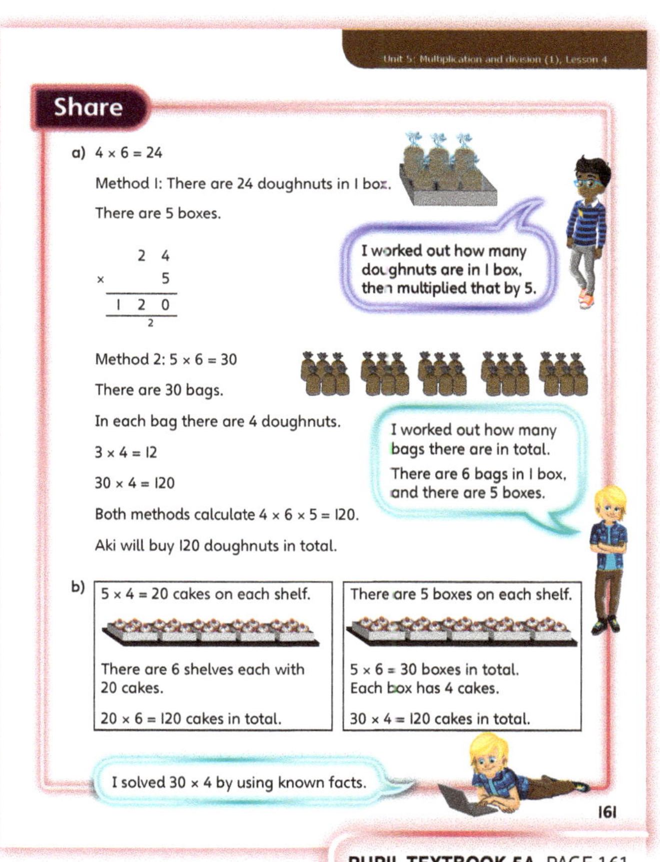

PUPIL TEXTBOOK 5A PAGE 161

Think together

WAYS OF WORKING Whole class teacher led (I do, We do, You do)

ASK

- Question ① a): *Which method would you have chosen to solve the problem? Why? What is similar and different about the two methods? Which method is more efficient? Why?*
- Question ②: *Is there more than one solution for each number? How does using the factor tree help break down a number into factors? Does it matter which factors you use first in the factor tree?*

IN FOCUS Question ① a) is important as it offers children a pictorial representation of both methods of solving the question posed. This will help them recognise the similarities and differences between the two methods and help scaffold their ability to explain the mathematics involved. This question also offers an opportunity to make children aware of the associative nature of multiplication (i.e. $2 \times 2 \times 7$ is 2 lots of 2×7 which is the same as 7 lots of 2×2).

STRENGTHEN If children are unsure which calculation they prefer in question ②, ask: *Can you represent each calculation in a similar way to question ①? How can these representations help you explain which calculation is more efficient?*

DEEPEN While solving question ② b), children could be encouraged to look for patterns and make generalisations about the numbers they are investigating. Ask: *Can you try some of your own numbers?*

ASSESSMENT CHECKPOINT Question ② offers a good opportunity to assess whether children can fluently and confidently use their understanding of factors to explain how they would go about solving a problem.

ANSWERS

Question ① a): $2 \times 2 \times 7 = 28$

Question ① b): $4 \times 7 = 2 \times 14$

Question ② a): $3 \times 5 \times 2 = 5 \times 2 \times 3 = 30$

Question ② b):

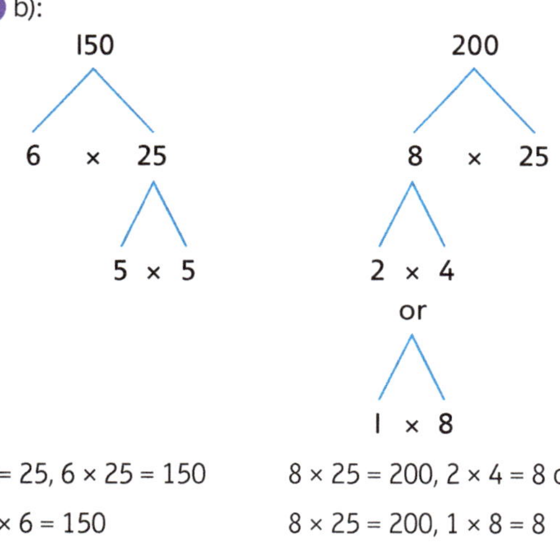

$5 \times 5 = 25$, $6 \times 25 = 150$
$5 \times 5 \times 6 = 150$

$8 \times 25 = 200$, $2 \times 4 = 8$ or
$8 \times 25 = 200$, $1 \times 8 = 8$
$2 \times 4 \times 25 = 200$ or
$1 \times 8 \times 25 = 200$

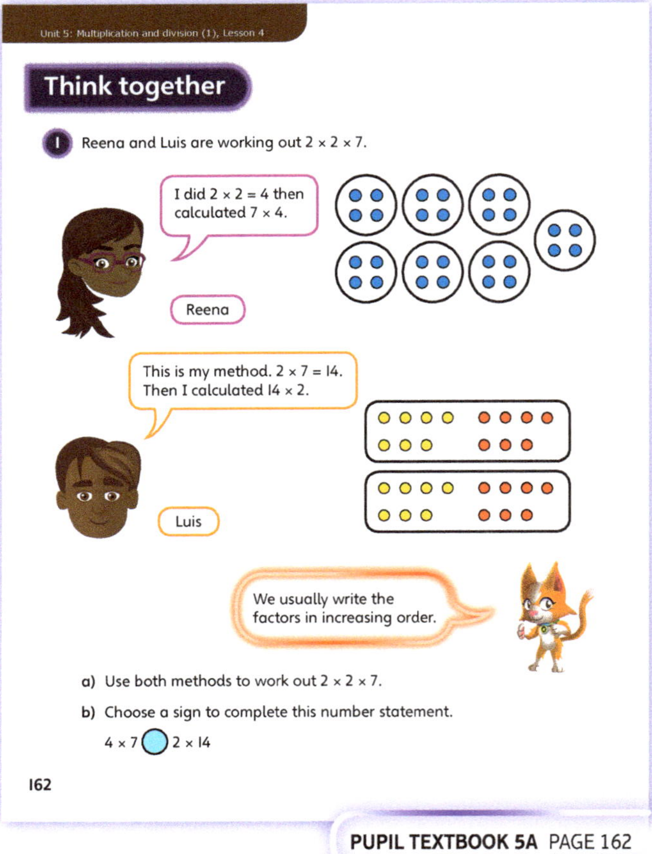

PUPIL TEXTBOOK 5A PAGE 162

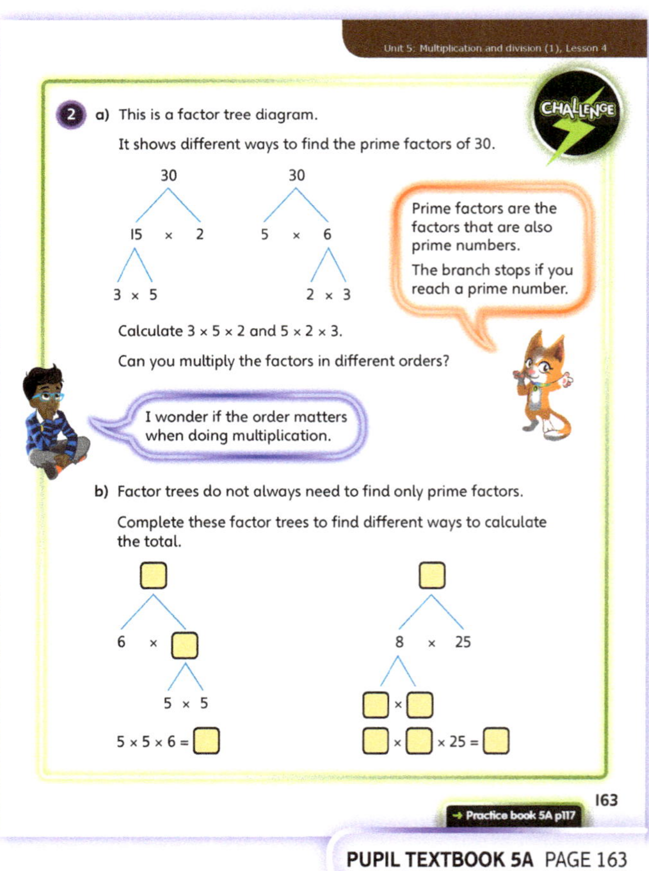

PUPIL TEXTBOOK 5A PAGE 163

Practice

WAYS OF WORKING Independent thinking

IN FOCUS Question ➊ helps children link their understanding with concrete representations of the calculations. It will also elicit children's understanding of the associative properties of multiplication (that 3 lots of 2 × 2 is the same as 2 lots of 3 × 2). Look for their recognition of this in their explanation.

Question ➌ develops children's flexibility by ensuring they look for more than one way of solving the same problem. As in question ➊, it also raises children's awareness of the associative law of multiplication.

STRENGTHEN If children have difficulty completing the factor trees in question ➍, ask: *What multiplication table is X in? Is there another pair of factors that could go in the factor tree?*

DEEPEN While solving question ➍, deepen children's reasoning about the factors they are using by asking:
- *Why have you chosen that pair of factors for the factor tree?*
- *Why do you think that is the easiest pair to solve? Explain.*

ASSESSMENT CHECKPOINT Question ➌ will offer the opportunity to assess whether children can apply their knowledge of factors independently to a problem in context. Look for children explaining how the calculation they can see relates to the 'story' of the problem to show their understanding of the link between the mathematics they know and the question.

Question ➎ will assess children's ability to apply their understanding of factors alongside their understanding of prime numbers to solve a mathematical puzzle.

ANSWERS Answers for the **Practice** part of the lesson appear in the separate **Practice and Reflect answer guide**.

Reflect

WAYS OF WORKING Independent thinking

IN FOCUS This final question is important as it gives an opportunity for children to apply their understanding from the lesson. It will be interesting for children to compare their ideas, discussing why they have chosen certain factors instead of others. Ask: *Why do you think someone may have chosen different factors to you?*

ASSESSMENT CHECKPOINT Look for children's recognition that using factors that they are fluent with can help them to solve multiplication calculations that would otherwise be more complex. Children should recognise that there are multiple ways of achieving this.

ANSWERS Answers for the **Reflect** part of the lesson appear in the separate **Practice and Reflect answer guide**.

After the lesson ⏸

- Were children able to confidently recognise that order of multiplication does not matter (the associative property of multiplication)?
- How confident were children at applying their understanding of factors? What barriers to their learning did they face?
- How will you account and prepare for these barriers next time you teach this lesson?

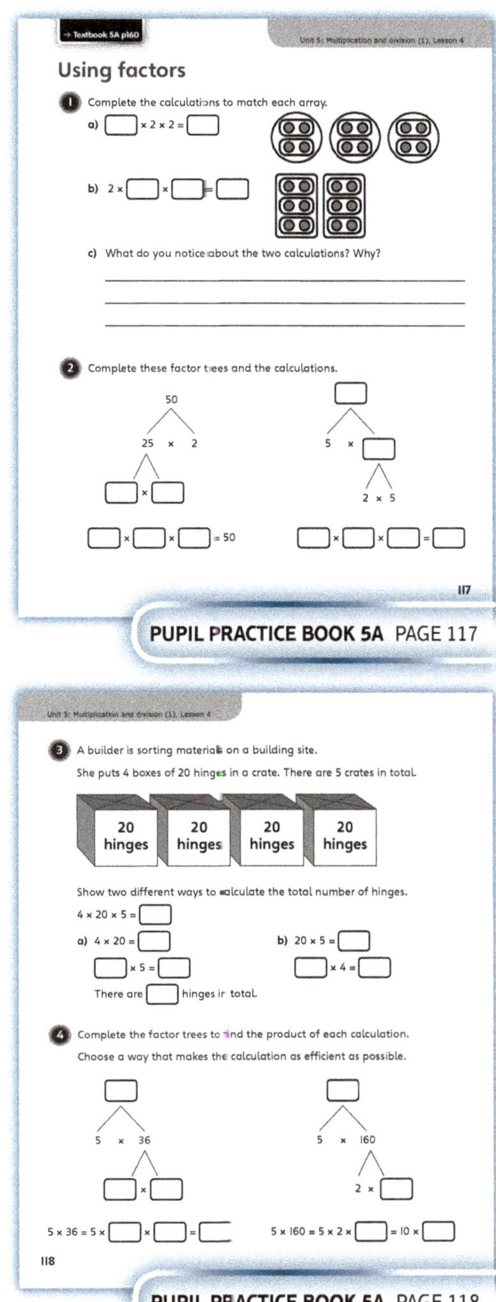

PUPIL PRACTICE BOOK 5A PAGE 117

PUPIL PRACTICE BOOK 5A PAGE 118

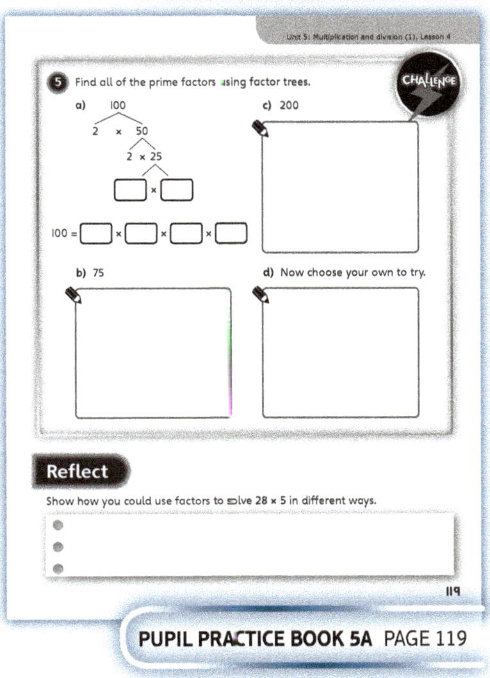

PUPIL PRACTICE BOOK 5A PAGE 119

Squares

Learning focus

In this lesson, children will learn about square numbers and how to recognise and represent square numbers pictorially before linking this to using notation, including squared (^{2}). Children will find square numbers in the multiplication grid and use them to solve calculations and problems.

Small steps

→ Previous step: Using factors
→ **This step: Squares**
→ Next step: Cubes

NATIONAL CURRICULUM LINKS

Year 5 Number – Multiplication and Division
- Recognise and use square numbers and cube numbers, and the notation for squared (^{2}) and cubed (^{3}).
- Solve problems involving multiplication and division including using their knowledge of factors and multiples, squares and cubes.

ASSESSING MASTERY

Children can understand the term 'square' and how it relates to number as well as shape, and use the correct notation when making calculations using squares. They can reliably find square numbers and can explain how the number links to the properties of the shape.

COMMON MISCONCEPTIONS

As finding a square number requires multiplying a number by itself, children may interpret this as X multiplied by 2, instead of X multiplied by X. Show children the appropriate square and ask:
- *If you look at this square as an array, what multiplication does it represent? How is that different to X × 2?*

STRENGTHENING UNDERSTANDING

Children could be encouraged to draw different squares using squared paper. Discuss the properties of squares and how each side is the same length. These properties can then be linked to arrays. Ask: *What squares have you made? What array does your square look like? Explain your answer.*

GOING DEEPER

Children can investigate relationships between prime numbers and square numbers. Ask:
- *Is it always true, sometimes true or never true that the sum of two primes is a square number?*

KEY LANGUAGE

In lesson: square number, squares, x^2, factors

Other language to be used by the teacher: multiply, multiplied, times, dimensions

STRUCTURES AND REPRESENTATIONS

arrays, multiplication grid

RESOURCES

Optional: multilink cubes, counters, squared paper, chessboard

 In the eTextbook of this lesson, you will find interactive links to a selection of teaching tools.

Before you teach ⏸

- How will you make sure that the concrete resources you use make the link between shape and number explicit?
- Are children sufficiently confident with factors and multiples?
- How will you ensure the two curriculum aims are approached?

Discover

WAYS OF WORKING Pair work

ASK

- Question **1** a): *What would be the best way to count the squares?*
- Question **1** a): *How many squares can you see?*
- Question **1** b): *How will you know you have found all of the squares?*

IN FOCUS Question **1** a) is important as it gives children their first opportunity to count the squares within a square. Discuss how children went about counting them – was there an efficient way?

Question **1** b) is a good opportunity to discuss other squares and begin children's pattern recognition in square numbers.

PRACTICAL TIPS Children could be encouraged to make different squares using different resources. Discuss the dimensions of the squares and how this relates to the area inside. To make this point especially clear, use multilink cubes or squared paper to allow children to count the squares inside the shape.

ANSWERS

Question **1** a): There are 64 small squares on the chessboard altogether.

Question **1** b): Possible squares:
1×1
2×2
3×3
4×4
5×5
6×6
7×7
8×8

Share

WAYS OF WORKING Whole class teacher led

ASK

- Question **1** a): *How did you count the squares?*
- Question **1** a): *Have you seen the square sign before? Where have you seen it used?*
- Question **1** b): *Why do you think these numbers are called 'square' numbers?*
- Question **1** b): *Can you draw the larger square numbers on squared paper?*

IN FOCUS Question **1** a) is important as it introduces children to the vocabulary and mathematical notation. It will be important for children to have plenty of opportunity to practise using both of these.

Question **1** b) provides a key opportunity to link children's understanding of the square shapes with their new understanding of how square numbers work.

PUPIL TEXTBOOK 5A PAGE 164

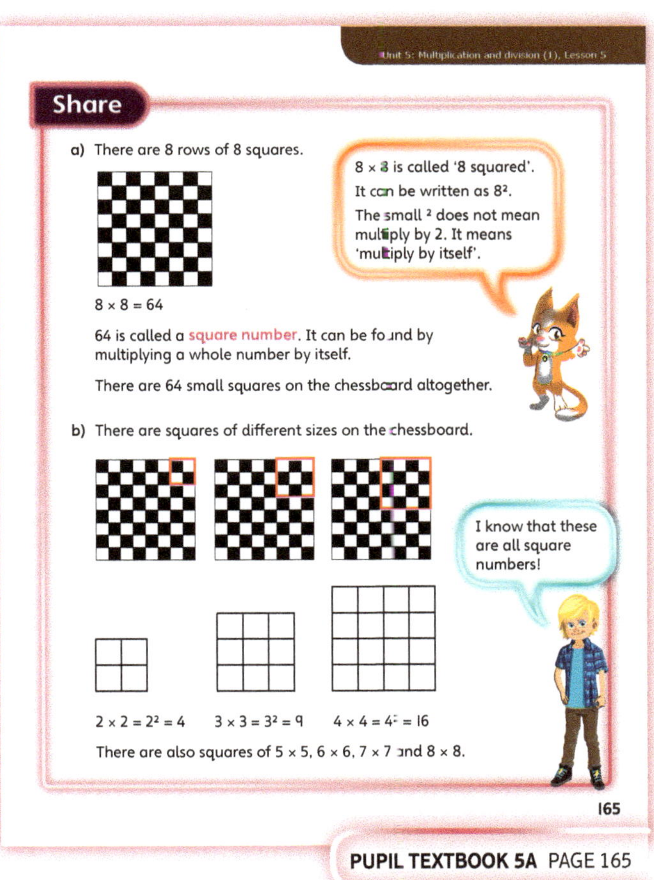

PUPIL TEXTBOOK 5A PAGE 165

Think together

WAYS OF WORKING Whole class teacher led (I do, We do, You do)

ASK

- Question **1**: *How are the arrays similar and different to a square? What is the most efficient way of finding how many dots there are?*
- Question **2**: *Is there a pattern in the square numbers you have found? Can you explain it?*
- Question **3**: *What is wrong with Jamilla's square? What is the clearest way of representing a square number? Why?*

IN FOCUS Question **2** offers a good opportunity to observe the patterns in square numbers. It provides an excellent opportunity for children to begin predicting the sequence of square numbers beyond 12 × 12.

Question **3** helps address the possible misconception that the squares around the outside of a square are all that need counting to find square numbers, instead of the inside as well. Question **3** b) aims to make children second guess whether 16 is a square number or not as the counters are arranged as two groups of 8, potentially causing a cognitive conflict where they will assume a mistake has been made.

STRENGTHEN To strengthen understanding in question **3**, ask: *Can you make a complete square with 12 cubes? Explain why not. How has Jamilla managed to make a square? Where has she gone wrong?*

DEEPEN Question **2** can be deepened by asking children to spot and explain the pattern that emerges in the squared numbers. Ask: *What pattern can you spot? Is there a pattern in how the numbers increase? How might this pattern help you predict the next square numbers in the sequence?*

ASSESSMENT CHECKPOINT Question **3** will offer the opportunity to assess whether children can reliably identify a square number. Look for children demonstrating their understanding of the concrete representations of square numbers and applying this to their explanations.

ANSWERS

Question **1**: $5^2 = 5 \times 5 = 25$

   10 squared is 100.

Question **2**: 1, 4, 9, 16, 25, 36, 49, 64, 81, 100, 121, 144

Question **3** a): Jamilla is incorrect as she has not made a complete, solid square.

Question **3** b): 16 is a square number. Look for children making or drawing a 4 by 4 square.

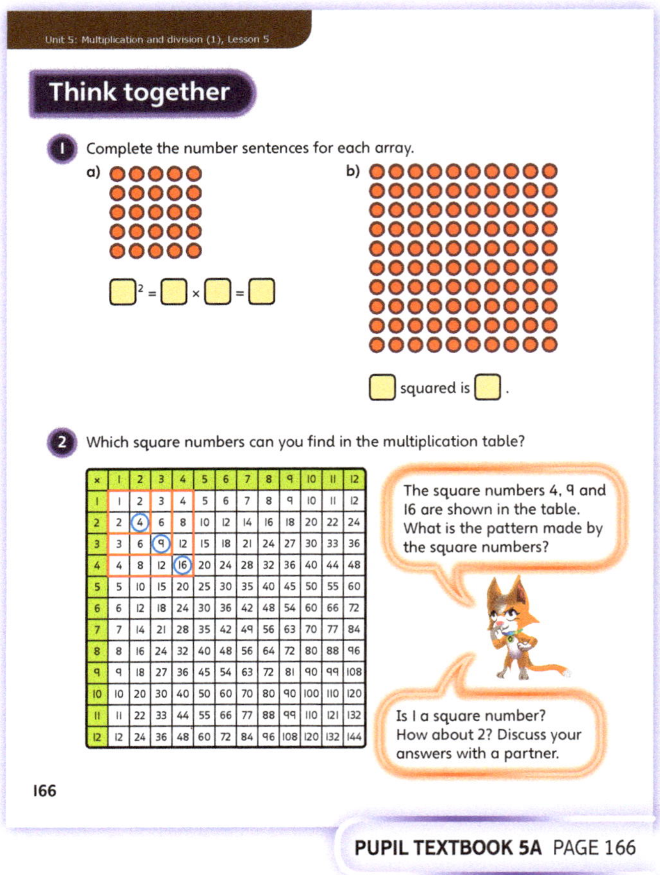

PUPIL TEXTBOOK 5A PAGE 166

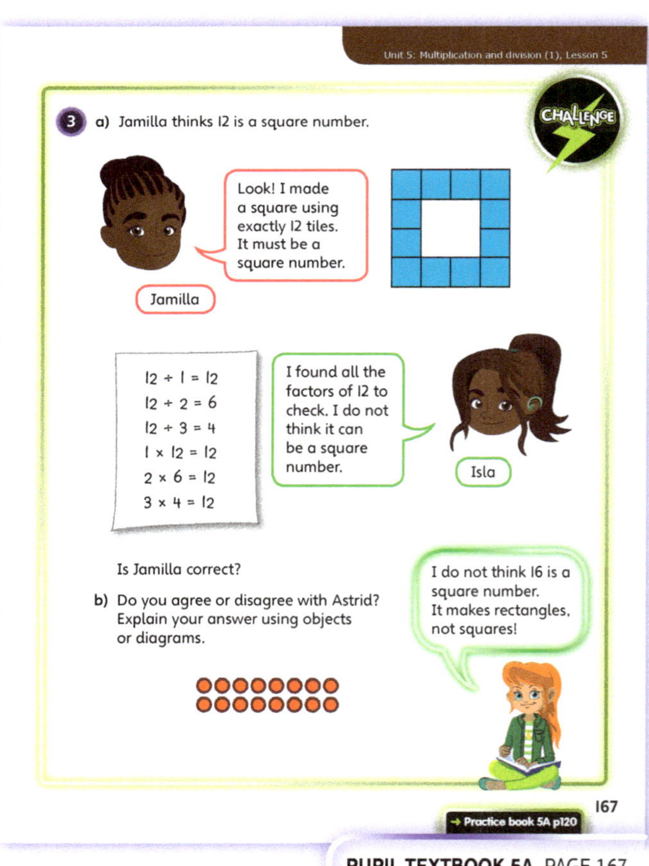

PUPIL TEXTBOOK 5A PAGE 167

Practice

WAYS OF WORKING Independent thinking

IN FOCUS Question ❶ scaffolds children's independent thinking about square numbers. Offering the pictures alongside the written representation will enable children to visualise the square numbers in later questions.

Questions ❷, ❸ and ❺ allow children to create pictorial representations of square numbers. Be sure that children link these to the sentences below so that they become fluent with the more efficient abstract way of recording.

Question ❹ offers an opportunity for children to develop and demonstrate their understanding of the squared (^{2}) notation. Be sure that children understand that 4^2 means 4×4, and not 4×2.

STRENGTHEN To support children in the **Think Differently** question, provide them with counters or cubes and ask: *What arrays can you make with those numbers? Are any of your arrays square? What does that mean?*

DEEPEN While solving question ❼, ask: *Do you agree with Isla? Can you find evidence that supports your ideas and proves them?*

THINK DIFFERENTLY This is a key question as it removes the scaffold of the square shape and presents children with square numbers in different contexts and representations. This will develop children's awareness that a square number is found through multiplication and that a square is just one representation of this among many.

ASSESSMENT CHECKPOINT Question ❻ will offer an opportunity to assess children's ability to recognise and identify square numbers with no scaffolding. Look for children using their understanding of arrays, number patterns and multiplication tables when identifying which of the numbers are squares.

ANSWERS Answers for the **Practice** part of the lesson appear in the separate **Practice and Reflect answer guide**.

Reflect

WAYS OF WORKING Independent thinking

IN FOCUS This question will offer an opportunity to assess children's fluency with square numbers. Children should be able to record some squares from memory by this point and know efficient methods of finding ones they have not committed to memory.

ASSESSMENT CHECKPOINT Look for which square numbers children have committed to memory. Use this opportunity to assess which children still rely on drawing square arrays and which children have moved into more abstract recording.

ANSWERS Answers for the **Reflect** part of the lesson appear in the separate **Practice and Reflect answer guide**.

After the lesson ⏸

- How many children moved beyond the concrete and pictorial representations of squared numbers and showed fluency with the abstract concepts, particularly the mathematical notation of squared (^{2})?
- Were children confidently able to identify square numbers in the multiplication grid and recognise the sequence of square numbers in the grid?
- How will you support those children who still need help with fully understanding the abstract concepts and using the mathematical notation?

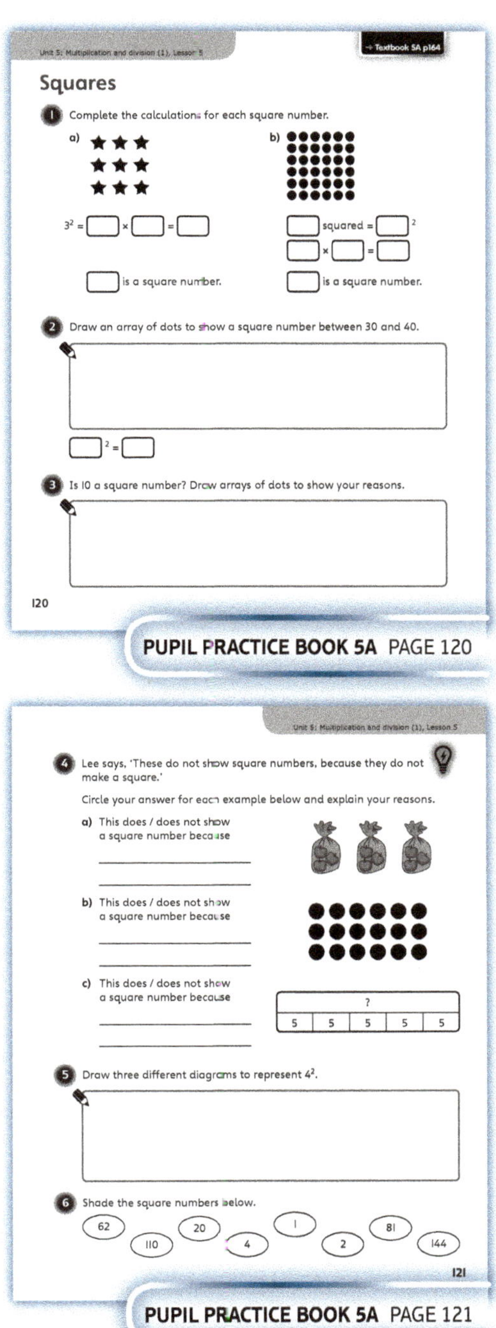

PUPIL PRACTICE BOOK 5A PAGE 120

PUPIL PRACTICE BOOK 5A PAGE 121

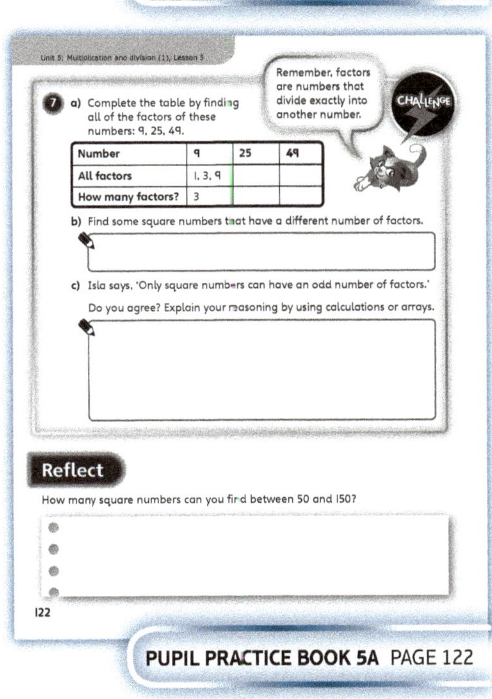

PUPIL PRACTICE BOOK 5A PAGE 122

Cubes

Learning focus

In this lesson, children will learn how to recognise and represent cube numbers pictorially before linking this to using notation, including cubed (^{3}). They will learn how to find cube numbers and use them to solve calculations and problems.

Small steps

→ Previous step: Squares
→ **This step: Cubes**
→ Next step: Inverse operations

NATIONAL CURRICULUM LINKS

Year 5 Number – Multiplication and Division

- Recognise and use square numbers and cube numbers, and the notation for squared (^{2}) and cubed (^{3}).
- Identify multiples and factors, including finding all factor pairs of a number, and common factors of two numbers.
- Solve problems involving multiplication and division including using their knowledge of factors and multiples, squares and cubes.

ASSESSING MASTERY

Children can understand the term 'cube' and how it relates to number as well as shape. They can reliably find cube numbers and use the notation for cube numbers (^{3}), and can explain how the number links to the properties of the shape, particularly as regards to dimensions.

COMMON MISCONCEPTIONS

As finding a cube number requires multiplying a number by itself three times, children may interpret this as X multiplied by 3, instead of X multiplied by X multiplied by X. Show children the appropriate cube and ask:

- *How is a cube different to a square? How many dimensions does a cube have? How many times will you need to multiply this number? What does the cube sign mean?*

STRENGTHENING UNDERSTANDING

Children should be given the opportunity to create cubes of different dimensions, prior to the lesson. Discuss the properties of the cube and how it was made. Ask: *How many smaller cubes were needed to make your big cube? How could you describe the dimensions of your cube?*

GOING DEEPER

Ask: *Is it always true, sometimes true or never true that the sum of 4 cube numbers is a cube number?*

KEY LANGUAGE

In lesson: cube number, cubes, multiplied

Other language to be used by the teacher: multiply, dimension

RESOURCES

Mandatory: cubes

Optional: base 10 equipment, puzzle cube

 In the eTextbook of this lesson, you will find interactive links to a selection of teaching tools.

Before you teach 

- Do children understand the mathematical notation for square?
- For children who found the abstract notation tricky, how will you offer support in this lesson?

Discover

 Pair work

- Question **1** a): *How could you test how many small cubes were used?*
- Question **1** a): *How many small cubes would be needed to make the next biggest cube?*
- Question **1** b): *Why does Isla think 2 × 2 × 2 = 6? What is her mistake?*
- Question **1** b): *Does what you learnt about the square sign apply in a similar way to the cube sign?*

 Question **1** b) is important as it approaches the misconception that the cube notation means 'multiply by 3'. This will be a key piece of learning during the lesson and so it is important to focus on this point during the class discussion.

 Children should be encouraged to make the cubes they see in the picture using multilink cubes. This will help secure their understanding of why a number is cubed when discussing it later. If children investigate different sized cubes, then a record of these, and their related multiplication facts, could be displayed in the classroom for future use.

Question **1** a): 8 small cubes make up the puzzle cube.

Question **1** b): Isla saw 2 multiplied 3 times, so did 2 × 3 = 6 (2 + 2 + 2 rather than 2 × 2 × 2).

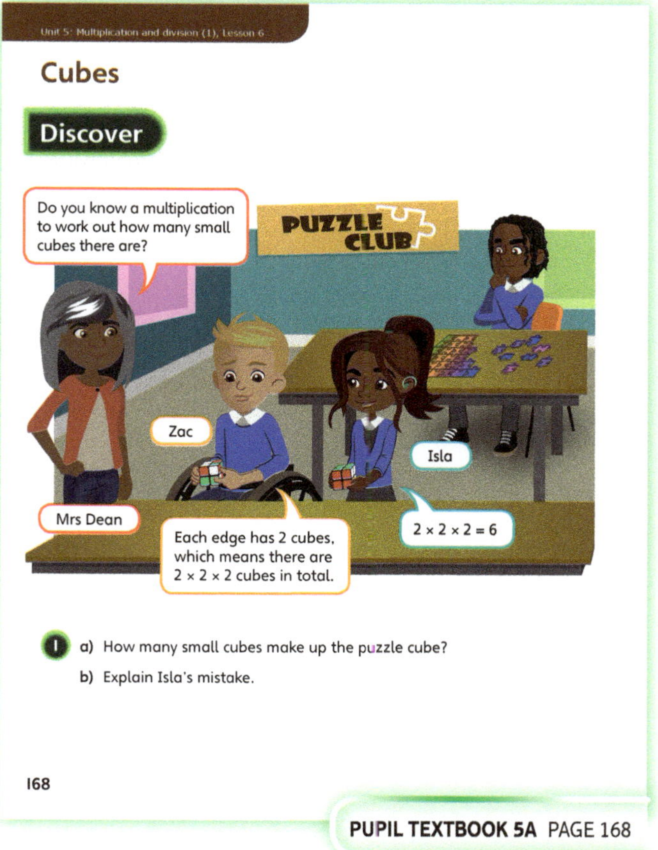

PUPIL TEXTBOOK 5A PAGE 168

Share

 Whole class teacher led

- Question **1** a): *How did you visualise the cube?*
- Question **1** a): *Did it make a difference to the answer how the cube was visualised? Why?*
- Question **1** a): *How many dimensions does a cube have?*
- Question **1** a): *How is the cube sign similar and different to the square sign?*
- Question **1** a): *How does the cube sign relate to the dimensions of the cube?*
- Question **1** b): *What would you say to Isla to help her correct her mistake?*

 Question **1** a) gives an excellent opportunity to link the concrete representation of a cube with the abstract concept of cubing numbers. If they haven't already, children should be encouraged to practically recreate what is shown in the picture to help secure their understanding.

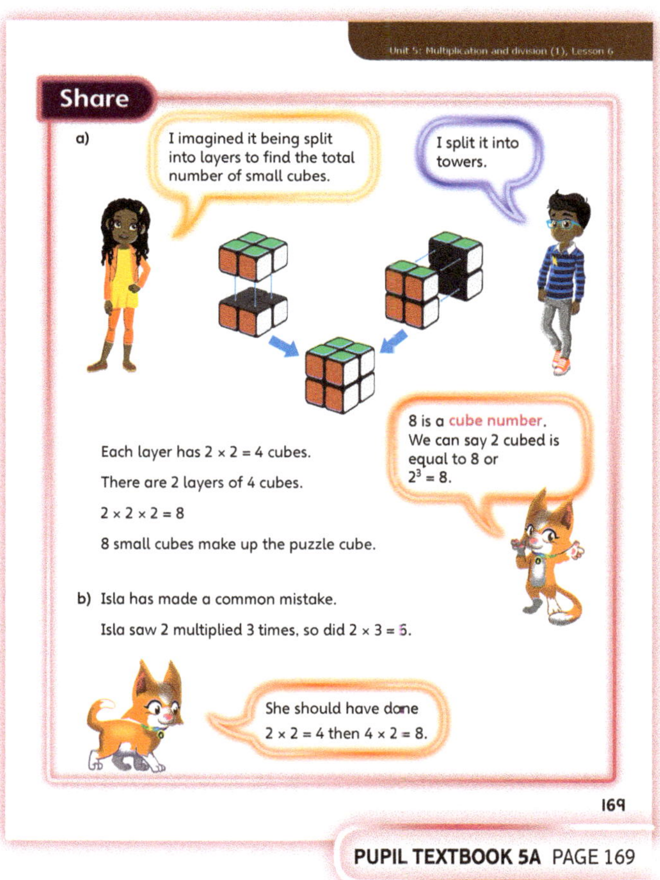

PUPIL TEXTBOOK 5A PAGE 169

Think together

WAYS OF WORKING Whole class teacher led (I do, We do, You do)

ASK

- Question ❶: *How could you test what cubes are necessary for the largest possible cube? In what ways can you record your findings? How could you make cubing 6 easier? Why does Amelia think the answer is 36?*
- Question ❷ b): *Why does Zac think 2 is a cube number? Could it be one?*
- Question ❸ a) and b): *How will you investigate these calculations? What resources or methods will you use?*

IN FOCUS Question ❶ allows children the opportunity to find cube numbers within a given number. It is also important as it scaffolds children's abstract recording of the calculations. It will be important to discuss Amelia's idea that the answer is 36 to help children differentiate between square and cube numbers.

Question ❷ a) gives children the opportunity to investigate what happens when you cube 1. This can be compared to what happens when 10 is cubed, providing an interesting opportunity for children to discuss patterns and make generalisations about place value and cubing.

Question ❷ b) deals with the misconception that a cube number is the starting number rather than the product.

STRENGTHEN To help children visualise cube 10 in question ❷ a), provide them with base 10 equipment. Ask: *Can you use the base 10 equipment to show 10^3? Can you explain how the equipment has shown 10 cubed?*

DEEPEN Question ❸ will deepen children's understanding by helping them to link their previous learning about the distributive properties of multiplication to their new understanding of cubes. Ask: *How can partitioning the numbers help make the multiplication easier?*

ASSESSMENT CHECKPOINT Question ❸ a) and b) will assess not only children's ability to cube numbers reliably, but also their ability to use their fluency in number and calculations to help them cube bigger numbers.

ANSWERS

Question ❶: The largest cube she can make is a 3 × 3 × 3 cube. This will need 27 smaller cubes.

$3 \times 3 \times 3 = 27$

$3^3 = 27$

27 is a cube number.

Question ❷ a): $1^3 = 1 \times 1 \times 1 = 1$

$10^3 = 10 \times 10 \times 10 = 1,000$

Question ❷ b): Zac is incorrect. The cube number is 8 (the product) not 2.

Question ❸ a): $5^3 = 125$

Question ❸ b): $6^3 = 216$

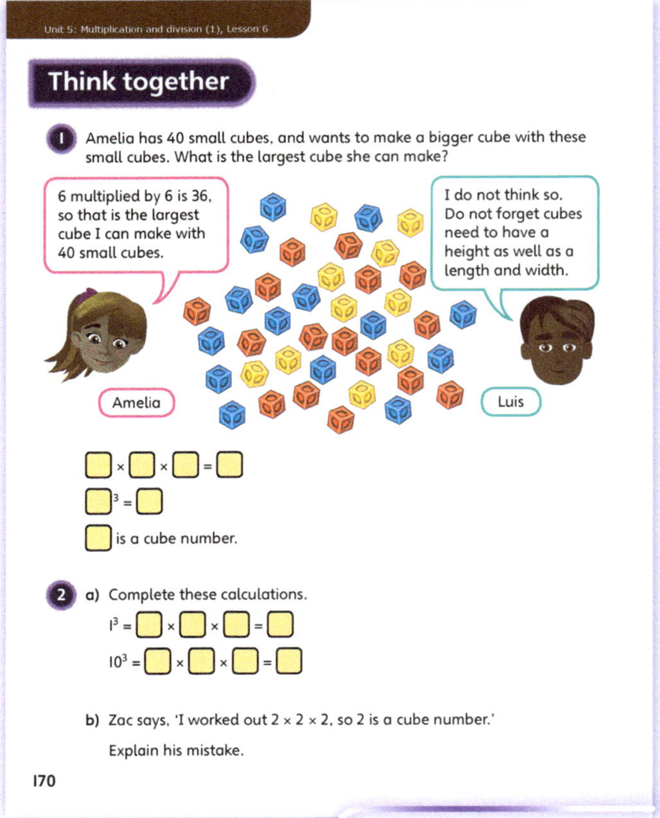

PUPIL TEXTBOOK 5A PAGE 170

PUPIL TEXTBOOK 5A PAGE 171

Practice

WAYS OF WORKING Independent thinking

IN FOCUS Question ① will help children to differentiate between correctly cubing and the potential misconceptions that can arise when trying to cube a number. It also shows pictorial and concrete representations of the concept of cubing and how it is different to multiplying by 3 or adding the same number three times.

Question ③ helps children link the distributive law to their understanding of cubing numbers. This is important as it will help them more successfully cube larger numbers in the future.

STRENGTHEN To strengthen children's understanding of how to approach each calculation in question ③, ask: *Can you build what is shown in the picture? Explain what you are building as you do so. How would you record each step?*

DEEPEN Expand question ③ by providing children with cubes to work out 6^2. Encourage them to split the cubes in similar ways to Luis to break down the calculation and write out their workings in full.

THINK DIFFERENTLY This question allows children to demonstrate their secure understanding by diagnosing and correcting the mistakes. Each mistake covers a different misconception, so it will be important to observe if any examples are found to be trickier across the class. This will show if an area of teaching needs revisiting.

ASSESSMENT CHECKPOINT Question ② will allow for the assessment of whether children can reliably calculate cube numbers. Look for their ability to link the pictorial representation with the abstract calculation.

Question ③ offers a good opportunity to assess children's fluency with number when finding cube numbers.

ANSWERS Answers for the **Practice** part of the lesson appear in the separate **Practice and Reflect answer guide**.

Reflect

WAYS OF WORKING Independent thinking

IN FOCUS This question allows for an overall assessment of children's progress. Children should be able to confidently find the first 5 cube numbers, explaining how they did so. It would be interesting to compare children's methods in a class discussion to highlight the different ways the cube numbers can be reached.

ASSESSMENT CHECKPOINT Look for children confidently cubing numbers using the concrete, pictorial and abstract representation used in the lesson. Children should be able to use the distributive law to help them with larger numbers.

ANSWERS Answers for the **Reflect** part of the lesson appear in the separate **Practice and Reflect answer guide**.

After the lesson ⏸

- Did children recognise how the distributive law is useful when cubing larger numbers?
- Could you have made this more explicit? If so, how?
- How confident are children with the abstract mathematical notation? Could they link it to their learning about squares?

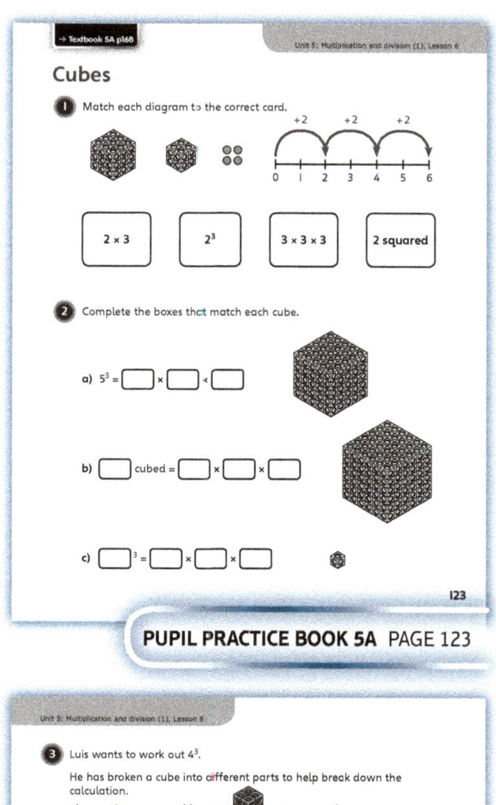

PUPIL PRACTICE BOOK 5A PAGE 123

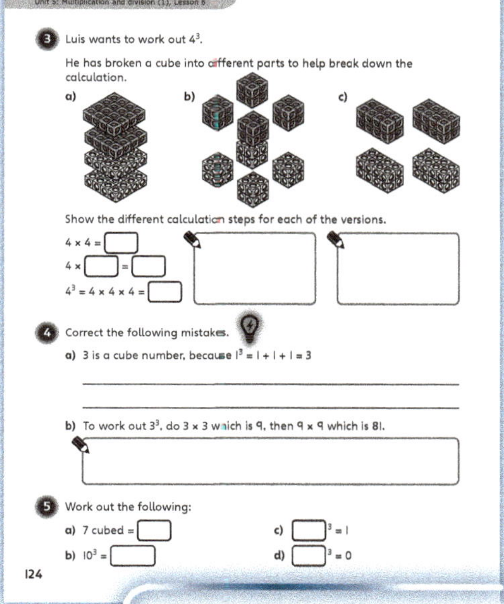

PUPIL PRACTICE BOOK 5A PAGE 124

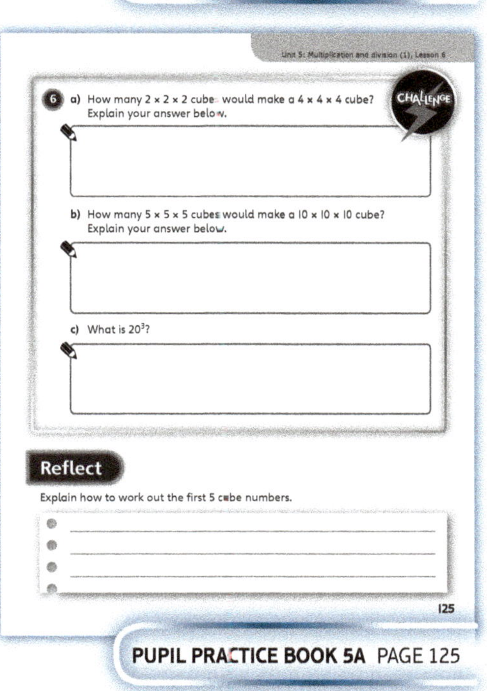

PUPIL PRACTICE BOOK 5A PAGE 125

Inverse operations

Learning focus

In this lesson, children will learn to recognise how knowing and using the inverse operation can help to check and solve problems, with a focus on using multiplication and division as inverse operations. There is also the opportunity for children to begin to develop proportional reasoning by using simple scaling.

Small steps

→ Previous step: Cubes
→ **This step: Inverse operations**
→ Next step: Multiplying whole numbers by 10, 100 and 1,000

NATIONAL CURRICULUM LINKS

Year 5 Number – Multiplication and Division

Solve problems involving multiplication and division, including scaling by simple fractions and problems involving simple rates.

ASSESSING MASTERY

Children can fluently identify the inverse of an operation and use this to check its accuracy, identifying where using the inverse operation can help them find mistakes or solve a problem. In addition, children can use inverse operations to help them work backwards through a problem. Children can also use simple scaling to solve problems.

COMMON MISCONCEPTIONS

In inverse operations, children may transpose numbers incorrectly or think that division is associative. For example, when recording the inverse of 5 × 6 = 30, children may record 5 ÷ 6 = 30 or 5 ÷ 30 = 6. Ask:
• *Can you show me a picture of that calculation? Does your picture match your solution? Explain your answer. Does the order in which you divide numbers matter?*

STRENGTHENING UNDERSTANDING

Children could be encouraged to build arrays to strengthen understanding of this lesson. Discuss with them the multiplication and division calculations they represent. Ask: *Can you list the multiplication and division calculations this array represents? Can you explain how these calculations are similar and different?*

GOING DEEPER

Encourage children to create their own 'Find my number' challenges as seen in question 6 of the **Practice Book**. This would create two levels of challenge. The first step would be to create a chain of operations that result in a given number. Children would then need to reverse this, ready to challenge a partner.

KEY LANGUAGE

In lesson: inverse operation, multiples, multiplication fact, division fact, remainder

Other language to be used by the teacher: addition, add, subtract, subtraction, minus, multiply, times, multiplication, divide, division

STRUCTURES AND REPRESENTATIONS

arrays, factor trees, number lines

RESOURCES

Mandatory: counters, cubes

 In the eTextbook of this lesson, you will find interactive links to a selection of teaching tools.

Before you teach ⏸

• Are children confident with all four operations?

Discover

 Pair work

ASK

- Question **1** a): *How many stars are on each flag? How many stars are needed in total?*
- Question **1** a): *How will you go about finding out how many flags are needed?*
- Question **1** b): *How will you test Miss Hall's conjecture?*
- Question **1** b): *How many buttons will they need to use? Can you prove it?*

IN FOCUS Question **1** a) gives children an opportunity to begin working with the four operations in context. It will be important to note that the four stars need to be multiplied by an unknown number to create the product of 60. This will allow children to begin considering the inverse operation.

PRACTICAL TIPS The scenario shown in the picture could be easily replicated in class with paper bunting triangles and stickers. Alternatively, counters will work as well as stickers.

ANSWERS

Question **1** a): They will make 15 flags.

Question **1** b): Miss Hall is not correct: there cannot be 43 buttons, because as 43 is not a multiple of 3. They will need 15 × 3 = 45 buttons.

PUPIL TEXTBOOK 5A PAGE 172

Share

  Whole class teacher led

ASK

- Question **1** a): *How was the inverse used to help solve the question?*
- Question **1** a): *Can you explain how the array demonstrates the explanation?*
- Question **1** a): *Which is the most efficient method? Explain why.*
- Question **1** b): *How did you prove that Miss Hall could not be correct?*
- Question **1** b): *How many buttons did they need? How did you prove it?*

IN FOCUS It will be important to discuss the efficiency of using the inverse operation when studying question **1** a). Children should be encouraged to see that, while it is possible to solve this by calculating every multiple of 4, it is much quicker to use the inverse and divide 60 by 4.

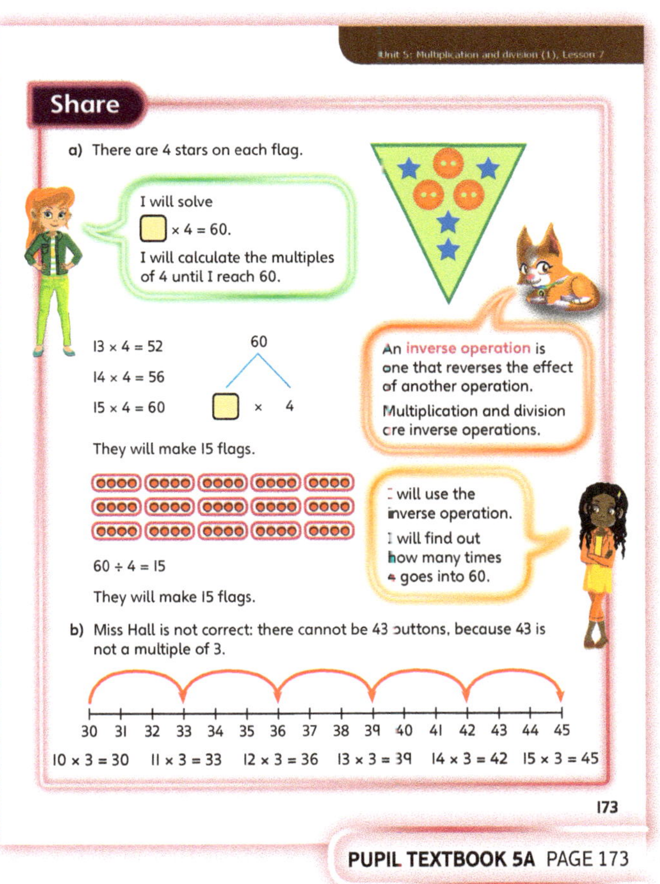

PUPIL TEXTBOOK 5A PAGE 173

Think together

WAYS OF WORKING Whole class teacher led (I do, We do, You do)

ASK

- Question ❶: *How many groups of 2 and 5 will you need? How do you know? How can there be two inverse operations for one multiplication calculation?*
- Question ❷: *How will you calculate how many flags she has made? Is it possible to solve the second part of this question before the first?*
- Question ❸: *What strategies were most useful when solving these calculations? How can you use the answer to a) to give an answer to c)? Can you multiply the factor tree by a multiple to help you?*

IN FOCUS Question ❶ helps children to link the inverse operations together. Discuss how each multiplication calculation has two related inverse division calculations. Encourage children to discuss why this is useful to remember.

STRENGTHEN When discussing the calculations in question ❸, it may help children's thinking to ask: *What do you know about the calculation? What facts are missing? How have you used the inverse operations in previous questions to rearrange a calculation? Can you use those strategies here?*

DEEPEN Extend question ❶ by asking children to draw flags of their own with squares and circles, then writing out multiplication and division facts for these.

ASSESSMENT CHECKPOINT Question ❸ thoroughly assesses children's grasp on how to recognise and use the correct inverse operation to solve a problem. Look for their fluency and reasoning about which inverse calculation to use and how they know it will work. Question ❸ also starts to develop proportional reasoning, since children might see that they can multiply a previous answer by a scale factor to give an answer to a later question by doubling or tripling.

ANSWERS

Question ❶ a): $2 \times 8 = 16$
$16 \div 8 = 2$
$16 \div 2 = 8$

Question ❶ b): $5 \times 8 = 40$
$40 \div 8 = 5$
$40 \div 5 = 8$

Question ❷ a): $36 \div 2 = 18$. She has made 18 flags.

Question ❷ b): $18 \times 6 = 108$. She had 108 buttons.

Question ❸ a): $12 \div 3 = 4$
$12 \div 4 = 3$
$4 \times 3 = 12$
$36 \div 3 = 12$

Question ❸ b): $22 \div 11 = 2$
$22 \div 2 = 11$
$44 \div 2 = 22$
$44 \div 22 = 2$

Question ❸ c): Lexi is thinking of 18 ($5 \times 3 = 15$, so a remainder of 3 would mean starting with 18)

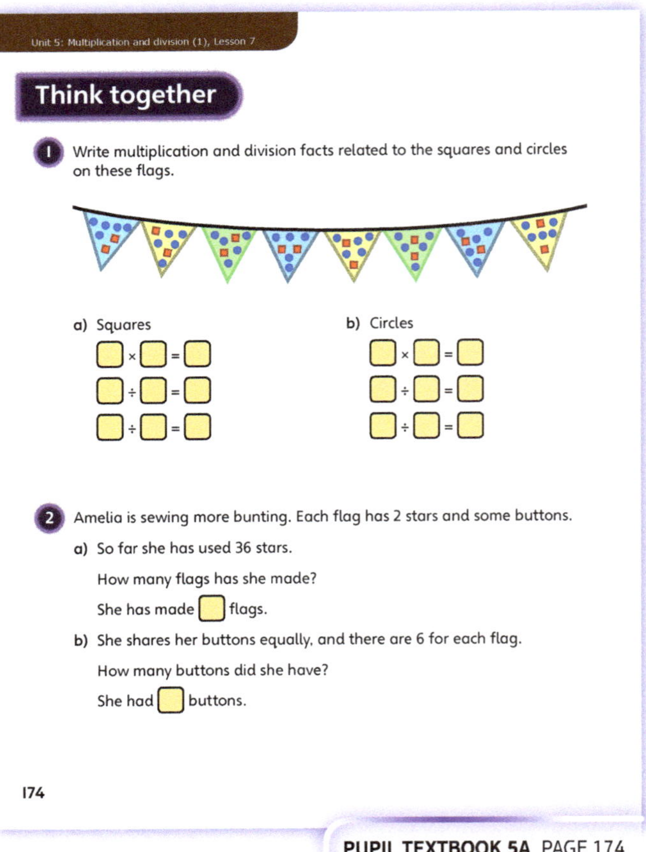

PUPIL TEXTBOOK 5A PAGE 174

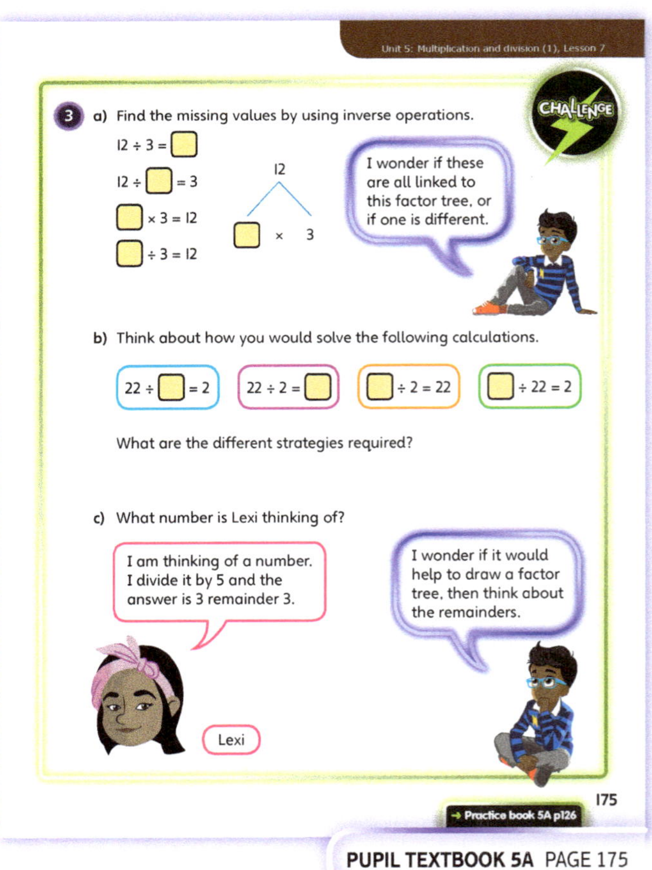

PUPIL TEXTBOOK 5A PAGE 175

Practice

WAYS OF WORKING Independent thinking

IN FOCUS Question ❶ scaffolds children's independent use of inverse operations by offering them concrete and pictorial representations of the calculations. This will help children link the two operations together and help them find the inverse more fluently.

Question ❷ scaffolds children's use of the inverse to solve contextual word problems. Children are given a cloze procedure for the abstract calculations that will be needed to solve the problem. Question ❸ gives children the opportunity to solve a similar word problem without the scaffold. Also, question ❸ prepares children for understanding proportional reasoning which will be fully introduced in Year 6 (using a ratio of 3 : 2 to represent the red and white roses, respectively, and a scale of 6).

Question ❺ approaches the misconception where children transpose the numbers in a calculation incorrectly when finding the inverse operation. It will be important to observe children's responses to this as it will indicate the likelihood that this misconception will be made or avoided in the class.

STRENGTHEN For question ❷, it may help to provide children with resources that allow them to act out the problem. Ask: *Can you use these resources to set up the problem as it is written? Can you show me how you would solve it?*

DEEPEN In question ❺, deepen children's reasoning by asking: *What advice would you give Bella to help her recognise her mistake? What resources or pictures would be best to use to help demonstrate the correct way of working?*

ASSESSMENT CHECKPOINT Question ❹ will assess children's fluency with recognising the patterns and process when finding the inverse of a calculation. There is also the opportunity to further explore the use of proportional reasoning, since children will recognise that numbers are being increased by simple scales.

Question ❻ will assess children's reasoning and problem-solving ability when using inverse operations.

ANSWERS Answers for the **Practice** part of the lesson appear in the separate **Practice and Reflect answer guide**.

Reflect

WAYS OF WORKING Independent thinking

IN FOCUS This question will be a good opportunity for a final assessment of children's understanding of the strategies used to find the missing values in the given calculations.

ASSESSMENT CHECKPOINT Look for children's ability to link the given calculations with their inverse. Can they use the inverse fluently to fill in the missing number?

ANSWERS Answers for the **Reflect** part of the lesson appear in the separate **Practice and Reflect answer guide**.

After the lesson ⏸

- How will you encourage children to use the problem-solving methods in this lesson in future lessons?
- How will you encourage children to use inverse operations to check their calculations?
- Do children recognise the usefulness of being able to find the inverse calculation? How will you support and develop their understanding moving forward?

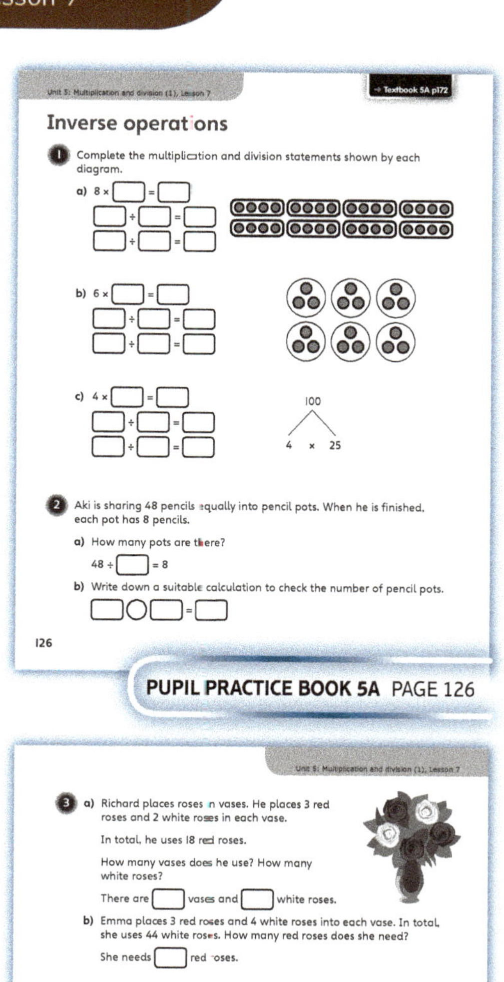

PUPIL PRACTICE BOOK 5A PAGE 126

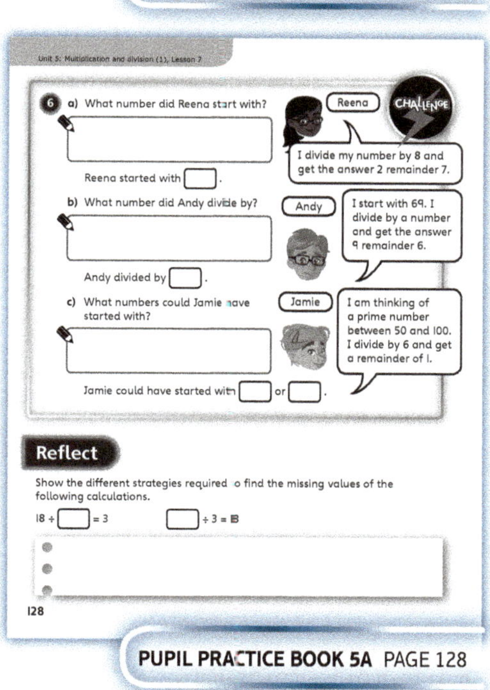

PUPIL PRACTICE BOOK 5A PAGE 127

PUPIL PRACTICE BOOK 5A PAGE 128

Multiplying whole numbers by 10, 100 and 1,000

Learning focus

In this lesson, children will use their understanding of place value to develop their ability to fluently multiply whole numbers by 10, 100 and 1,000.

Small steps

→ Previous step: Inverse operations
→ **This step: Multiplying whole numbers by 10, 100 and 1,000**
→ Next step: Dividing whole numbers by 10, 100 and 1,000

NATIONAL CURRICULUM LINKS

Year 5 Number – Multiplication and Division

Multiply and divide whole numbers and those involving decimals by 10, 100 and 1,000.

ASSESSING MASTERY

Children can reliably multiply whole numbers by 10, 100 and 1,000, link their understanding of place value to their calculations and confidently represent their thinking using concrete, pictorial and abstract representations.

COMMON MISCONCEPTIONS

A common misconception when multiplying by 10, 100 and 1,000 is to just 'add a zero on to the end'. This is a damaging habit as this will cause errors when multiplying by 100 and 1,000, and later on, decimals. Ask:

• *Show me what you mean by 'add a zero'. What does 'add a zero' actually mean? Make your original number, and your final number, using base 10 equipment. How are they similar and different?*

STRENGTHENING UNDERSTANDING

Children could be encouraged to count in 10s, 100s and 1,000s before this lesson to secure their fluency with the patterns in those number sequences. This could also be approached through chants, rhymes and songs.

GOING DEEPER

Encourage children to create contextual word problems to multiply 10, 100 or 1,000. Challenge children to make them as real to life as possible so that they are required to consider the real-life application of this skill.

KEY LANGUAGE

In lesson: multiply, multiplication, ten, one hundred, one thousand, 10, 100, 1,000

Other language to be used by the teacher: place value

STRUCTURES AND REPRESENTATIONS

place value grid, number lines

RESOURCES

Mandatory: base 10 equipment, place value counters

Optional: printed place value grids, toy car parts

 In the eTextbook of this lesson, you will find interactive links to a selection of teaching tools.

Before you teach 

• How confident are children with place value and using base 10 equipment?
• How confident are children already when multiplying whole numbers by 10, 100 and 1,000?

Discover

 Pair work

ASK

- Question **1** a): *How many wheels are on each car?*
- Question **1** a): *How can you test how many wheels would be needed for 10 cars?*
- Question **1** a) and b): *What patterns can you spot in the numbers?*

IN FOCUS Questions **1** a) and b) will give children their first opportunity to multiply single digit numbers by 10 and 100. This will give a good opportunity to initially assess children's confidence before starting the main part of the lesson.

PRACTICAL TIPS This introduction would lend itself well to junk modelling. Children could be given the parts to make one junk model car. The questions posed in the **Discover** section could then be posed about their car parts.

ANSWERS

Question **1** a): 40 wheels are needed for 10 cars.

400 wheels are needed for 100 cars.

Question **1** b): 20 lamps are needed for 10 cars.

200 lamps are needed for 100 cars.

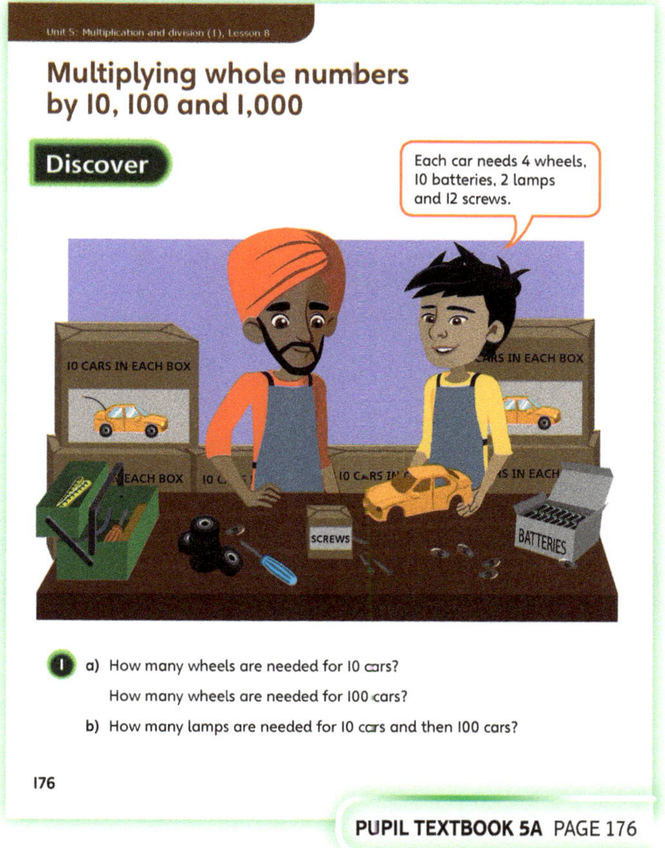

PUPIL TEXTBOOK 5A PAGE 176

Share

 Whole class teacher led

ASK

- Question **1** a): *How does the array and number line make the multiplication clear?*
- Question **1** a): *What would be different if the car needed 6 wheels?*
- Question **1** a): *Which method is more efficient, counting in 4s or 10s? Why?*
- Question **1** b): *How do the numbers change and stay the same as you multiply by 10 and 100?*
- Question **1** b): *Can you use the base 10 equipment to explain how the numbers change?*

IN FOCUS It will be important when focusing on these two questions to make sure children recognise the patterns in the numbers. Question **1** a) also reinforces the proportionality of each multiple of 4. This should be supported with the use of the concrete resources, so children can manipulate and experience the difference between 1 four, 10 fours and 100 fours.

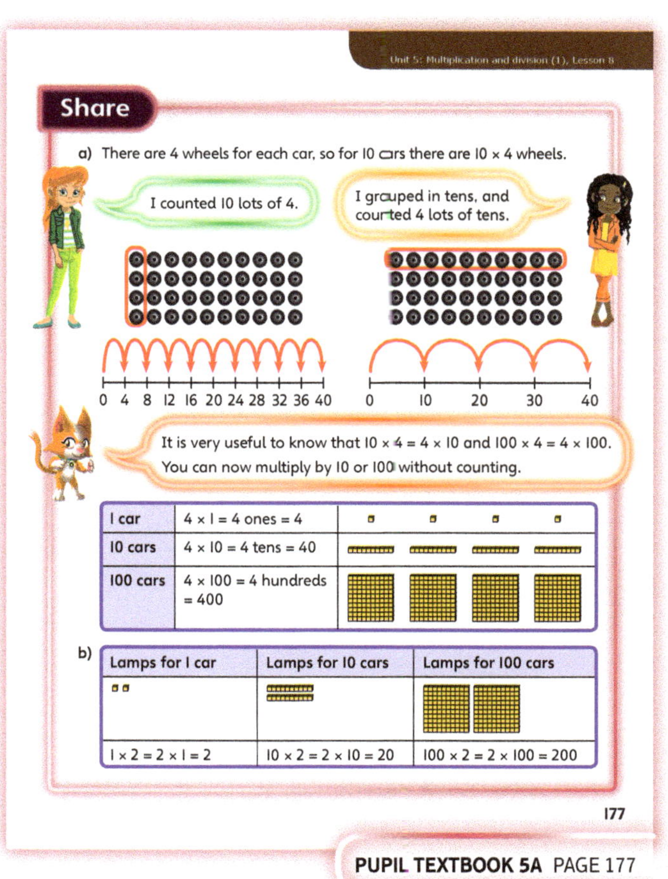

PUPIL TEXTBOOK 5A PAGE 177

Think together

ASK

- Question **1** a): *How does using the base 10 equipment make the multiplication clear? What is similar and different about each calculation?*
- Question **1** b): *How does the number line show the multiplication clearly?*
- Question **2**: *How would you explain Aki's mistake to him?*
- Question **3** a): *What has happened to the digit '3'? How has the exchange altered the value of the digit?*
- Question **3** c): *What happens when you repeatedly multiply by 10?*

IN FOCUS Questions **2** and **3** are important as they tackle the misconception of 'add a zero'. Question **3** a) uses exchanging to show that a number moved from the ones column to the tens column is being multiplied by 10, rather than 'just adding a zero'.

STRENGTHEN If children are finding it difficult to diagnose Aki's misconception in question **2**, they could build the calculations using concrete resources.

DEEPEN Question **3** deepens children's understanding of how multiplying by 10 is linked to place value. Ask: *How can you use multiplying by 10 to multiply by 100 and 1,000? How could you use this to multiply by greater numbers?*

ASSESSMENT CHECKPOINT Questions **1** a) and b) will assess children's ability to more fluently multiply whole numbers by 10, 100 and 1,000.

Question **3** will assess children's fluency with how multiplying by 100 and 1,000 is linked to multiplying by 10. It will also highlight misconceptions about place value and 'just adding a zero' when multiplying by 10.

ANSWERS

Question **1** a): $10 \times 10 = 100$
$10 \times 100 = 1,000$
$10 \times 1,000 = 10,000$

Question **1** b): $12 \times 10 = 120$
$12 \times 100 = 1,200$
$12 \times 1,000 = 12,000$

Question **2**: Aki has solved the first calculation correctly, $23 \times 100 = 2,300$
Aki has solved the second calculation incorrectly by multiplying 20 by 10, instead of 100.

Question **3** a): Each set of 10 unit counters are exchanged for a ten place value counter, giving 3 tens counters in the tens column.
$10 \times 3 = 30$

Question **3** b): $3 \times 10 = 30$
$17 \times 10 = 170$
The numbers move to next column in the place value grid when multiplying by 10.

Question **3** c): $3 \times 10 \times 10 = 300$
$17 \times 10 \times 10 = 1,700$
Multiplying by 10 and then 10 again is the same as multiplying by 100.

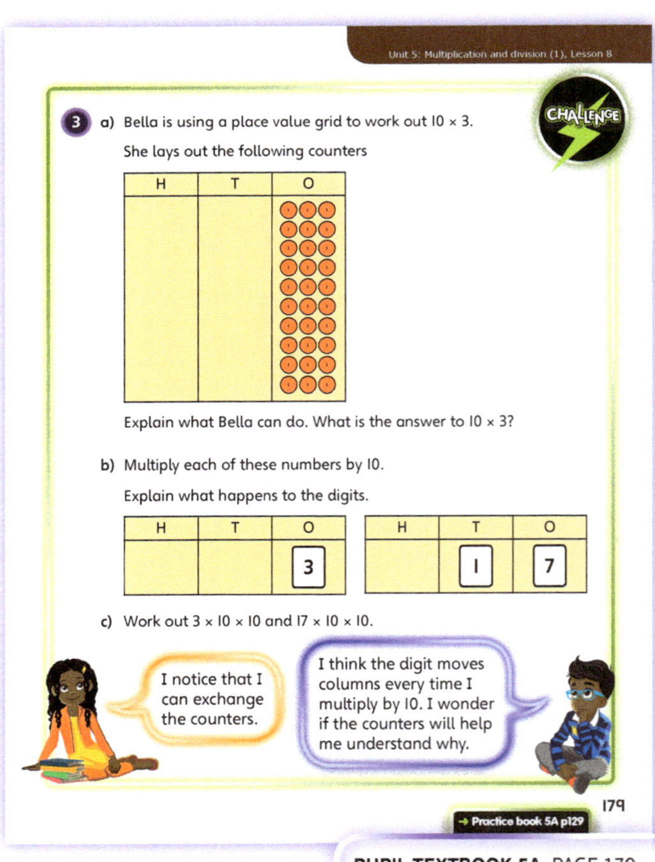

PUPIL TEXTBOOK 5A PAGE 178

PUPIL TEXTBOOK 5A PAGE 179

Practice

WAYS OF WORKING   Independent thinking

IN FOCUS   Question **1** develops children's ability to multiply by 10, 100 and 1,000 while scaffolding their understanding by using the pictorial representation of drawn place value counters.

Question **2** requires focuses on the different ways that multiplying by 10, 100 and 1,000 can be represented.

Question **3** requires children to use abstract written calculations. As children progress through the question, more of it is left blank, requiring more input. This will develop children's ability to link the concrete and pictorial representations they have been experiencing to the abstract written methods they need to begin using.

STRENGTHEN   If children are finding it tricky to complete the calculations in question **6**, encourage them to build the calculations first, and then begin spotting patterns that will help them complete the rest. Ask: *Can you build the first and second calculations? Can you spot any patterns in the numbers that will help you complete the next two? How can you apply what you have noticed to the next sequences?*

DEEPEN   Question **7** is an excellent opportunity for children to develop their ability to reason and justify their opinions. Deepen this by encouraging two children that have different ideas about who is correct to convince each other of their ideas. Ask: *Can you convince your partner that you are correct?*

THINK DIFFERENTLY   Question **4** requires children to diagnose another child's mistakes. Develop children's written reasoning ability by asking them to describe the mistakes and write their advice on how to avoid the mistakes next time.

ASSESSMENT CHECKPOINT   Question **5** links multiplication to the place value grid and gives an opportunity to check for any misconceptions with 'adding a zero' and place value in general.

ANSWERS   Answers for the **Practice** part of the lesson appear in the separate **Practice and Reflect answer guide**.

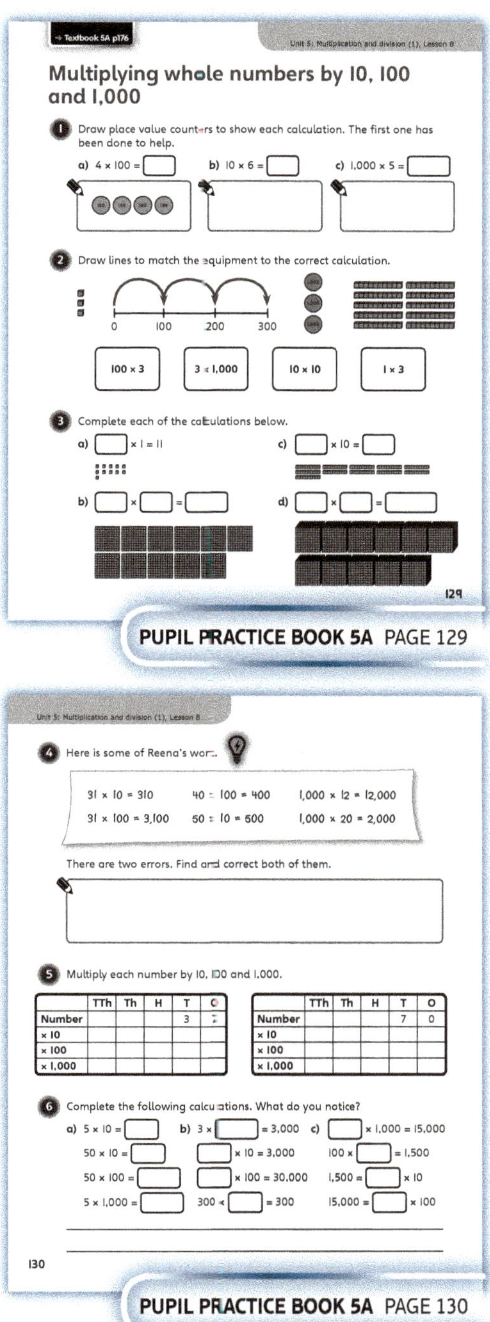

PUPIL PRACTICE BOOK 5A PAGE 129

PUPIL PRACTICE BOOK 5A PAGE 130

Reflect

WAYS OF WORKING   Independent thinking

IN FOCUS   This question also allows children to use their learning from the previous lesson. Children should confidently recognise the multiplication facts they have focused on today and they should be able to use their knowledge of inverse operations to begin considering the related divisions.

ASSESSMENT CHECKPOINT   Look for children fluently recognising and interpreting the concrete representations of a number being multiplied by 100. Children should be able to confidently link the representation to an appropriate calculation.

ANSWERS   Answers for the **Reflect** part of the lesson appear in the separate **Practice and Reflect answer guide**.

After the lesson

- Did children recognise the usefulness of distributive properties when multiplying by 100 and 1,000?
- What concrete manipulatives were most effective in this lesson? Why?

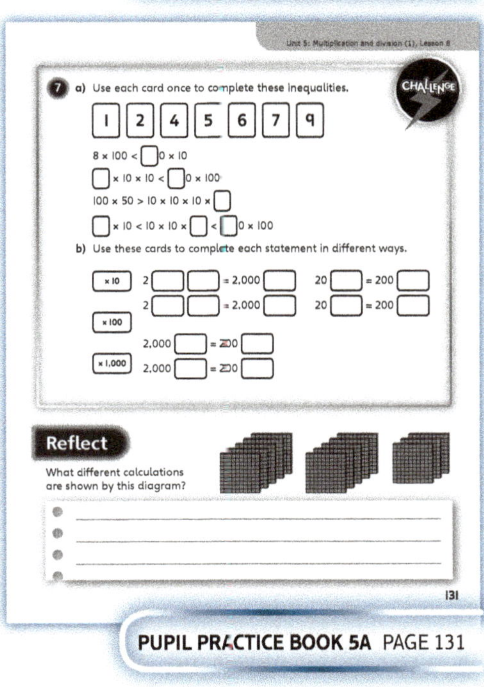

PUPIL PRACTICE BOOK 5A PAGE 131

Dividing whole numbers by 10, 100 and 1,000

Learning focus

In this lesson, children will use their understanding of place value to develop their ability to fluently divide whole numbers by 10, 100 and 1,000.

Small steps

→ Previous step: Multiplying whole numbers by 10, 100 and 1,000
→ **This step: Dividing whole numbers by 10, 100 and 1,000**
→ Next step: Multiplying and dividing by multiples of 10, 100 and 1,000

NATIONAL CURRICULUM LINKS

Year 5 Number – Multiplication and Division

- Multiply and divide whole numbers and those involving decimals by 10, 100 and 1,000.
- Solve problems involving multiplication and division, including scaling by simple fractions and problems involving simple rates.

ASSESSING MASTERY

Children can reliably divide whole numbers by 10, 100 and 1,000, link their understanding of place value to their calculations and represent their thinking using concrete, pictorial and abstract representations.

COMMON MISCONCEPTIONS

Children may recognise dividing by 10, 100 and 1,000 as just 'taking away a zero'. Ask:
- *What does it mean to 'take away a zero'? How are dividing by ten and taking away zero similar/different?*

STRENGTHENING UNDERSTANDING

Recap division and factors with children before this lesson, if necessary. Give children opportunities to experience this concept by sharing objects into groups and linking this to the division calculation. Make arrays, noting the related division and multiplication facts and that multiplication and division are inverse operations.

GOING DEEPER

Give children this statement to investigate: *Is it always true, sometimes true or never true that when you divide a 3-digit whole number by 10, 100 or 1,000 it will result in a number greater than 1?*

KEY LANGUAGE

In lesson: divide, division, ten (10), one hundred (100), one thousand (1,000), inverse operation

Other language to be used by the teacher: place value, ones, tens, hundreds, thousands, tens of thousands, share

STRUCTURES AND REPRESENTATIONS

factor trees, bar models, place value grids, number lines

RESOURCES

Mandatory: place value counters, base 10 equipment

Optional: printed place value grids

 In the eTextbook of this lesson, you will find interactive links to a selection of teaching tools.

Before you teach 

- Are children confident multiplying by 10, 100 and 1,000?
- How will you strengthen understanding of inverse operations?

Discover

 Pair work

ASK

- Question **1** : *Is it better to come first, second or third in this competition? Explain why.*
- Question **1** : *How much do you win for coming first, second or third?*

IN FOCUS Question **1** a) and b) are important as children may assume that the people who come second win more as the shared prize pot is far greater. The conversation from these questions will give children their first opportunity to understand the effect of dividing a whole number by 10, 100 and 1,000.

PRACTICAL TIPS To help engage children in this area of learning, you could create a small competition or race where the first 10 children to succeed are able to share a prize pot. This activity, while difficult to achieve for 100 and 1,000 participants, will give children hands-on, contextual experience with the main concept in the lesson. It may also give an opportunity to recap that division is not commutative, and the order of operations matters.

ANSWERS

Question **1** a): Each 1st prize winner will receive £38.

Question **1** b): Each 2nd prize winner will receive £12.

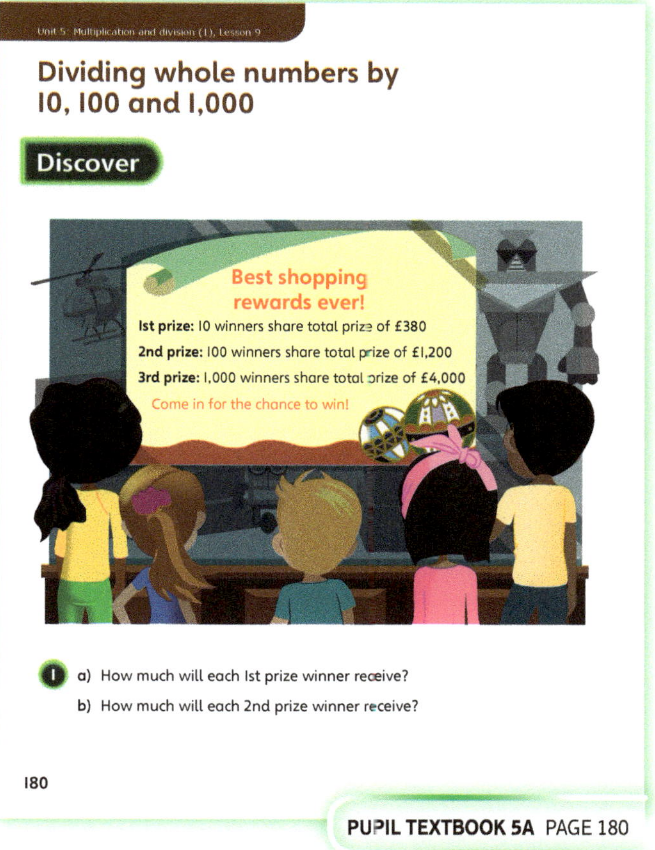

PUPIL TEXTBOOK 5A PAGE 180

Share

 Whole class teacher led

ASK

- Question **1** a): *What calculation does the bar model show? Explain how.*
- Question **1** a): *What would you need to multiply 10 by to have a product of 380?*
- Question **1** a): *Can you use your knowledge of inverse operations to help you solve this?*
- Question **1** b): *How much would each person have won if the prize pot was £15,000?*
- Question **1** b): *At what amount would it be that second prize winners were winning more than first prize winners? Explain how you know.*

IN FOCUS Both questions **1** a) and **1** b) are important to focus on as they show how children may solve the calculations in multiple ways. Making this clear will help develop children's fluency when solving similar calculations in the future.

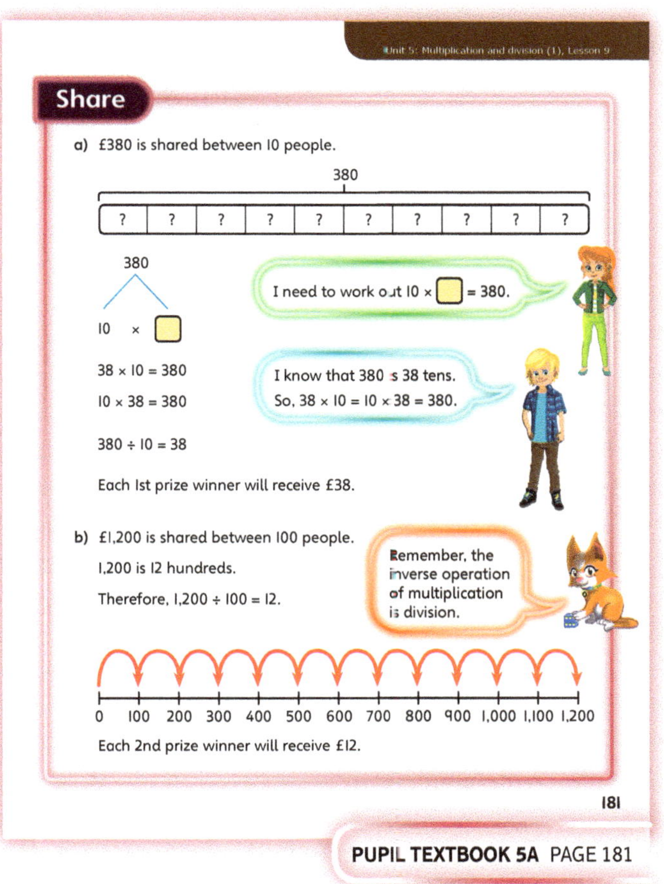

PUPIL TEXTBOOK 5A PAGE 181

Think together

WAYS OF WORKING Whole class teacher led (I do, We do, You do)

ASK

- Question **1**: *How can the factor tree help you solve this?*
- Question **2**: *What patterns can you spot in these calculations? What can you say about what happens every time you divide by 10, 100 or 1,000?*
- Question **3**: *How does your knowledge of base 10 equipment help you with this question?*

IN FOCUS Question **2** offers an excellent opportunity for children to begin generalising about dividing by 10, 100 and 1,000. It will be important to discuss the patterns they see and link what they know of place value and dividing to the quotients found in the division calculations.

Question **3** is important as it requires children to use place value when dividing by 10, 100 and 1,000, to complete multi-step calculations. It will be important to discuss how, similarly to multiplying by 10 and 10 again, dividing by 10 and then by 10 again is equivalent to dividing by 100 (as an example).

STRENGTHEN If children need support with question **3**, encourage them to use base 10 equipment to develop the concept of exchange in place value. This should help them visualise the problem more clearly.

DEEPEN When solving question **2**, deepen children's understanding by encouraging them to generalise. Ask: *What patterns do you notice? What is similar and different about the calculations? How can this help you solve similar calculations more quickly in the future?*

ASSESSMENT CHECKPOINT Question **3** will provide good evidence of children's understanding of the concepts in this lesson, and their ability to link them to their learning from the previous lesson and about place value. Look for children's recognition that, for example, dividing by 10, then by 10 again, is equivalent to dividing by 100.

ANSWERS

Question **1**: 4,000 is 4 thousands.
$4 \times 1,000 = 4,000$
$4,000 \div 1,000 = 4$
Each third prize winner receives £4.

Question **2** a): $30 \div 10 = 3$, $300 \div 100 = 3$, $3,000 \div 1,000 = 3$

Question **2** b): $310 \div 10 = 31$, $3,100 \div 100 = 31$, $31,000 \div 1,000 = 31$

Question **2** c): $300 \div 10 = 30$, $3,000 \div 100 = 30$, $30,000 \div 1,000 = 30$
Each set of calculations gives the same answer.

Question **3** a): $4,000 \div 10 = 400$; $4,000 \div 10 = 40$
$3,200 \div 10 = 320$; $3,200 \div 10 = 32$

When you divide the numbers by 10 or 100, the digits stay the same but move right in the place value grid.

Question **3** b): Max is correct, it is the same as dividing by 1,000.

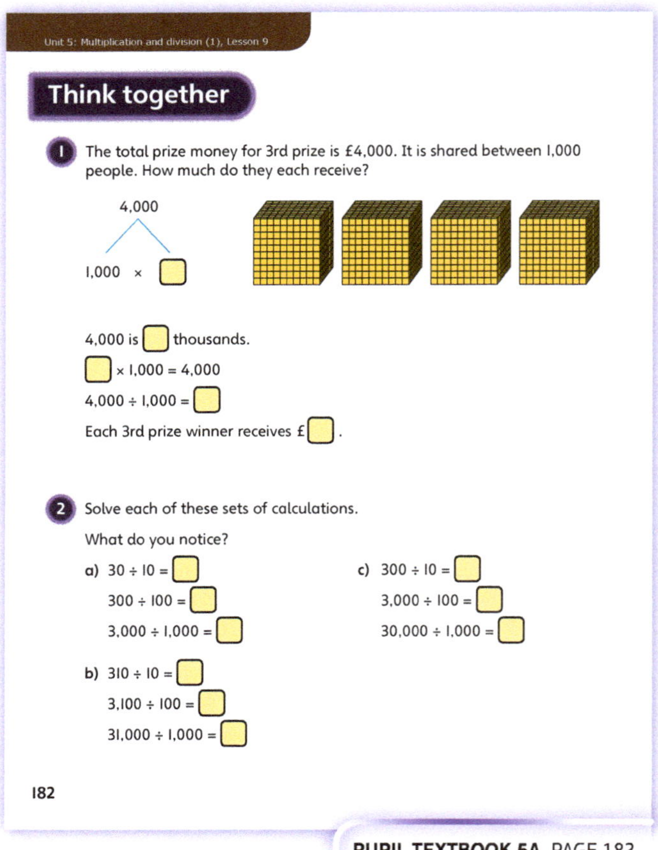

PUPIL TEXTBOOK 5A PAGE 182

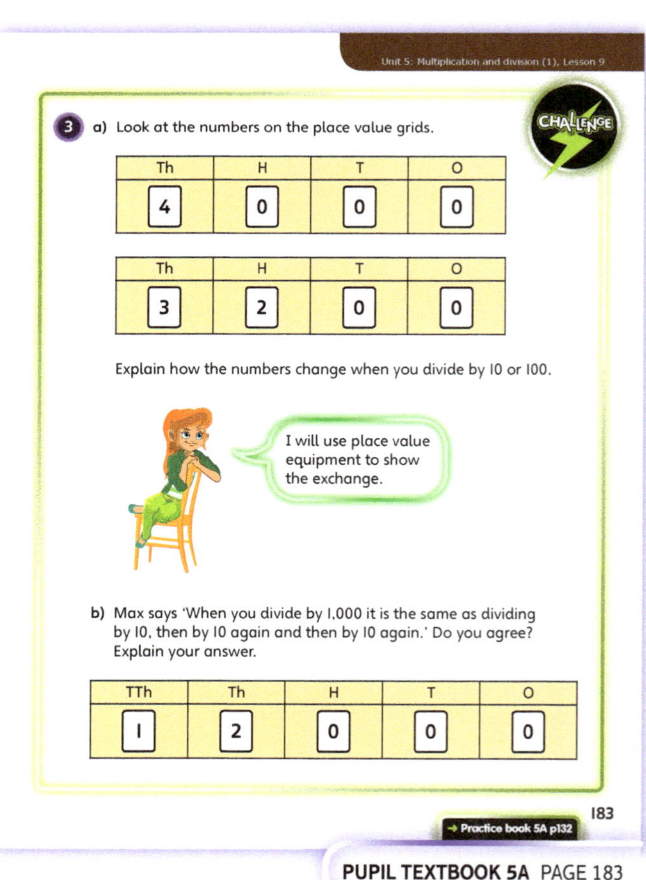

PUPIL TEXTBOOK 5A PAGE 183

Practice

WAYS OF WORKING Independent thinking

IN FOCUS Questions ❶ and ❷ are important as they offer children scaffolded activities where they can begin to independently make the link between their understanding of factor trees and their new understanding of dividing by 10, 100 and 1,000. As children progress, there is more for them to complete.

Question ❸ begins to show children how their learning can be used in real life contexts.

When solving question ❹, remind children of the patterns they noticed and the generalisations they made earlier in the lesson. How can their thinking and ideas help them?

Question ❻ is important as it encourages children to think algebraically, considering what they know about each side of the equality and how knowing one number will affect the other unknown number.

STRENGTHEN In question ❻, ask children to consider using a factor tree. Ask: *How can you use the factor tree to begin finding out what the triangle will be worth? What number do you need to multiply or divide by 10?*

DEEPEN Question ❺ deepens children's fluency, reasoning and problem-solving by putting their learning into a context. It also tries to catch children out by requiring them to divide and multiply by 10. This is not initially clear in the question and will only become clear once children have considered the problem carefully. Ask: *How did you find out how many marbles were in each jar? What did you do to find out how many jars were needed?*

ASSESSMENT CHECKPOINT Question ❻ will assess children's ability to use their understanding of multiplying and dividing by 10, 100 and 1,000 to solve a mathematical problem. Look for their ability to apply the patterns and solutions they have found to other examples when finding more solutions with the same relationship.

ANSWERS Answers for the **Practice** part of the lesson appear in the separate **Practice and Reflect answer guide**.

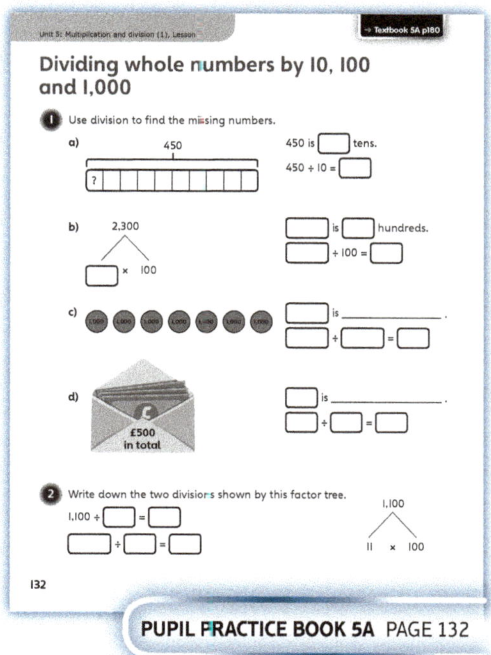

PUPIL PRACTICE BOOK 5A PAGE 132

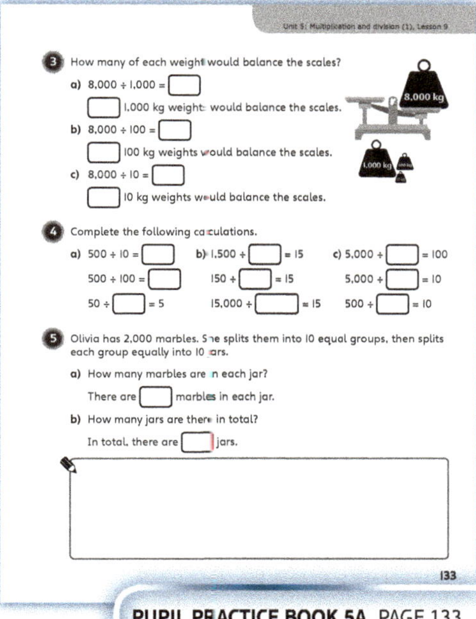

PUPIL PRACTICE BOOK 5A PAGE 133

Reflect

WAYS OF WORKING Independent thinking

IN FOCUS This question will help assess whether children have fully understood the effect that dividing by 10, 100 and 1,000 has on the place value of digits in a number.

ASSESSMENT CHECKPOINT Look for fluent explanations of which calculation is incorrect and why, and accurate use of place value. Children should use multiplication and division and demonstrate their understanding of their inverse relationship to check their calculations.

ANSWERS Answers for the **Reflect** part of the lesson appear in the separate **Practice and Reflect answer guide**.

After the lesson 

- Are children equally confident with dividing by 10, 100 and 1,000 as they were with multiplying?
- Are children confident in using the inverse relationship between multiplication and division?
- Can children confidently use factor trees to aid their thinking?

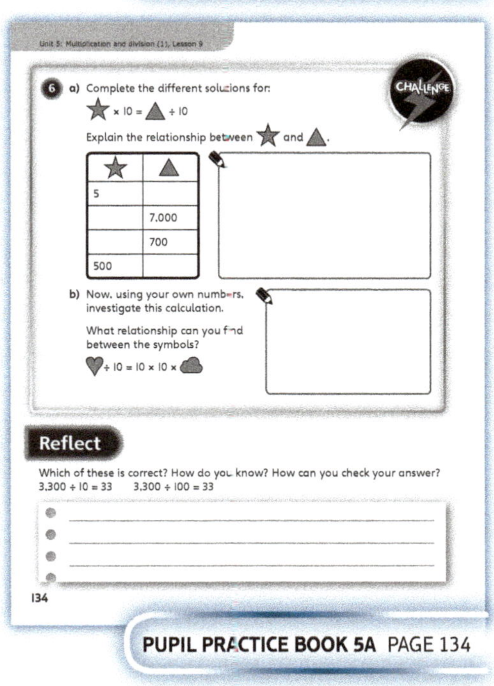

PUPIL PRACTICE BOOK 5A PAGE 134

Multiplying and dividing by multiples of 10, 100 and 1,000

Learning focus

In this lesson, children will use their knowledge and understanding of multiplying and dividing by 10, 100 and 1,000 to reliably multiply numbers by multiples of 10, 100 and 1,000 using known multiplication facts.

Small steps

→ Previous step: Dividing whole numbers by 10, 100 and 1,000
→ **This step: Multiplying and dividing by multiples of 10, 100 and 1,000**
→ Next step: Measuring perimeter

NATIONAL CURRICULUM LINKS

Year 5 Number – Multiplication and Division

Multiply and divide whole numbers and those involving decimals by 10, 100 and 1,000.

ASSESSING MASTERY

Children can use their understanding of multiplying and dividing by 10, 100 and 1,000, and their understanding of multiplication facts, to reliably multiply and divide given numbers by multiples of 10. They can use this ability to solve mathematical problems, explaining their reasoning and giving clear evidence of their thinking.

COMMON MISCONCEPTIONS

Children may solve a calculation such as 180 ÷ 6 by solving 18 ÷ 6, recording the answer as 3 in error. Ask:
• *Can you show me both of those calculations using a concrete or pictorial representation?*

Children may assume that 20 × 50 = 100 (an equal number of zeros on either side of the equals sign). Ask:
• *What is 2 × 5? What is 20 × 5? Can you represent 20 × 5 and 20 × 50 using resources or a picture?*

STRENGTHENING UNDERSTANDING

Develop children's fluency with multiples of 10 by counting through a times-table and linking it to the same times-table multiplied by 10, 100 or 1,000. Ask: *Can you count up in the 4 times-table? What will be different and what will stay the same if you count in the 40 times-table? 400? 4,000? What patterns can you find?*

GOING DEEPER

Encourage children to create their own lesson or tutorial on how to multiply and divide with multiples of 10, 100 and 1,000. Ask: *How will you make sure your explanation is clear to someone who does not yet understand?* Children could then teach each other and discuss how and where they would improve the lessons.

KEY LANGUAGE

In lesson: multiple, multiply, multiplication, divide, division, ten (10), hundred (100), thousand (1,000), metres (m), grams (g)

Other language to be used by the teacher: place value, ones, tens, hundreds, thousands, tens of thousands

STRUCTURES AND REPRESENTATIONS

factor trees, bar models, number lines

RESOURCES

Mandatory: base 10 equipment, place value counters

Optional: multiplication squares, calendar

 In the eTextbook of this lesson, you will find interactive links to a selection of teaching tools.

Before you teach 

• Are there any misconceptions from the previous two lessons that will inhibit learning in this lesson?

Discover

  Pair work

ASK

- Question **1** a): *How many days are there in April?*
- Question **1** a): *What type of calculation do you need to do here?*
- Question **1** b): *How could knowing how many words Emma will learn in 10 days help you solve this question?*

IN FOCUS Question **1** a) offers children their first opportunity to begin calculating with multiples of 10.

Question **1** b) is important as it encourages children's fluency and flexibility when calculating with the types of numbers used in the lesson.

PRACTICAL TIPS Children could be encouraged to set their own target for a 30-day month. They could be encouraged to set three levels of challenge for themselves; a low, middle and high number of words to enable them to investigate the effect of multiplying different numbers by a multiple of 10.

ANSWERS

Question **1** a): Emma plans to learn 150 words in April.

Question **1** b): Ebo knows that $10 \times 30 = 300$. So, he knows 5×30 must be half of 300, which is 150.

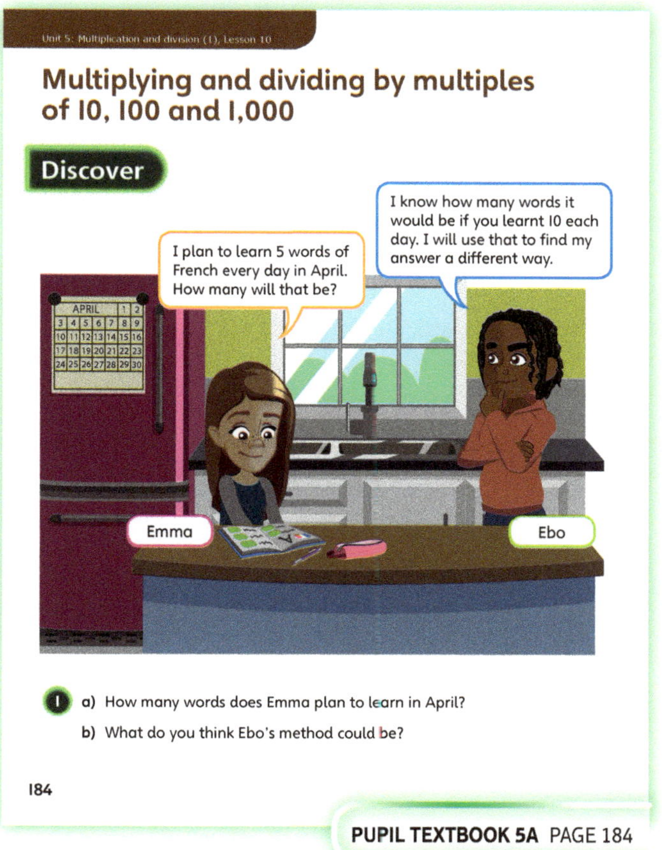

PUPIL TEXTBOOK 5A PAGE 184

Share

  Whole class teacher led

ASK

- Question **1** a): *How does the base 10 equipment make the multiplication clear?*
- Question **1** a): *How is the base 10 equipment representation and the number line similar? How are they different?*
- Question **1** a): *How is 5×3 similar to 5×30? How can you use multiplication knowledge to help you solve a trickier calculation like this?*
- Question **1** b): *How does the bar model make Ebo's way of calculating clear?*
- Question **1** b): *Did you find another way of solving this?*

IN FOCUS Questions **1** a) and **1** b) both offer the opportunity to scaffold children's understanding of the concepts covered in the questions by helping them make links between the different representations and methods of the same calculation.

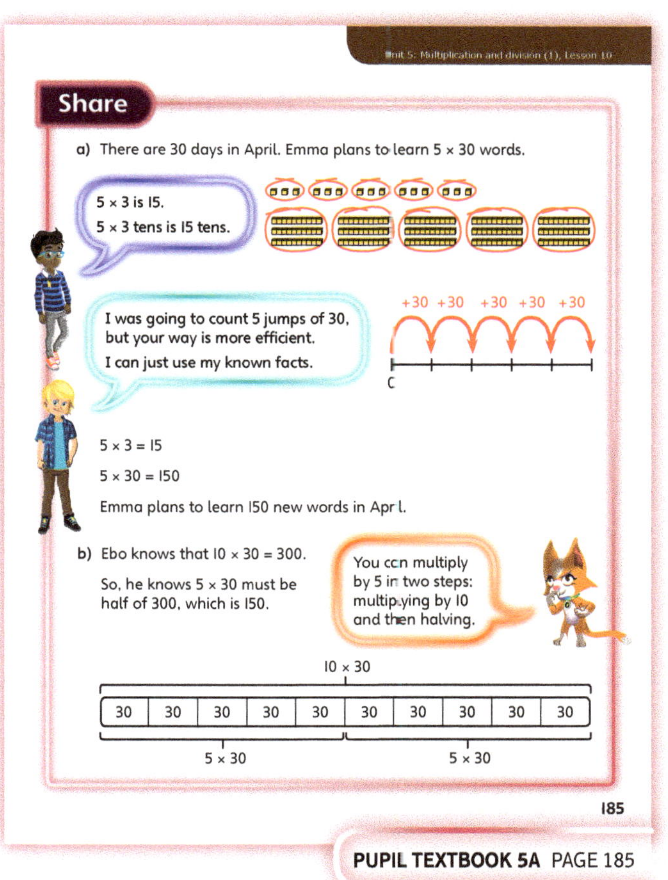

PUPIL TEXTBOOK 5A PAGE 185

Think together

WAYS OF WORKING Whole class teacher led (I do, We do, You do)

ASK

- Question ① : *How will you represent the problem? Which method is most efficient? Why?*
- Question ② : *What is similar and different about the two arrays of place value counters?*
- Question ③ a): *Can you draw the rest of the bar model that would represent this problem?*
- Question ③ b): *What number facts are useful to know when solving these calculations?*

IN FOCUS Question ① reinforces the methods shown with concrete resources. Discuss which method children prefer and why. Both methods encourage children to recognise that 180 ÷ 30 can be more easily solved as 18 ÷ 3.

Question ② reinforces the learning from question ① and represents it in the more abstract way of place value counters. Discussing what is similar and different about the two sets of place value counters will be key to ensuring children's understanding.

STRENGTHEN While solving questions ①, ② and ③, give children access to the concrete resources pictured in the questions. If children need help ask: *Can you make the calculation using the resources you have used before? What does your concrete representation show?*

DEEPEN When solving question ③ b), deepen children's ability to recognise generalisations by asking:
- *What do you notice about all of the calculations?*
- *How are the calculations similar and different?*
- *Can you come up with a rule that would help you calculate with any multiple of 10, 100 or 1,000?*

ASSESSMENT CHECKPOINT Questions ① and ② will assess children's ability to recognise the similarities between calculations, such as 4 × 300 and 4 × 3. Look for their recognition that, having identified and understood the pattern, they can use their simpler multiplication and division knowledge to solve trickier calculations.

Question ③ b) assesses children's fluency when recognising other multiplication and division facts that can help them solve trickier calculations more efficiently.

ANSWERS

Question ① : 180 ÷ 30 = 6, so Isla needs to learn 6 new words each day.

Question ② a): 4 × 3 = 12; 4 × 300 = 1,200

Question ② b): 24 ÷ 6 = 4; 2,400 ÷ 600 = 4

Question ③ a): 800 × 12 = 600 × 16
Athlete B will have to train for 16 days before she has run as far as Athlete A.

Question ③ b): 9 × 3,000 = 90 × 300
4,000 × 4 = 8 × 2,000
35,000 ÷ 7,000 = 35 ÷ 7
2,400 ÷ 120 = 1,200 ÷ 60

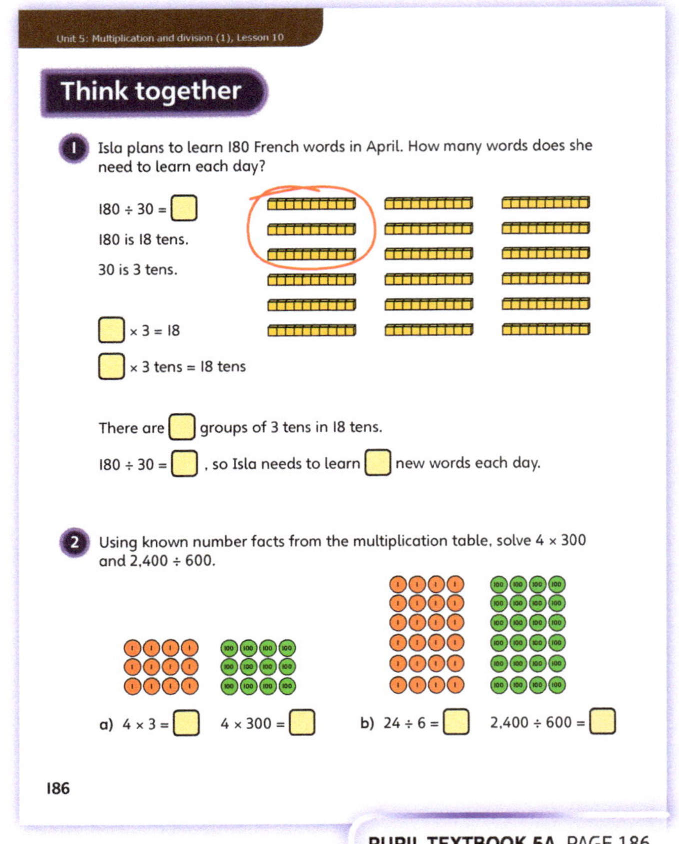

PUPIL TEXTBOOK 5A PAGE 186

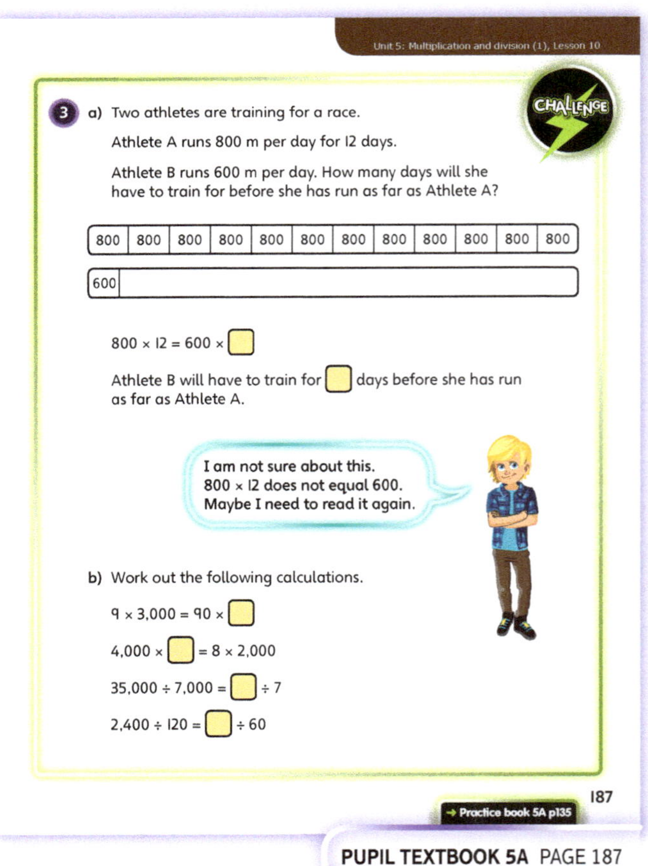

PUPIL TEXTBOOK 5A PAGE 187

Practice

WAYS OF WORKING Independent thinking

IN FOCUS Question ❶ helps children's understanding of how the pictorial and concrete representations link to the calculations. It also scaffolds children's understanding that larger multiples of 10, 100 and 1,000 can be read as single digit numbers for the purposes of calculating easily.

Question ❷ encourages children to link their understanding of the calculations to their own drawn pictorial evidence. Give children access to the concrete place value counters to help support their thinking here.

Question ❸ helps children link 300 × 6 and 30 × 60, and encourages use of known multiplication facts to solve problems involving more than 2 digits.

STRENGTHEN When solving question ❸, encourage children to see the link between the calculations. Ask: *Once you have found all of the calculations for one of the factor trees, how easy will it be to find the calculations for the other?*

In question ❹, it may help to have a multiplication square at hand. Ask: *Can you find any similar multiplication or division facts on this multiplication square? How can these facts help you solve these calculations?*

DEEPEN In question ❻, encourage children to write their own contextual word problems for their partner to solve.

THINK DIFFERENTLY Question ❺ helps tackle the misconception of counting the zeros in a calculation and simply making sure the answer has the same amount. See if children know where Ambika has applied her procedural understanding, and not shown full understanding of the concept of place value and multiplying by 10, 100 and 1,000.

ASSESSMENT CHECKPOINT Questions ❶ and ❷ will assess children's ability to link the abstract calculations to the concrete and pictorial representations they have been using throughout the unit. Look for children making clear the links between the representations in their verbal explanations.

Question ❸ will assess children's ability to recognise the link between multiples of 1, 10, 100 and 1,000.

ANSWERS Answers for the **Practice** part of the lesson appear in the separate **Practice and Reflect answer guide**.

Reflect

WAYS OF WORKING Pair work

IN FOCUS This question will offer a chance to briefly assess whether children in the class are able to apply their learning from the previous three lessons. It will be interesting for children to compare their calculations, discussing the similarities and differences they observe.

ASSESSMENT CHECKPOINT Look for children recognising how their two calculations can be linked. Children may want to show this using factor trees or one of the concrete or pictorial representations they have used in the lesson.

ANSWERS Answers for the **Reflect** part of the lesson appear in the separate **Practice and Reflect answer guide**.

After the lesson ⏸

- Were children able to apply their learning from the previous two lessons fluently to this lesson?

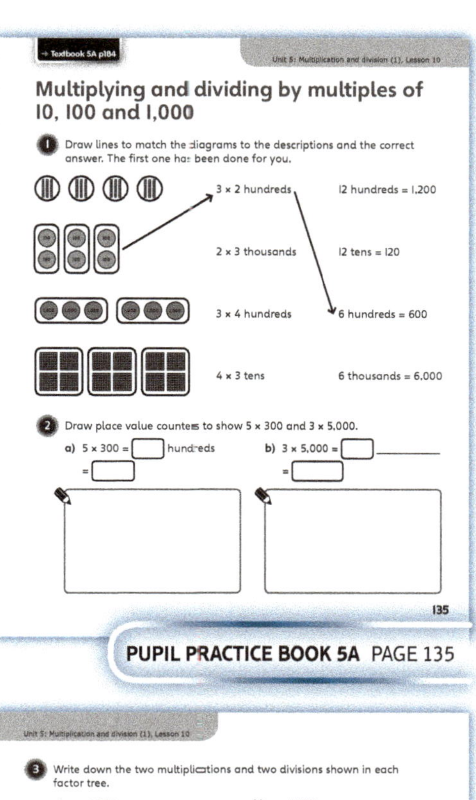

PUPIL PRACTICE BOOK 5A PAGE 135

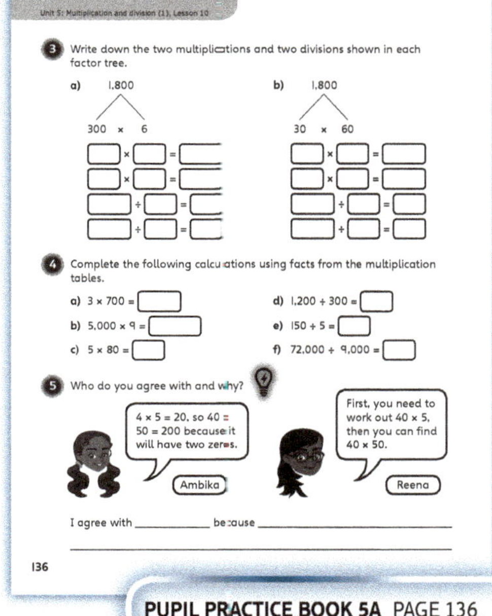

PUPIL PRACTICE BOOK 5A PAGE 136

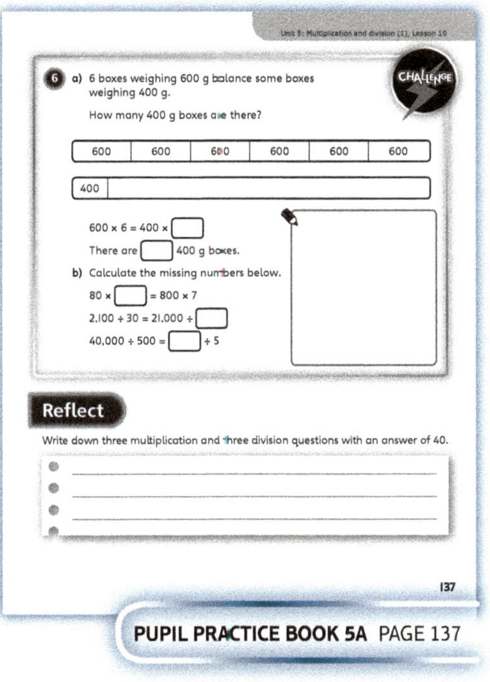

PUPIL PRACTICE BOOK 5A PAGE 137

End of unit check

Don't forget the *Power Maths* unit assessment grid on p26.

WAYS OF WORKING Group work adult led

IN FOCUS Question **1** assesses children's ability to recognise and find factors and multiples.

- Question **2** assesses children's recognition and understanding of square numbers.
- Question **3** assesses children's ability to recognise multiplication facts they know and use them to solve more complex calculations.
- Question **4** assesses children's ability to work out prime numbers.
- Questions **5** and **6** assess children's fluency with multiplying and dividing whole numbers by multiples of 10, 100 and 1,000.
- Question **6** is a SATs-style question and assesses children's recognition of prime numbers.
- Question **7** is a problem-solving question working with a multiple of 10.

ANSWERS AND COMMENTARY Children who have mastered the concepts in this unit will reliably find multiples and factors of given numbers. They will be able to explain the unique properties of prime, square and cube numbers. They will be able to reliably multiply and divide whole numbers by 10, 100 and 1,000, and they will be able to confidently use their understanding of these concepts to multiply and divide whole numbers by multiples of 10, 100 and 1,000.

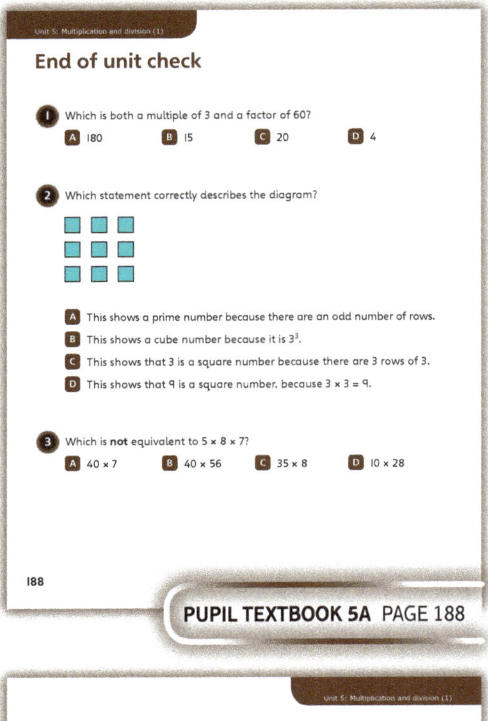

PUPIL TEXTBOOK 5A PAGE 188

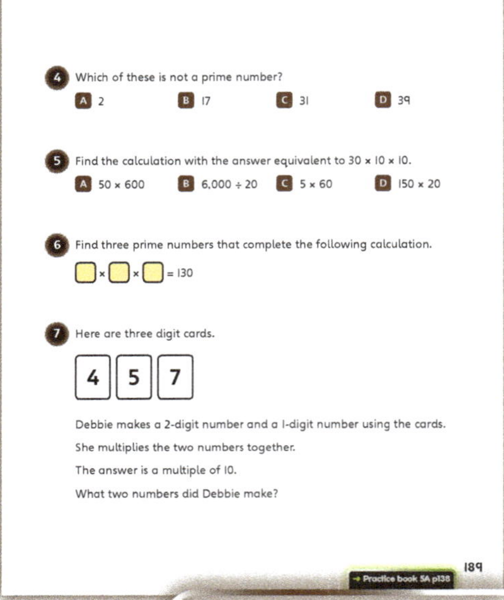

PUPIL TEXTBOOK 5A PAGE 189

Q	A	WRONG ANSWERS AND MISCONCEPTIONS	STRENGTHENING UNDERSTANDING
1	B	A or C may show that children have muddled factors and multiples.	Encourage children to build or draw the numbers and calculations shown in the question using the representations they are most confident with. Discuss what their representation shows them and how they can apply that to the question they are solving.
2	D	B indicates a misconception in how to read of X^3. C suggests that the starting number is the cube number.	
3	B	A, C or D suggests that children made an error or have gaps in their multiplication table knowledge.	
4	D	A, B or C suggests children do not understand the features of a prime number.	
5	D	A, B or C suggests that chidren are finding inverses tricky, or that they are less confident with multiples of 10, 100 or 1,000.	
6		$2 \times 5 \times 13 = 130$ Children may neglect to consider 2 as a prime number as it is even.	
7		75 and 4 or 74 and 5	

My journal

WAYS OF WORKING Independent thinking

ANSWERS AND COMMENTARY

Children may record answers such as:
- I know 250 isn't a square number because 15 squared is 225 and 16 squared is 256.
- 2,500 is a square number because 50×50 is 2,500.
- I know 2,500 is going to be square because 5×5 is 25. If I multiply both 5s by 10 then the answer must be multiplied by 100. $25 \times 100 = 2,500$.

If children are finding it difficult to unpick the reasoning in the question, ask:
- *What resources or representations have helped you find square numbers before?*
- *What calculation do you use to find a square number?*

Power check

WAYS OF WORKING Independent thinking

ASK

- *How do you feel about the new concepts you learnt in this unit?*
- *Could you teach someone about square and cube numbers?*
- *What is useful about multiplying and dividing by 10, 100 and 1,000?*

Power puzzle

WAYS OF WORKING Pair work

IN FOCUS This Power puzzle activity will assess whether children can find factors and recognise prime numbers. Ask:
- *How will you know when you have found a prime number?*
- *Do some prime numbers appear more than others? Why do you think this might be?*

ANSWERS AND COMMENTARY

Prime factors of 90 = $2 \times 3 \times 3 \times 5$

Prime factors of 210 = $2 \times 3 \times 5 \times 7$

Children should be able to reliably find the factors of each given number, factorising until they reach the prime factors. If children are not sure how to begin finding factors, ask:
- *Can the number be divided by 2? Why would this be helpful?*

If children need extra support to recognise prime numbers, ask:
- *What is a prime number? Can you show me what a prime number looks like using an array?*
- *What are the first prime numbers to 20? How might knowing this help you?*

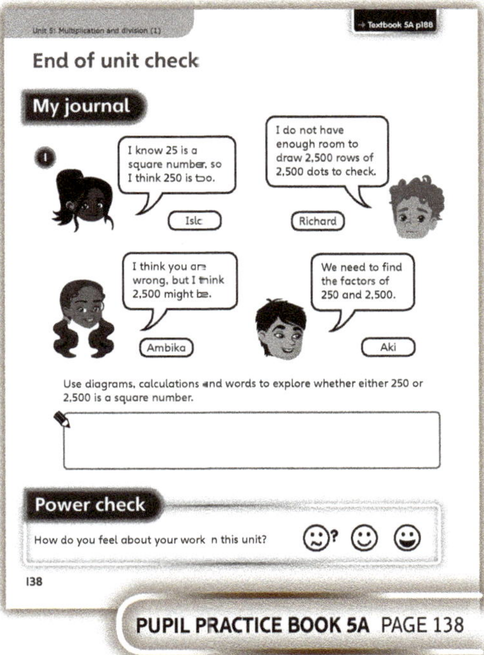

PUPIL PRACTICE BOOK 5A PAGE 138

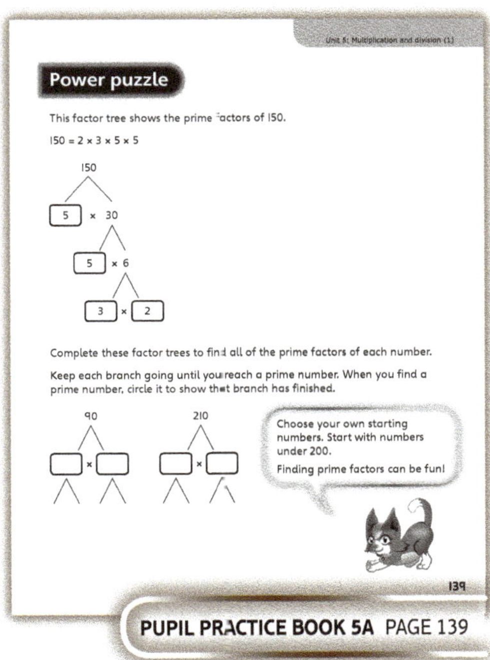

PUPIL PRACTICE BOOK 5A PAGE 139

After the unit ⏸

- Are children confident explaining what a prime number is and finding factors?
- Do children understand how to use the inverse operation to check their calculations?

Strengthen and **Deepen** activities for this unit can be found in the *Power Maths* online subscription.

Unit 6
Measure – area and perimeter

Don't forget to watch the Unit 6 video!

WHY THIS UNIT IS IMPORTANT

This unit provides children with numerical strategies to calculate area and perimeter of rectangles and squares. They will be introduced to simple formulae such as *perimeter = 2 × (length + width)* and *Area = length × width*. Application of these methods will include working inversely and using a systematic approach to find rectangles with a given perimeter or area.

WHERE THIS UNIT FITS

→ Unit 5: Multiplication and division (1)
→ **Unit 6: Measure – area and perimeter**
→ Unit 7: Multiplication and division (2)

This unit builds on the concepts of area and perimeter learned in Year 4. Previous methods (including the use of concrete representations and squares) will be used as a starting point to derive numerical strategies.

Before they start this unit, it is expected that children:
- can define the concepts of area and perimeter
- can find the perimeter of shapes when all side lengths are given
- can find the area of rectilinear shapes drawn on squared paper by counting squares.

ASSESSING MASTERY

Children who have mastered this unit will be able to find the perimeter of rectilinear shapes by accurately measuring a shape's sides and by using the given dimensions, including examples where only some of the shape's dimensions are supplied and the unknowns need to be derived. They can apply their knowledge inversely to suggest the dimensions of a shape given its perimeter. Children can calculate the area of rectangles by multiplying length by width and use the *length × width = area* formula correctly. They can apply their knowledge of area inversely to suggest the dimensions of a shape when given its area. Children can estimate the area of irregular shapes, explaining which squares are to be counted.

COMMON MISCONCEPTIONS	STRENGTHENING UNDERSTANDING	GOING DEEPER
Children may think that every side of a shape needs to be labelled with a measurement in order to calculate its perimeter.	Provide children with paper shapes. Encourage them to bend these shapes to prove that combinations of side lengths are equivalent (and so they can derive missing lengths).	Provide children with more complex unlabelled rectilinear shapes. Ask them to label as few sides as possible so that the perimeter can still be calculated. Which sides are essential?
Children may overly rely on counting squares to find area instead of using the *length × width* formula. They may therefore have difficulty finding the area of those rectangles drawn without squares.	Present children with the dimensions of a rectangle and ask them to represent this using an array. Children could measure and draw their rectangle on plain paper, then check by using a transparent overlay of squared paper.	Provide children with further opportunities to estimate area. Give them irregular shapes to draw around on squared paper. When identifying the whole squares, encourage children to use the *l × w* relationship to find the area, rather than counting each individual square.

WAYS OF WORKING

Use these pages with the whole class to revise the concepts of area and perimeter covered in Year 4. What do children remember about arrays and their link with multiplication? Which multiplication fact does Ash's diagram show? (4 × 7). Discuss how the rectangles are presented, explaining that the class is going to learn how to work out the areas and perimeters of rectangles and squares using a rule (formula) rather than by counting squares.

STRUCTURES AND REPRESENTATIONS

2D rectilinear shapes represented on squared paper: This model allows children to count the number of side lengths around a shape and the number of squares that fit inside a shape.

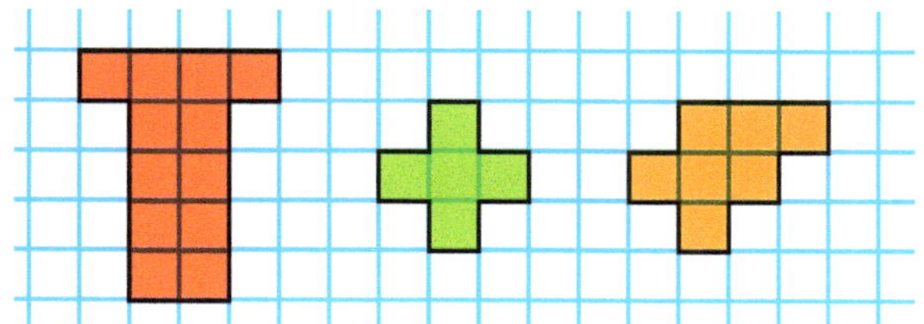

Array: This model is an essential link between the idea of multiplying two numbers to find a total and the concept of length × width to calculate a rectangle's area.

KEY LANGUAGE

There is some key language that children will need to know as part of the learning in this unit:

→ perimeter, distance, area, space

→ scale, actual area/actual size, convert

→ centimetres (cm), metres (m), square centimetres (cm²), square metres (m²)

→ rectangle, square, rectilinear shape, sides, length, width

→ measure, combine, brackets, total, double, estimate, array

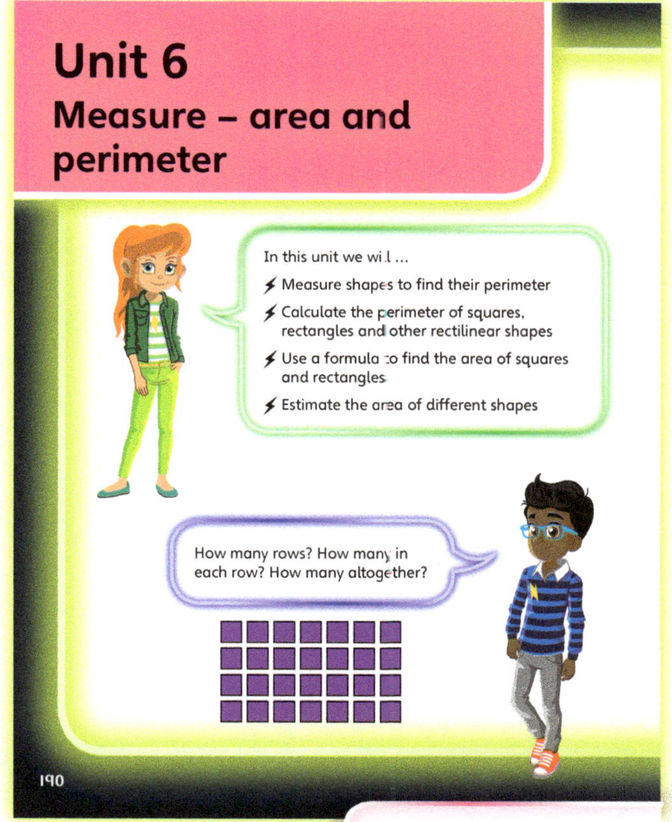

PUPIL TEXTBOOK 5A PAGE 190

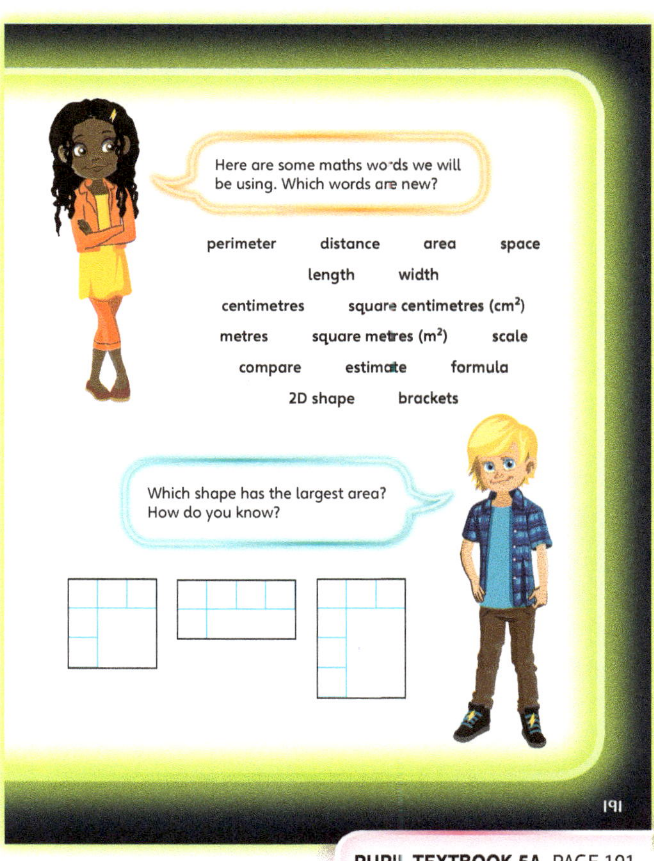

PUPIL TEXTBOOK 5A PAGE 191

Measuring perimeter

Learning focus

In this lesson, children will find the perimeter of rectilinear shapes through measurement in centimetres.

Small steps

→ Previous step: Multiplying and dividing by multiples of 10, 100 and 1,000
→ **This step: Measuring perimeter**
→ Next step: Calculating perimeter (1)

NATIONAL CURRICULUM LINKS

Year 5 Measurement

Measure and calculate the perimeter of composite rectilinear shapes in centimetres and metres.

ASSESSING MASTERY

Children can accurately find the perimeter of rectilinear shapes by measuring with a ruler. Children recognise that they only need to measure two sides of a rectangle and one side of a square to find their perimeters.

COMMON MISCONCEPTIONS

Children may think that they need to measure all of the sides of a rectilinear shape in order to find its perimeter. Stop children after measuring one or two side lengths. Ask:
• *What is special about the opposite sides of a rectangle / the sides of a square? Do you have enough information to find the perimeter?*

STRENGTHENING UNDERSTANDING

To strengthen understanding, provide practical opportunities for children to measure the perimeter of larger shapes in metres (using trundle wheels). Working at a large scale can consolidate children's understanding of perimeter being the distance around a shape. Reduce the number of sides of the rectangular shapes children can measure from four down to three then to two. Ask: *Can you still find the perimeter if you only measure three sides? Two sides?*

GOING DEEPER

Provide children with different-sized rectangles and squares (playing cards, business cards, coffee mats, tiles). Challenge children to take any two or three objects and place them together to form a rectilinear shape. Ask: *What is the perimeter of your shape? How could you arrange the objects in a different way to make a shape with a larger perimeter?*

KEY LANGUAGE

In lesson: perimeter, 2D, measure, centimetres (cm), rectangle, square, double, equal sides, length, width, distance, combine

Other language used by the teacher: rectilinear, composite, opposite sides

STRUCTURES AND REPRESENTATIONS

2D shapes

RESOURCES

Mandatory: rulers/measuring tapes (cm)

Optional: trundle wheels, metre rulers, paper strips, straws

 In the eTextbook of this lesson, you will find interactive links to a selection of teaching tools.

Before you teach 

• This lesson involves practical measurement activities. Are there any adaptations you can make to this lesson to link these activities to your own school context?
• How can you improve the teaching of problem solving through this lesson?

Discover

WAYS OF WORKING Pair work

ASK

- Question ❶ a): *What is the same and what is different about the shapes of the two stickers?*
- Question ❶ a): *Can you put each question into your own words without using the word 'perimeter'?*
- Question ❶ a): *Can you use your finger to trace what is meant by the perimeter of each shape?*
- Question ❶ b): *There are no measurements labelled on these shapes, so how can you find the perimeter?*

IN FOCUS When considering questions ❶ a) and b) link the practical activity described below to the concept of the distance around each sticker shown in the picture. Ask children to trace around the shapes with their finger to show the perimeter. As no dimensions are given in the picture, encourage children to realise that they will need to take physical measurements.

PRACTICAL TIPS Give children copies of rectilinear shapes on paper and ask them to cut them out. Ask: *Who had to cut furthest around their shape? How could you find this out?* Draw one continuous line by rotating shapes and drawing along each side so that they show the cumulative distance (this could be colour-coded). Alternatively, rotate the shape along a ruler so that each new side length adds to the cumulative distance. Ask: *What is the total distance you cut?*

ANSWERS

Question ❶ a): The perimeter of the red stripy sticker is 28 cm.

Question ❶ b): The perimeter of the blue dotty sticker is 26 cm.

Share

WAYS OF WORKING Whole class teacher led

ASK

- Question ❶ a): *Can you describe more than one way to measure the perimeter of the red stripy sticker?*
- *Does it matter where you start measuring from?*
- Question ❶ b): *How can you use doubling to help find the perimeter of the blue dotty sticker?*
- Question ❶ b): *How many sides would you need to measure if the blue dotty sticker was square?*
- *Do you think shapes with more sides will always have a longer perimeter than those with fewer sides?*

IN FOCUS Ensure that children can explain why doubling works as a strategy for finding the perimeter in question ❶ b). (Rectangles have two pairs of opposite, equal sides.) Ask: *Can you explain how to use this doubling method with any rectangle?* Children should recognise that they can either double the total of the length and width or double each measurement separately before adding them.

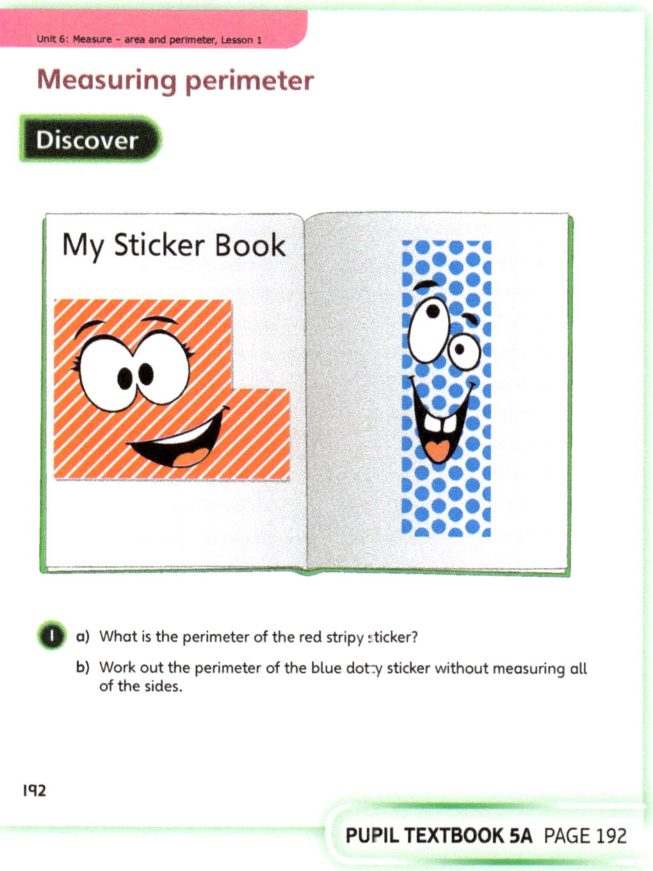

PUPIL TEXTBOOK 5A PAGE 192

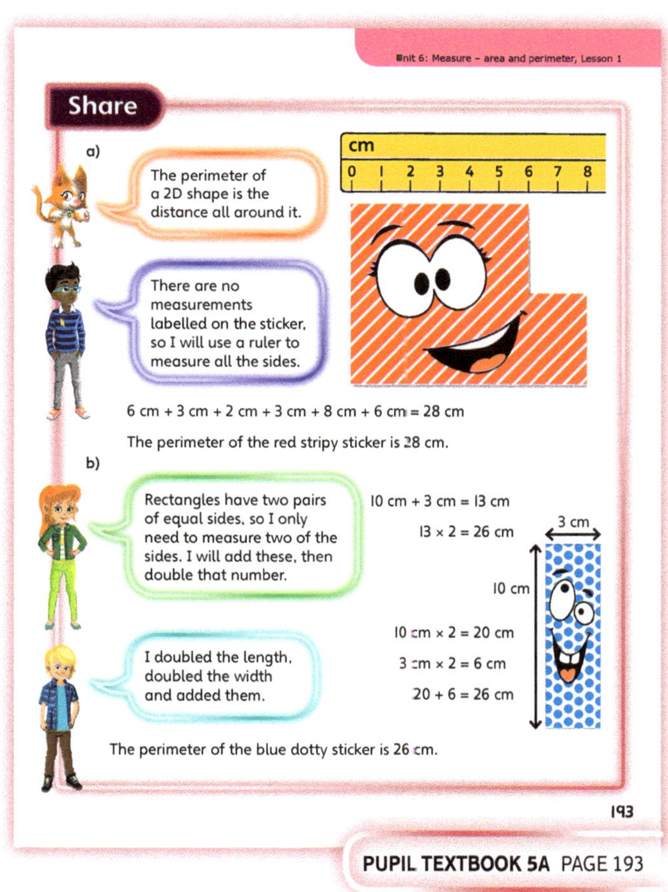

PUPIL TEXTBOOK 5A PAGE 193

Think together

WAYS OF WORKING Whole class teacher led (I do, We do, You do)

ASK

- Question **1**: *How would you describe this shape? Can you see any numbers that combine to help you add them quickly? Can you spot a way to find the answer more quickly than measuring all eight sides?*
- Question **2**: *Although this sticker has been ripped, what information do you know? What properties of rectangles can help you answer this question? Can you think of two ways to work out its perimeter?*

IN FOCUS Question **3** focuses on identifying the two measurements (width and length) of an irregular rectilinear shape that will enable children to quickly find the perimeter without having to measure all of the sides.

STRENGTHEN To support understanding of question **2**, provide small rectangular pieces of paper of different sizes to model the problem concretely. Ask: *What do you need to know to work out a rectangle's perimeter?* Rip your rectangle once so that you still have enough information to work out the perimeter. Encourage children to find the perimeter twice: by doubling the length and width and then checking by putting the rectangle back together and measuring and adding all of the sides.

DEEPEN For question **3**, provide children with thin strips of paper or paper straws, encouraging them to recreate the shape shown by cutting six strips to match the lengths shown. Challenge them to move their strips around to explore the equivalences between different side lengths. Ask: *What do you notice? Which two straws could you measure to help calculate the perimeter of the whole shape? Why does this work?*

ASSESSMENT CHECKPOINT Children should be able to find the perimeter of rectilinear shapes by measuring sides and adding accurately. Children should be beginning to realise that not all sides need to be measured, correctly identifying which sides to measure to find the overall perimeter. Some children will be able to use doubling.

ANSWERS

Question **1**: $5 + 6 + 3 + 3 + 1 + 3 + 1 + 6 = 28$ cm

The perimeter is 28 cm.

Question **2**: The perimeter of the sticker was 32 cm. Explanation should mention the fact that the missing length and width are the same as the visible ones so the measurable ones can be doubled to find the perimeter.

Question **3**: The two sides children should point to are the horizontal base (the shape's width) and the right-hand vertical side (the shape's length). The remaining sides are equal to either the width or the length, so they simply need to measure the length and the width and then double them to find the perimeter. The perimeter of the shape is 62 cm.

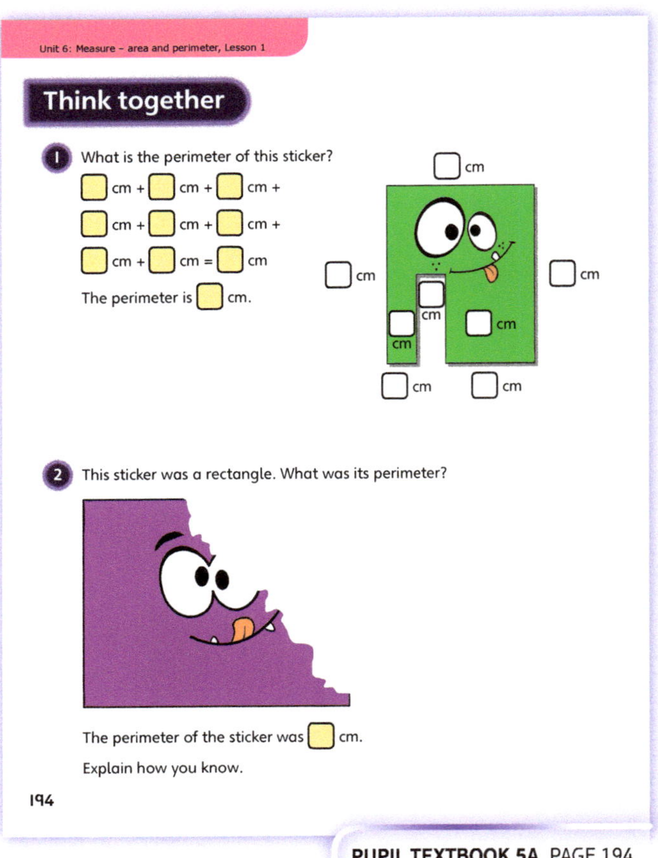

PUPIL TEXTBOOK 5A PAGE 194

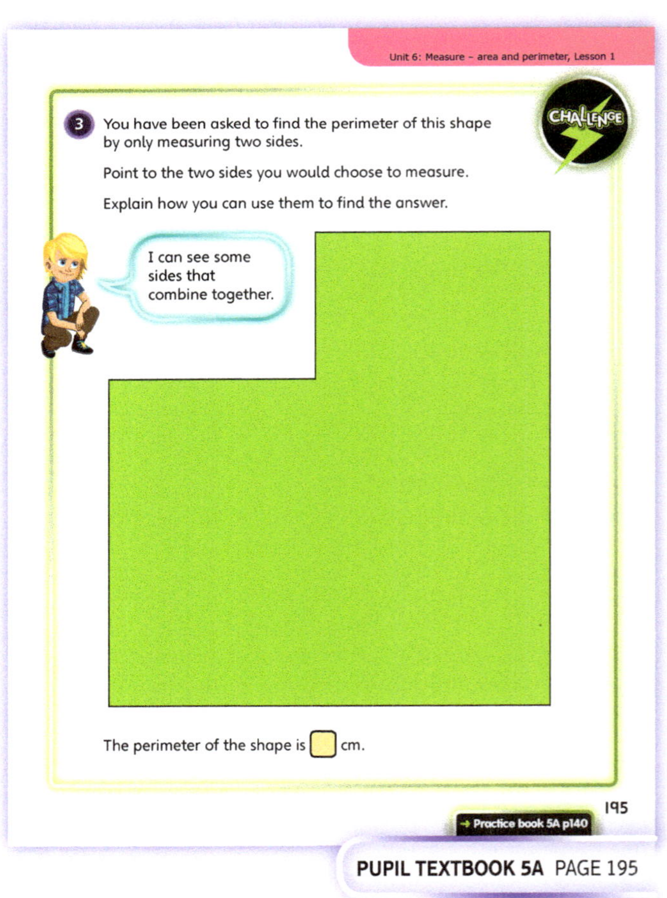

PUPIL TEXTBOOK 5A PAGE 195

Practice

WAYS OF WORKING Independent thinking

IN FOCUS Use questions ② and ③ to address two misconceptions about perimeter: i) larger shapes have a larger perimeter (question ②) and ii) when two shapes are put together, the resulting composite shape's perimeter will be the total of the two separate perimeters (question ③).

STRENGTHEN To strengthen children's understanding of perimeter as being the total of all the side lengths, provide children with lengths of string. They can rotate these around each shape, marking off each side length as they do so. The perimeter will be displayed as the final length, which children can then measure using a ruler. This strategy will also enable children to compare perimeters.

DEEPEN Question ⑤ challenges children to decide whether statements about perimeter are true or false, giving their reasons. Place emphasis on the reasoning part of this question, encouraging children to share their responses with peers. Extend this question by asking children to write their own true or false statements.

THINK DIFFERENTLY Question ③ explores the perimeter of a composite shape made from two rectangles. Can children predict whether or not Lee is correct? Can children explain why Lee is incorrect? Encourage children to see that the lengths of the sides that are joined are not part of the perimeter – they are inside the shape not outside it.

ASSESSMENT CHECKPOINT Children should be able to calculate perimeter by measuring the sides of rectilinear shapes. Can children identify which sides need to be included in a shape's perimeter, particularly in composite shapes, where some sides are inside the shape? Do they also understand that not every side needs to be measured?

ANSWERS Answers for the **Practice** part of the lesson appear in the separate **Practice and Reflect answer guide**.

Reflect

WAYS OF WORKING Independent thinking

IN FOCUS This question reveals children's understanding of strategies for measuring the perimeter of rectilinear shapes. The aim of the question is for children to generalise about how to find a perimeter without having a specific perimeter to find.

ASSESSMENT CHECKPOINT Look for children who describe finding the perimeter by measuring the sides. Note those children who include the fact that they need not measure every side and who use reasoning to explain which sides they would measure and how they would use them to help. Can children generalise about finding the perimeter of rectangles using doubling?

ANSWERS Answers for the **Reflect** part of the lesson appear in the separate **Practice and Reflect answer guide**.

After the lesson ⏸

- How could you provide children with practical opportunities to explore finding the perimeter of shapes in their school environment?
- Do children understand that they do not always need to measure every side of a rectilinear shape?
- Which mathematical ideas and processes did children develop during the lesson on measuring perimeter?

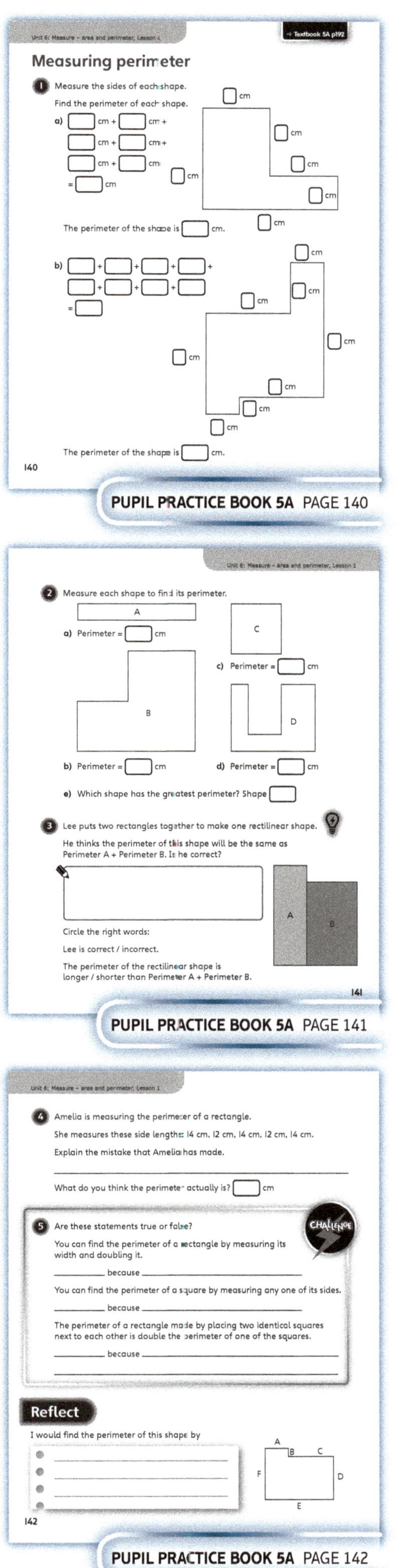

PUPIL PRACTICE BOOK 5A PAGE 140

PUPIL PRACTICE BOOK 5A PAGE 141

PUPIL PRACTICE BOOK 5A PAGE 142

Calculating perimeter ❶

Learning focus

In this lesson, children will calculate the perimeter of rectilinear shapes in centimetres and metres. They will also use a shape's perimeter to derive its dimensions.

Small steps

→ Previous step: Measuring perimeter
→ **This step: Calculating perimeter (1)**
→ Next step: Calculating perimeter (2)

NATIONAL CURRICULUM LINKS

Year 5 Measurement

Measure and calculate the perimeter of composite rectilinear shapes in centimetres and metres.

ASSESSING MASTERY

Children can calculate the perimeter of squares and rectangles, applying their understanding to solve related problems. They understand how to find the perimeter of a rectangle when given the length and the width and how to find the perimeter of a square when given one side length. Children can also apply their knowledge of perimeter to work backwards to suggest the dimensions of a shape by using its perimeter.

COMMON MISCONCEPTIONS

Children may think they need all the measurements of a rectilinear shape in order to find its perimeter. Ask:
• *What so you know about the properties of a square/rectangle? What is the perimeter of half of the rectangle? Which two sides are the same length as this side?*

STRENGTHENING UNDERSTANDING

Ensure children are confident in calculating the perimeter of squares and rectangles before exploring the perimeter of composite shapes. Allow children to make composite shapes using a combination of two or more squares/rectangles. Let them sketch and label their shape on squared paper before working out the perimeter. Ask: *Do any of the side lengths combine to equal the length of one of the other sides? Which sides of the rectangles are now not part of the perimeter?*

GOING DEEPER

Challenge children to devise their own perimeter problems for others to solve. Remind them to consider the information that is needed to solve the problem and to add a layer of reasoning where possible. For example, a problem may involve a shape made from squares where only the square dimensions are given, or it may involve a rectangle where only the length is given and a clue about how to find the width.

KEY LANGUAGE

In lesson: perimeter, length, width, sides, distance, **brackets**, operations, square, rectangle, doubling

Other language used by the teacher: rectilinear shape, composite shape, dimension

STRUCTURES AND REPRESENTATIONS

bar models, 2D shapes

RESOURCES

Mandatory: 2D squares and rectangles, rulers

Optional: paper straws, string, squared paper

 In the eTextbook of this lesson, you will find interactive links to a selection of teaching tools.

Before you teach 

• Do children understand what the perimeter of a shape is and know a way to work it out?
• Do children remember the properties of squares and rectangles?
• How can you improve the teaching of problem solving through the lesson?

Discover

 Pair work

ASK

- Question **1** a): *How can you tell what shape the pitch is just by reading the children's statements?*
- Question **1** a): *Do you have all the information you need to work out the perimeter?*
- Question **1** b): *What do you know about the sides of a square?*

IN FOCUS When discussing the picture, help children to visualise the shapes they are dealing with by encouraging them to imagine that they are looking down on the scene. What does the pitch look like from above? Challenge children to represent the football pitch and the playground pictorially. Ask: *What shapes will you need to draw? How will you label your diagram?*

PRACTICAL TIPS Prior to the lesson, give children practical experience of running around a rectangular pitch or court. Provide opportunities for them to measure the length and width using trundle wheels in order to calculate the perimeter.

ANSWERS

Question **1** a): The perimeter of the football pitch is 340 metres.

Question **1** b): The length of the playground is 50 m.

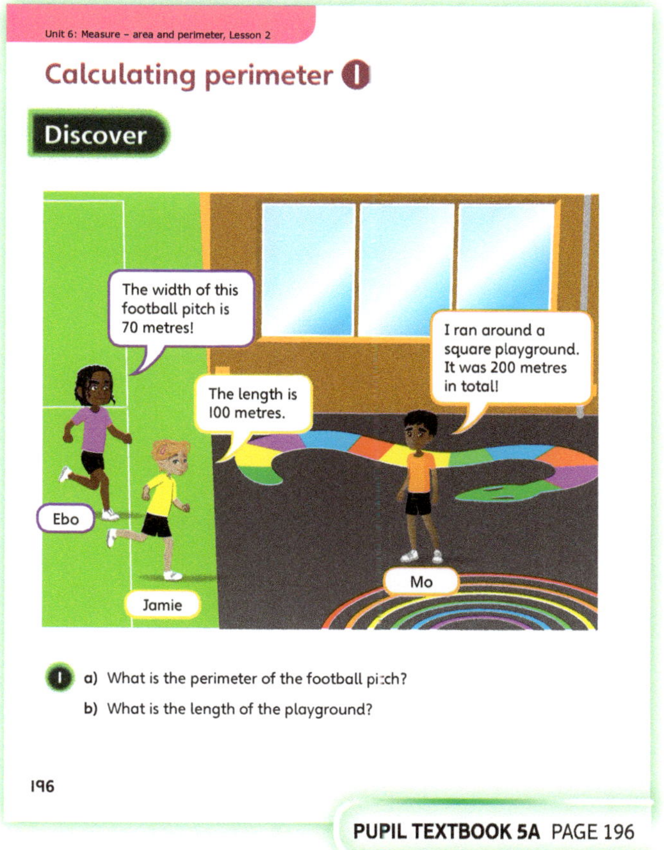

PUPIL TEXTBOOK 5A PAGE 196

Share

 Whole class teacher led

ASK

- Question **1** a): *What do you know about the sides of a rectangle? How many measurements do you need to be able to calculate the perimeter of a rectangle?*
- Question **1** a): *Can you think of more than one way to calculate the perimeter of the football pitch?*
- Question **1** a): *Can you see how the brackets help set out the calculation?*
- Question **1** b): *How many measurements do you need to be able to calculate the perimeter of a square?*
- Question **1** b): *How can you work backwards from a square's perimeter to find its length?*

IN FOCUS The focus in question **1** b) is to work backwards from a given perimeter to derive a square's dimensions. Use the bar model to ensure that children understand that a square's perimeter is made up of four equal sides. Alter the perimeter of the playground to allow children to suggest how they would find the answer for different perimeters. Ensure that children understand why this method works for squares, but not rectangles (i.e. because all four sides of a square are equal).

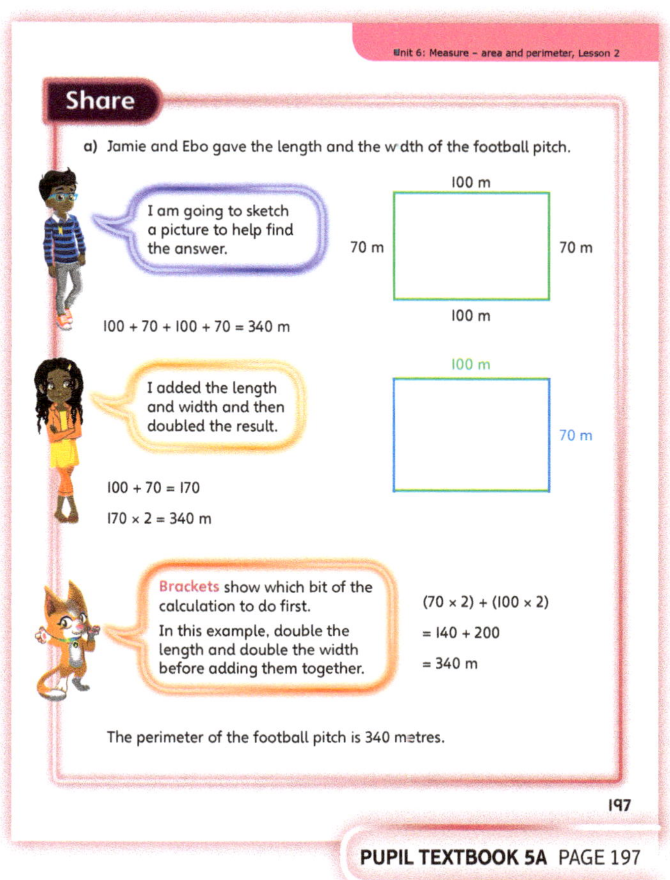

PUPIL TEXTBOOK 5A PAGE 197

Think together

WAYS OF WORKING Whole class teacher led (I do, We do, You do)

ASK

- Question ❶: *Do you have enough information to calculate the perimeter of the pitch? How is this multiplication method the same as doubling? Why is it helpful to use brackets in the calculation?*
- Question ❷: *What is the same and what is different about questions ❷ a) and ❷ b)? Why does part a) need multiplication to find the answer but part b) need division? Which times-table do you think it might be useful to know when working with the perimeter of the square?*

IN FOCUS The method shown for question ❶ a) uses brackets to denote the order of operations. Discuss this with children. Ask: *What would the calculation look like without brackets? Why do you think they are important?* Talk about the alternative doubling method and how brackets would be used here too: $(l + w) \times 2$. For further practice, write similar examples and challenge children to draw the brackets where they think they belong.

STRENGTHEN Provide children with paper straws to model the problems in questions ❶ and ❷ a). Ask: *How many paper straws do you need to model the shape in the question? So how can you use the length of one side to find the perimeter of a square?* Ask children to draw squares on squared paper, labelling the length of the sides and the perimeter. Ask: *What is the connection between the perimeter and the length of one side? How can you use the perimeter to find one side length of a square?*

DEEPEN Following on from question ❸, ask children to explore making shapes with four squares, finding and recording the perimeter of each different shape. Ask children to put the shapes in order of perimeter, shortest to longest. Can they reason about which shapes give the shortest/longest perimeters?

ASSESSMENT CHECKPOINT At this point, children should be able to confidently describe how to use the given dimensions of a square or rectangle to derive its perimeter. They should also be able to work from a given perimeter to find the side length of a square.

ANSWERS

Question ❶: $(120 \times 2) + (70 \times 2)$
= 240 + 140
= 380
The perimeter of the rugby pitch is 380 metres.

Question ❷ a): $40 \times 4 = 160$
The perimeter of the car park is 160 metres.

Question ❷ b): $80 \div 4 = 20$
The length of the lawn is 20 metres.

Question ❸: Lee is correct. The perimeter of the rectangle is the same as six times the length of the square. Two of the squares' sides are now inside the rectangle.

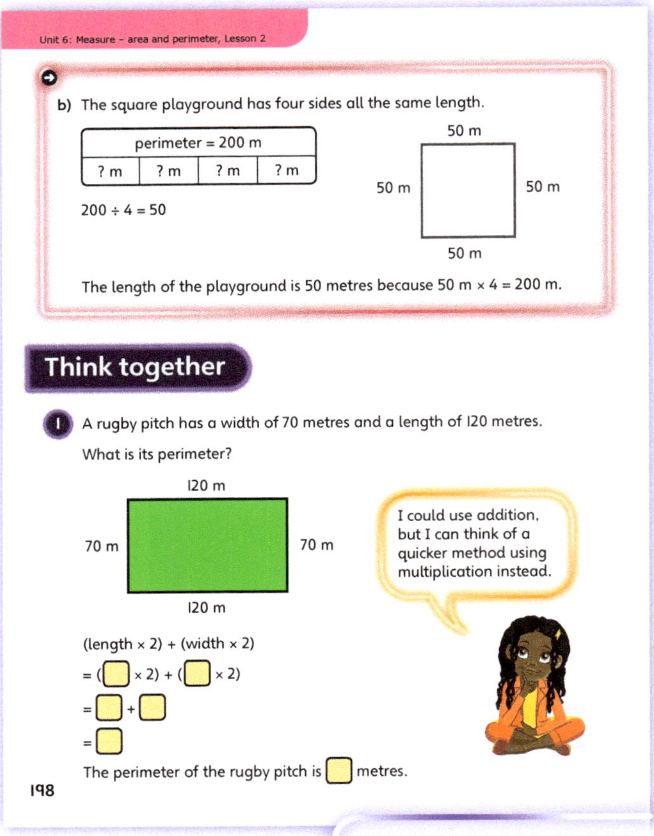

PUPIL TEXTBOOK 5A PAGE 198

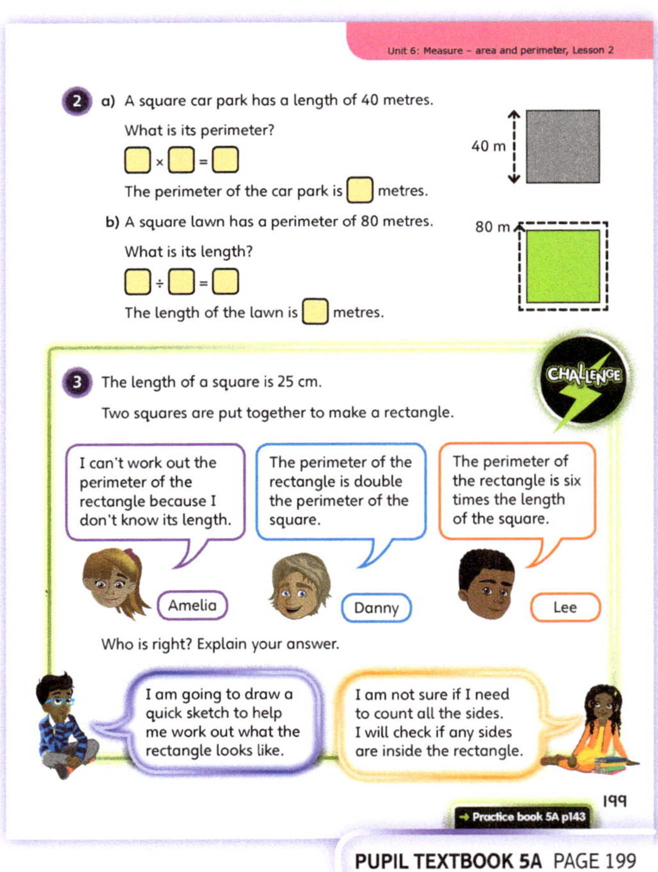

PUPIL TEXTBOOK 5A PAGE 199

Practice

WAYS OF WORKING Independent thinking

IN FOCUS Questions ❶ and ❷ focus on using doubling to work out the perimeters of rectangles by gradually removing the scaffolding and then presenting the shapes in different orientations.

STRENGTHEN To support children in question ❸, provide 10 cm × 10 cm square card tiles for them to physically make the shapes. Children should then draw around their tiles, possibly on 1 cm-squared paper, so that they are left with the composite shape. Ask: *How many sides do you need to count to work out the perimeter?* Ask them to label these sides 10 cm.

DEEPEN Ask children to explore other possible dimensions of a rectangle given its perimeter. Can they find all the whole number solutions? Make sure that you use even numbers and at least one multiple of 4 so that a square is a possible solution.

THINK DIFFERENTLY In question ❺, children are given a perimeter problem without the use of the word 'perimeter'. They are expected to recognise that the length of the piece of wire is equal to its perimeter. From this, they should apply their knowledge of the properties of squares and division to derive the length of one side.

ASSESSMENT CHECKPOINT Can children calculate the perimeter of composite rectilinear shapes where not all the dimensions are given and where the diagrams are not drawn to scale? Children should also be able to confidently use a shape's perimeter to find its dimensions or possible dimensions.

ANSWERS Answers for the **Practice** part of the lesson appear in the separate **Practice and Reflect answer guide**.

Reflect

WAYS OF WORKING Independent thinking

IN FOCUS The focus here is on showing that there are different methods of finding the perimeter of a rectangle. Discuss which method is the more efficient (using doubling). Ask: *Why is Jamie's method not correct? Can you think of another method that would also work? [A clue: it is similar to Max's method.]*

ASSESSMENT CHECKPOINT Can children identify both correct methods (adding all four sides; doubling the length, doubling the width and adding them together)? Can children explain which is the more efficient method?

ANSWERS Answers to the **Reflect** part of the lesson appear in a separate **Practice and Reflect answer guide**.

After the lesson ⏸

- How did children respond to the materials and approaches used? What would you do differently next time, particularly when modelling more complex perimeter problems in different ways?
- Can children work out the perimeter of everyday objects found around school (whiteboard, door, book, table), given the width and length? Do children use the correct unit (cm or m) of measure?

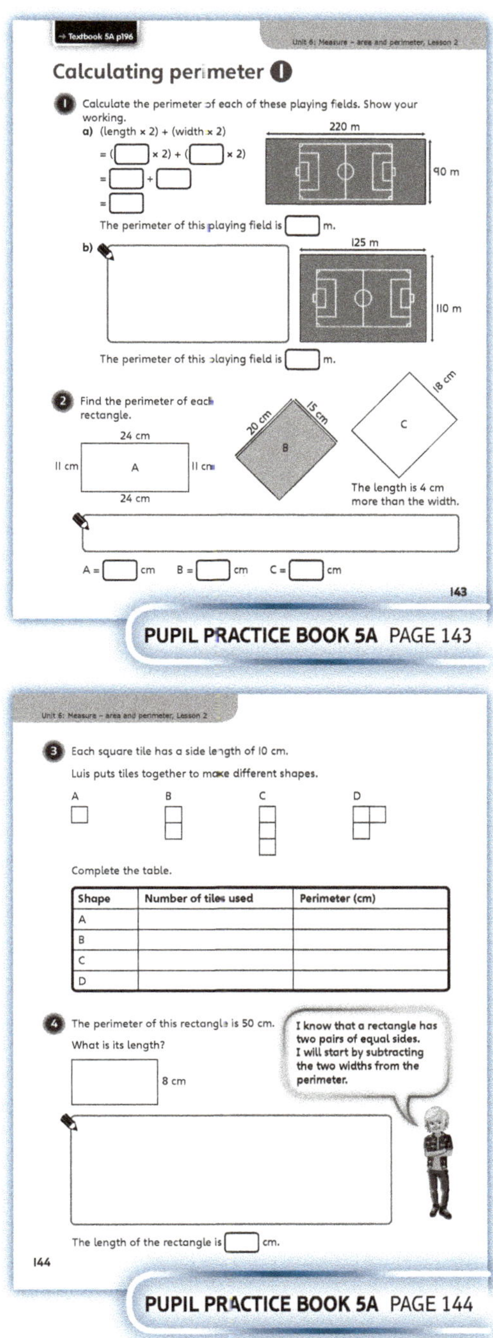

PUPIL PRACTICE BOOK 5A PAGE 143

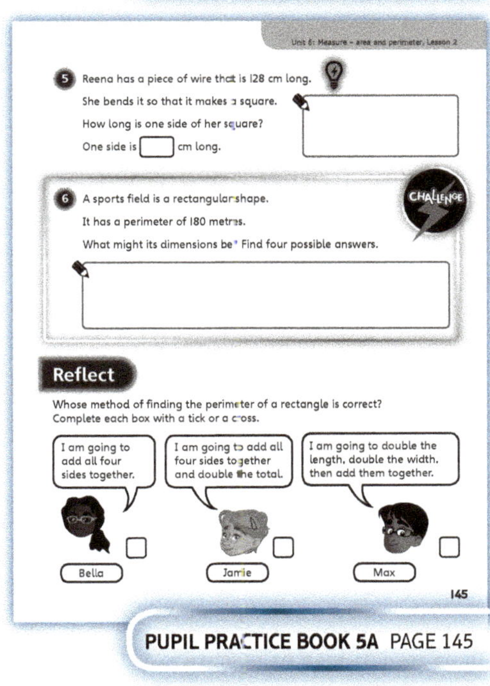

PUPIL PRACTICE BOOK 5A PAGE 144

PUPIL PRACTICE BOOK 5A PAGE 145

Calculating perimeter ❷

Learning focus

In this lesson, children will apply their knowledge of perimeter to solve problems, including calculating unknown lengths of composite rectilinear shapes.

Small steps

→ Previous step: Calculating perimeter (1)
→ **This step: Calculating perimeter (2)**
→ Next step: Calculating area (1)

NATIONAL CURRICULUM LINKS

Year 5 Measurement

Measure and calculate the perimeter of composite rectilinear shapes in centimetres and metres.

ASSESSING MASTERY

Children can use side lengths and perimeters they are given to derive unknown side lengths. They can apply their knowledge of the properties of composite shapes to make statements about their dimensions.

COMMON MISCONCEPTIONS

Children may look at a partially or sparsely-labelled rectilinear shape and think that they simply do not have enough information to find the length of an unknown side. Ask:

• *Have you been given enough information to find the length of this side or do you need more? Can you explain how to use the measurements you have been given to find the answer?*

STRENGTHENING UNDERSTANDING

Provide children with opportunities to model shapes concretely. For example, ask children to cut out rectilinear shapes and then fold them (either vertically or horizontally) to physically prove the equivalence of the different side lengths. Ask: *Which sides are the same length? If I could only tell you the length of one of these sides, which would you want to know? Why?*

GOING DEEPER

Challenge children to set their own problems using rectilinear shapes. Establish that their question may ask for a shape's perimeter or one (or more) of its unknown sides. Ask: *What is the least information you need to provide for your partner to be able find the answer?* Encourage children to use composite shapes using two identical rectangles where only the dimensions of the rectangle are given (as in **Think together** question ❸).

KEY LANGUAGE

In lesson: perimeter, length, width, sides, distance, square, rectangle, double, rectangular, dimensions, +, –, ×, ÷, =, calculate

Other language used by the teacher: rectilinear shape, composite shape, measurements, properties, derive

STRUCTURES AND REPRESENTATIONS

bar models, representations of 2D rectilinear shapes

RESOURCES

Mandatory: 2D squares and rectangles, rulers

Optional: paper straws (or similar), coloured pens, rectilinear shapes to cut out

 In the eTextbook of this lesson, you will find interactive links to a selection of teaching tools.

Before you teach 

• How will you help children to adapt their strategies when a problem's constraints are changed?
• What opportunities can you identify to make the problem-solving in this lesson as practical as possible?

Discover

WAYS OF WORKING Pair work

ASK

- Question **1** b): *How many corners would you have to turn to drive all around once? How many straight lengths of road would you have to drive along? What shape is the complete road?*
- Question **1** b): *Are any of the straight lengths of road the same distance?*

IN FOCUS Question **1** a) asks children to problem solve how to find the length of one side of the rectilinear shape by using the information they have been given in the picture. Children will need to identify which of the four operations will help them answer the question (subtraction). For question **1** b), children need to understand they are measuring the distance given by the orange outer line not the white dashed line.

PRACTICAL TIPS Mark out a large rectilinear shape with chalk on the ground outside and encourage children to model the problem by 'being' the cars. Challenge them to demonstrate the route Car A takes as it travels once around the road back to its driveway. Ask: *What is the name given to this distance?*

Discuss together children's experiences of walking around their neighbourhood. Are there any pathways or roads near where children live or around the school's perimeter that make a rectilinear shape if they walk around them? Walk around the school building or use a plan of the school to help children to visualise the problem.

ANSWERS

Question **1** a): The length of the queue of three vehicles is 35 metres.

Question **1** b): Car A travels 450 metres.

Share

WAYS OF WORKING Whole class teacher led

ASK

- Question **1** a): *What has been added to this diagram to help you answer the question?*
- Question **1** a): *Can you explain why you need to use subtraction to find the length of the queue of vehicles?*
- Question **1** b): *How do the different coloured lines added to this diagram help to answer the question?*
- Question **1** b): *Astrid mentions doubling. What needs to be doubled? Why?*

IN FOCUS It is important that children understand how they can use dimensions they are given to derive those that are missing. Ask: *Can you write an addition and a subtraction that both feature the missing side length? (95 – 60 = ? and 60 + ? = 95).* Point out how the colours of opposite sides and the arrows are used to show the connections between equal lengths.

PUPIL TEXTBOOK 5A PAGE 200

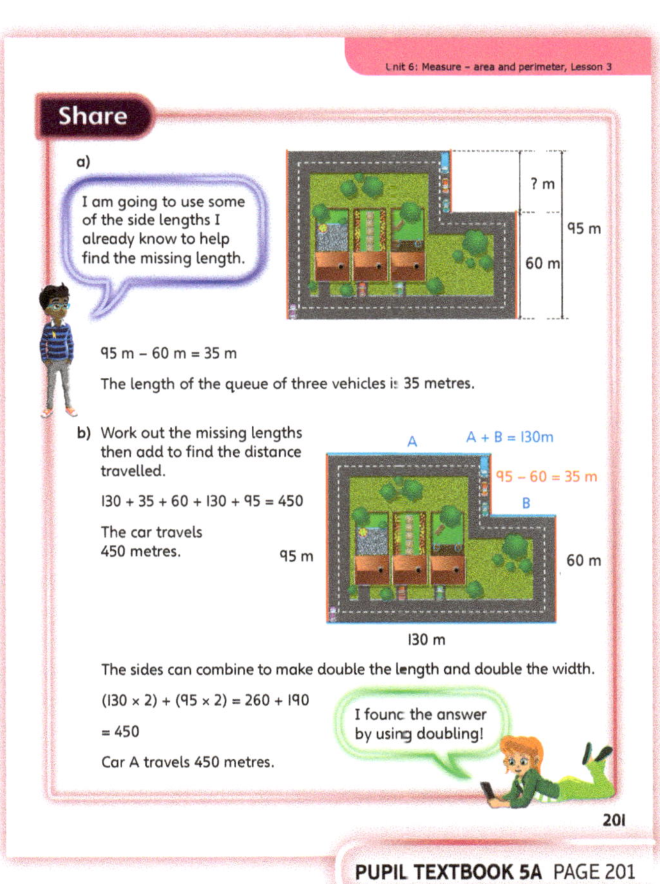

PUPIL TEXTBOOK 5A PAGE 201

Think together

WAYS OF WORKING Whole class teacher led (I do, We do, You do)

ASK

- Question **1**: *How is this shape the same/how is it different to the road in the **Discover** section?*
- Question **1**: *Without using a ruler, how could you prove which sides are the same length?*
- Question **2**: *Which sides of this shape are the same length? How can you find the missing lengths?*
- Question **3**: *Which side lengths do you know? Which one do you not know?*

IN FOCUS Question **2** gives children experience with more complex rectilinear shapes. Assure children that they have enough information to solve the problem. Consider each unknown side and ask: *Do you know what this length is?* Give children opportunities to discuss their reasoning in pairs before sharing with the class. The top horizontal lines could have various values but their total matches the bottom horizontal: 60 m.

STRENGTHEN It is important that children recognise that they do not need to find out each individual side length. Allow children to draw the shapes and colour code the vertical and horizontal lines to find equivalent lengths, as modelled in the **Share** section.

DEEPEN Ask children to explain their methods for answering question **3**. Did they think of the shape as a 6-sided rectilinear shape or as two separate rectangles? How did they work out the missing dimension? Ask children to explore putting together two or more identical rectangles in different arrangements and working out each perimeter. Ask: *What is the longest/shortest perimeter you can make with two/three identical rectangles?*

ASSESSMENT CHECKPOINT Can children derive unknown side lengths of rectilinear shapes given the information supplied? Can children explain how they know the value of a missing length or how to calculate a shape's perimeter from the limited information given?

ANSWERS

Question **1** a): A + C = 250 m
      B + D = 300 m

Question **1** b): (250 × 2) + (300 × 2)
      = 500 + 600
      = 1,100 m

Question **2**: The perimeter of the playground is 330 m.

Question **3**: The perimeter of the shape is 198 cm
      (or 1·98 m).

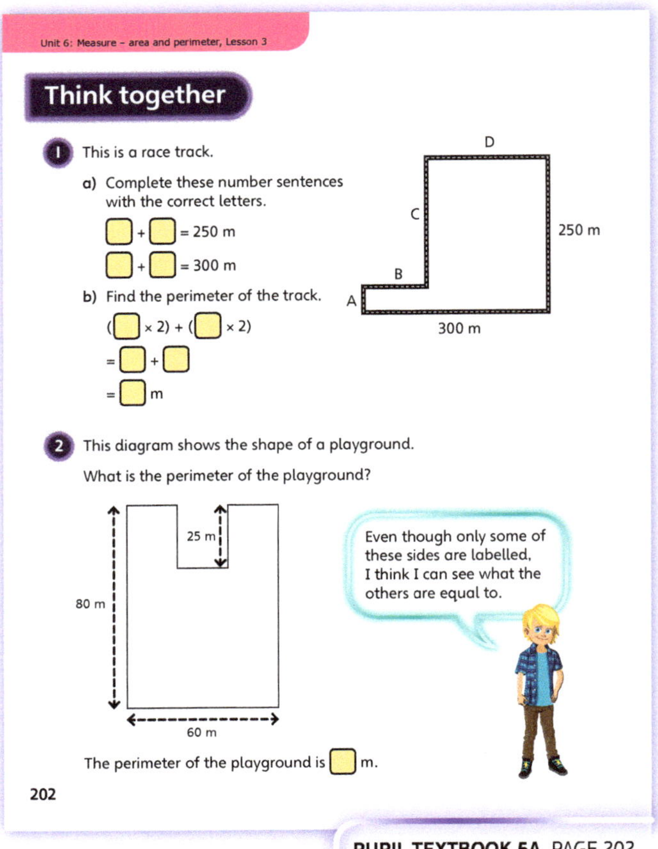

PUPIL TEXTBOOK 5A PAGE 202

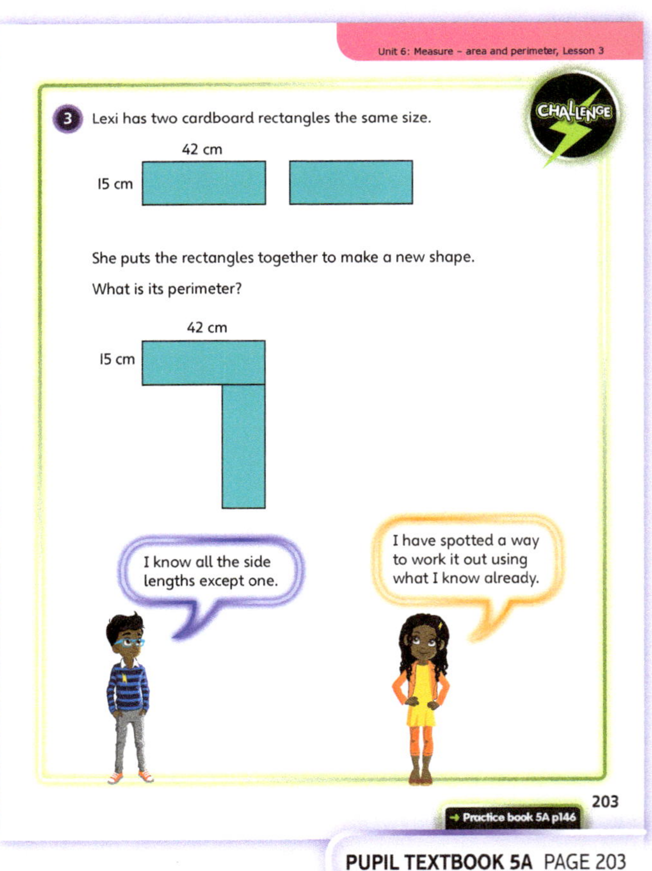

PUPIL TEXTBOOK 5A PAGE 203

Practice

WAYS OF WORKING Independent thinking

IN FOCUS All questions focus on deriving an unknown length from the information supplied in order to work out perimeters of rectilinear shapes. The support given is gradually reduced and the complexity increased. Discuss the worked example in question **3**, ensuring children understand the methodology and providing further examples where necessary. Highlighting the parallel/equal sides in different colours as before may help.

STRENGTHEN Encourage children to represent similar shapes concretely by cutting paper straws. Ask: *Can you colour code the sides of your shape so that if two side lengths equal another side length, they are all the same colour? Can you move the sides of your shape to prove that the sides are equal?* Encourage children to physically group their colour coded straws, making two lengths of one colour and two of another. Discuss how this illustrates the doubling method of finding perimeters of rectilinear shapes.

DEEPEN Ask children to extend question **5**: *How many different shapes can you make using exactly six square tiles? Ask: Is it always, sometimes or never true that shapes made from the same number of tiles will have the same perimeter? How many different perimeters can you make with exactly six square tiles? Which types of arrangement produce longer/shorter perimeters? Can you make a shape using eight square tiles that has a shorter perimeter than one you made with six tiles?*

ASSESSMENT CHECKPOINT Can children derive unknown side lengths of rectilinear shapes given the information supplied? Can children explain how they know the value of a missing length or how to calculate a shape's perimeter from the limited information given? Are children beginning to reason about the type of shapes that have longer/shorter perimeters?

ANSWERS Answers for the **Practice** part of the lesson appear in the separate **Practice and Reflect answer guide**.

Reflect

WAYS OF WORKING Independent thinking

IN FOCUS This question is different in that it gives children the four shortest side lengths and leaves the two longest sides unlabelled. Children should identify the pairs of sides that are equal to the unknown side lengths and use this to calculate the perimeter. Encourage them to annotate the diagram. Ask: *Could you always work out the perimeter, no matter what side lengths you were given?*

ASSESSMENT CHECKPOINT Can children explain how to derive the two missing side lengths in order to calculate the perimeter (which is 146 cm)? Do children then use addition or do they use doubling?

ANSWERS Answers for the **Reflect** part of the lesson appear in the separate **Practice and Reflect answer guide**.

After the lesson ⏸

- How well did the prompts and questions promote learning about finding perimeters of rectilinear shapes where not all sides are labelled?
- How well do you feel children have achieved the curriculum objective of being able to measure and calculate the perimeter of composite rectilinear shapes in centimetres and metres?
- What further opportunities to practise this concept could you provide in other curriculum areas, such as DT or Art?

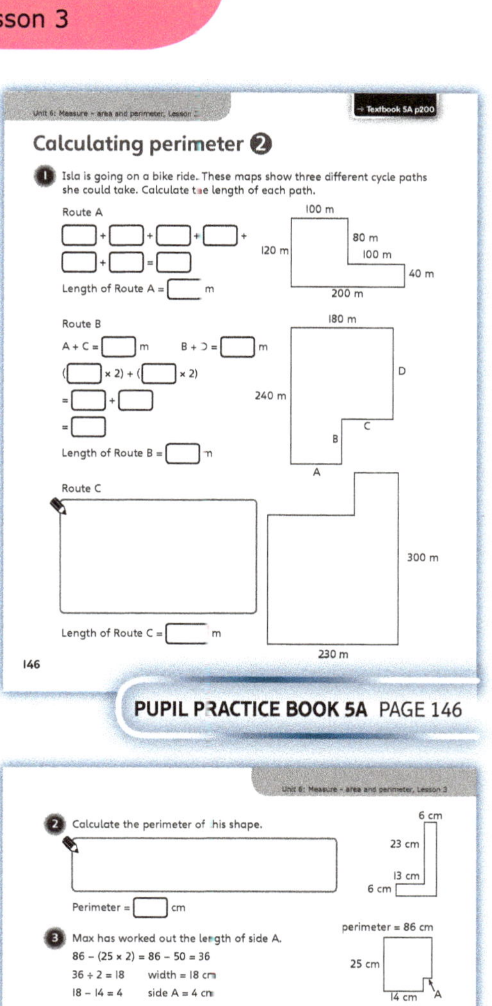

PUPIL PRACTICE BOOK 5A PAGE 146

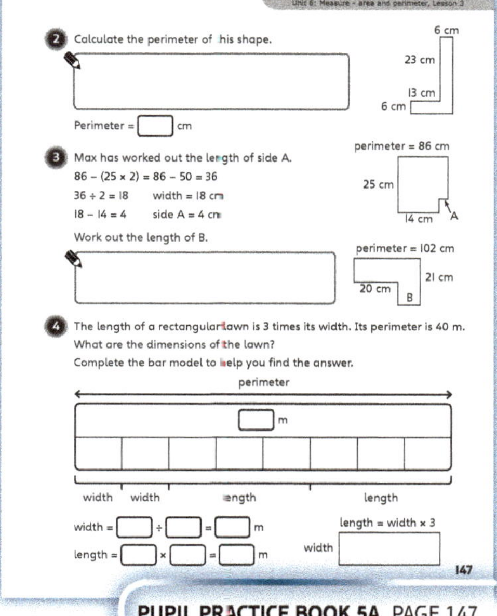

PUPIL PRACTICE BOOK 5A PAGE 147

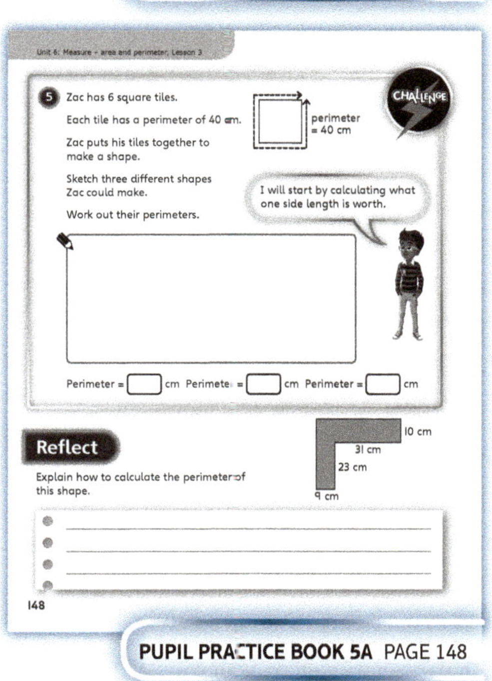

PUPIL PRACTICE BOOK 5A PAGE 148

Calculating area ❶

Learning focus

In this lesson, children will consolidate their knowledge of the area of rectangles by calculating area using square centimetres and square metres from scale drawings.

Small steps

→ Previous step: Calculating perimeter (2)
→ **This step: Calculating area (1)**
→ Next step: Calculating area (2)

NATIONAL CURRICULUM LINKS

Year 5 Measurement
- Calculate and compare the area of rectangles (including squares), and including using standard units, square centimetres (cm^2) and square metres (m^2) and estimate the area of irregular shapes.
- Solve problems involving multiplication and division, including scaling by simple fractions and problems involving simple rates.

ASSESSING MASTERY

Children can find the area of rectangles using cm^2 and m^2. Children are able to use scale drawings with confidence when finding area in square metres.

COMMON MISCONCEPTIONS

Children may ignore the scale information presented and simply count the squares to find the area and may assume that the measurement is always in cm^2. Ask:
- *What information does the scale tell you about the area of the rectangle? How can you use this to help you find the area? What unit of measurement is being used to show the area?*

STRENGTHENING UNDERSTANDING

Ask children, in small groups, to tape together pieces of newspaper to show 1 square metre (to fit inside a square made with four metre sticks) and to compare it with 1 square centimetre shown on 1 cm^2 grid paper. Ask children to show 3 m^2 using their newspaper square metres. Ask children to show 3 cm^2 on grid paper. Ask: *Does the scale matter?* Draw a 2 by 3 rectangle on squared paper and vary the keys: 1 square = 1 cm^2, 4 m^2, 9 cm^2. Work out each area ensuring children use the correct unit of measure and understand that changing the scale changes the overall area.

GOING DEEPER

Provide opportunities to explore the effect that a change in scale has, using online maps or a variety of physical maps (large and small scale, and preferably of the local area) so children can see how much detail each one shows. Draw attention to how the scale is shown on maps. Set children the challenge of rewriting some of the questions in their **Practice book**, altering the scale on each of them. Ask: *Will the answer be larger or smaller? Why?*

KEY LANGUAGE

In lesson: area, **square metres (m^2)**, **square centimetres (cm^2)**, scale, rectangle, square, actual area/size, rectangular

Other language used by the teacher: key, convert, calculate, compare, diagram, vary, unit of measure

STRUCTURES AND REPRESENTATIONS

2D shapes

RESOURCES

Mandatory: metre sticks, 1 cm-squared paper

Optional: maps of the local area, newspapers, base 10 equipment, card squares

 In the eTextbook of this lesson, you will find interactive links to a selection of teaching tools.

Before you teach 

- Are there any ways that you can adapt this lesson to link it to other lessons or curriculum work?
- Are there any additional strategies you could use to support children's understanding of scale?

Discover

WAYS OF WORKING Pair work

ASK

- Question ❶ a): *What shapes can you see on the map?*
- Question ❶ a): *What is the length and the width of the Top Secret HQ?*
- Question ❶ a): *How is the representation of area different to how you have seen it shown before?*
- Question ❶ a): *What is the actual area in real life of one square on the map? What about two squares? Or three squares?*

IN FOCUS The picture in the **Discover** section introduces a new idea that children may be familiar with from real life – namely the fact that objects and distances shown on maps are drawn to a scale. This means that we can represent their actual size without having to draw them at that size. Children might enjoy pondering the question: *What would happen if you could not draw maps to scale?*

PRACTICAL TIPS To help children visualise a square metre, ask them to use metre rulers to mark out a 1 m × 1 m square, discussing that this is a unit of measurement used to measure large areas. Children could be encouraged to imagine a huge piece of squared paper with square metres marked on it. Ask: *How is using square metres to measure area different from using square centimetres? Why do you need a key to help you?*

ANSWERS

Question ❶ a): The actual area of the Top Secret HQ is 270 m².

Question ❶ b): The actual area of the Hidden Vault is 150 m².

Share

WAYS OF WORKING Whole class teacher led

ASK

- Question ❶ a): *What does the term 'actual area' mean?*
- Question ❶ a): *In Year 4, you found area by counting squares. Is that how you should find the area here?*
- Question ❶ a): *What does it mean for the map to be drawn to 'scale'? Where have you seen this idea used in real life?*
- Question ❶ b): *How is it possible to work out the area of the Hidden Vault when part of the map is missing?*

IN FOCUS These questions are important because they deal with the concept of 'actual area'. It may be worth spending time considering maps based on real life (for example, a map of the school grounds on a squared grid). Ask: *Is this the actual size of the school?* Ensure children understand that each square on the map represents a much larger square in real life, so the actual area of the school depends on the scale value. Work through question ❶ a) using the methodology shown (counting squares, then multiplying by their scale).

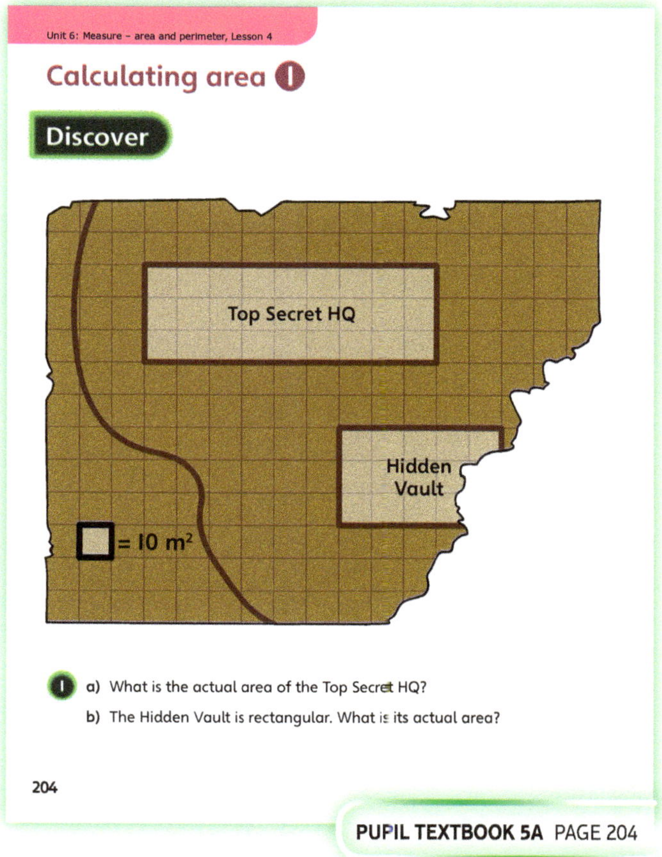

PUPIL TEXTBOOK 5A PAGE 204

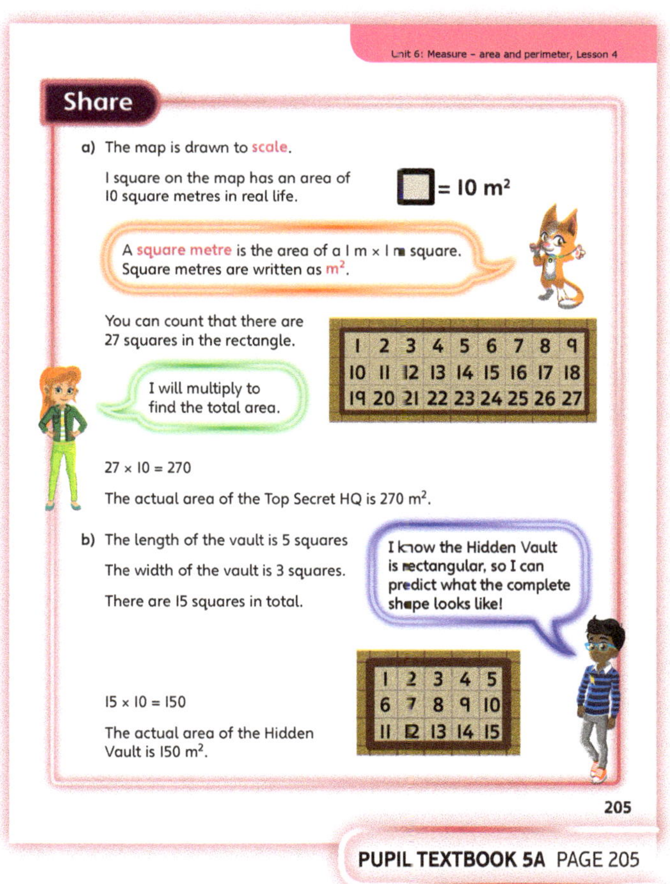

PUPIL TEXTBOOK 5A PAGE 205

Think together

WAYS OF WORKING Whole class teacher led (I do, We do, You do)

ASK

- Question ❶: *How many squares does the Underground Base take up on the map? Do you know a quick way to work this out? What is each square worth in real life?*
- Question ❷: *What are the two steps needed to calculate the area of each rectangle? Is it always, sometimes or never true that a larger areas drawn on this grid have an area that is a multiple of 4? Shape C does not have any squares inside it. How can you calculate its area?*

IN FOCUS Use question ❶ to take children through the series of steps needed to calculate the actual area from a scale diagram. Emphasise the scale and consolidate understanding of the difference between the pictorial representation of the area of the rectangle (measured in map squares) and the area in real life (measured in m² according to the key).

STRENGTHEN Use base 10 equipment 100 squares to illustrate question ❶: encourage children to draw around their 100 squares to make a large version of the 2 × 6 rectangle. Then ask them to place their 100 square inside one of the squares and explain that each square represents 100 square metres (100 m²). Ask: *How many lots of 100 will you need to cover the rectangle? How many square metres is this?* The same strategy can be used for question ❷, replacing the 100 squares with squares of card folded into quarters (representing 4 cm²).

DEEPEN Encourage children to discuss and write a sentence about how different scales affect the overall actual area. Can they draw a series of simple scale diagrams including a scale key to show whether this statement is always, sometimes or never true: *The largest number of squares shows the largest area.* (Sometimes.)

ASSESSMENT CHECKPOINT Can children describe the steps needed to calculate area: counting squares, then multiplying by the scale value? Do children understand that a pictorial shape showing a scale reflects a different value in real life?

ANSWERS

Question ❶: The area of the rectangle on the map is made up of 12 squares.
Each square is worth 100 square metres.
You can find the actual area by calculating 12 × 100.
The actual area of the Underground Base is 1,200 m².

Question ❷: a) Shape A = 24 squares × 4 cm² = 96 cm²
b) Shape B = 18 squares × 4 cm² = 72 cm²
c) Shape C = 20 squares × 4 cm² = 80 cm²

Question ❸: Shape A = 16 squares × 10 m² = 160 m²
Shape B = 18 squares × 8 m² = 144 m²
Shape C = 32 squares × 5 m² = 160 m²
Shape B has the smallest actual area;
A and C both have the same actual area.

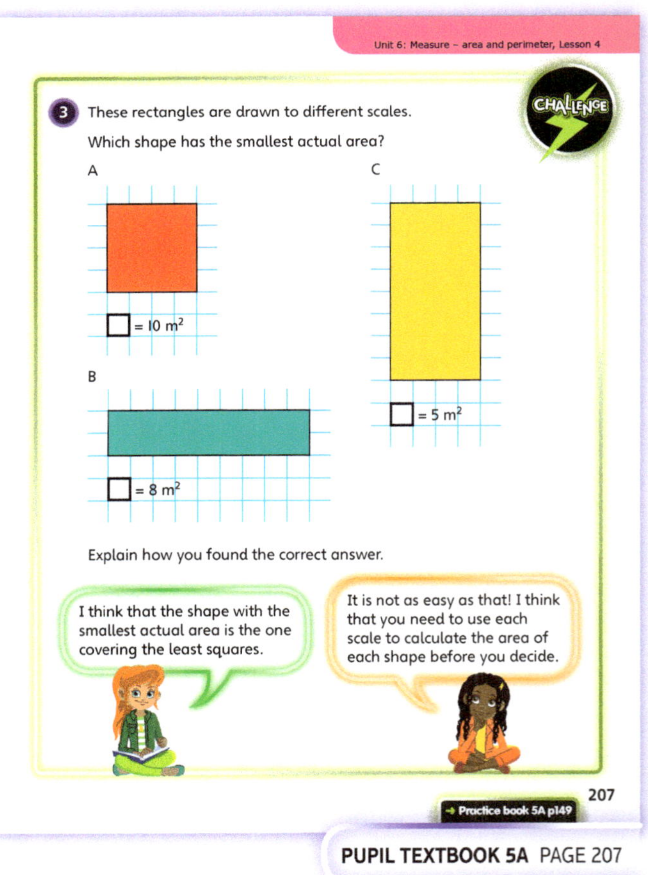

PUPIL TEXTBOOK 5A PAGE 206

PUPIL TEXTBOOK 5A PAGE 207

Practice

WAYS OF WORKING Independent thinking

IN FOCUS In question **3**, children are encouraged to explore the relationship between a rectangle's scale and its area. The rectangle has eight squares but the value of one square is varied. Children will also need to use division to work backwards from given areas to establish what one square is equal to.

STRENGTHEN Provide children with enlarged copies of the rectangles shown in the questions. Ask children to divide each rectangle into squares, writing the value of each square inside it or placing a plastic counter in each square to aid their counting. Children should tick each square or pick up a counter as they count in the relevant multiple.

DEEPEN Extend by giving children different areas and scales to explore, for example, 100 cm² with scales of 5 cm² and 10 cm². Ask: *Can you find more rectangles with a scale that gives an actual area of 100 cm²?* Where questions have the value of a square as 1, 4, 9 or 100 units (square numbers), can children work out what each side length is and the square's perimeter?

THINK DIFFERENTLY Question **4** encourages children to reverse the thinking done so far in this lesson and to work out the scale from a given area.

ASSESSMENT CHECKPOINT Can children find the area of rectangles by counting squares? Can children find the area of the actual object when presented with a scale diagram with a key?

ANSWERS Answers to the **Practice** part of the lesson appear in a separate **Practice and Reflect answer guide**.

Reflect

WAYS OF WORKING Independent thinking

IN FOCUS After children have discussed the scenario, ask them what is meant by it being a 'scale drawing' and how they know what each square is worth. Ask: *What method will you use to find the actual area of the room?*

ASSESSMENT CHECKPOINT Look for children who can explain how to find the rectangle's actual area, using the given scale to calculate in m². Watch for the misconception that all children need to do is count the squares (incorrectly answering 24 m²).

ANSWERS Answers to the **Reflect** part of the lesson appear in a separate **Practice and Reflect answer guide**.

After the lesson ⏸

- Did children understand how to use a given scale to find the actual area of rectangles in square metres?
- Are children still counting squares or using multiplication to find the number of squares in a rectangle?
- Are children ready to use the $l \times w$ formula for area presented in the next lesson?

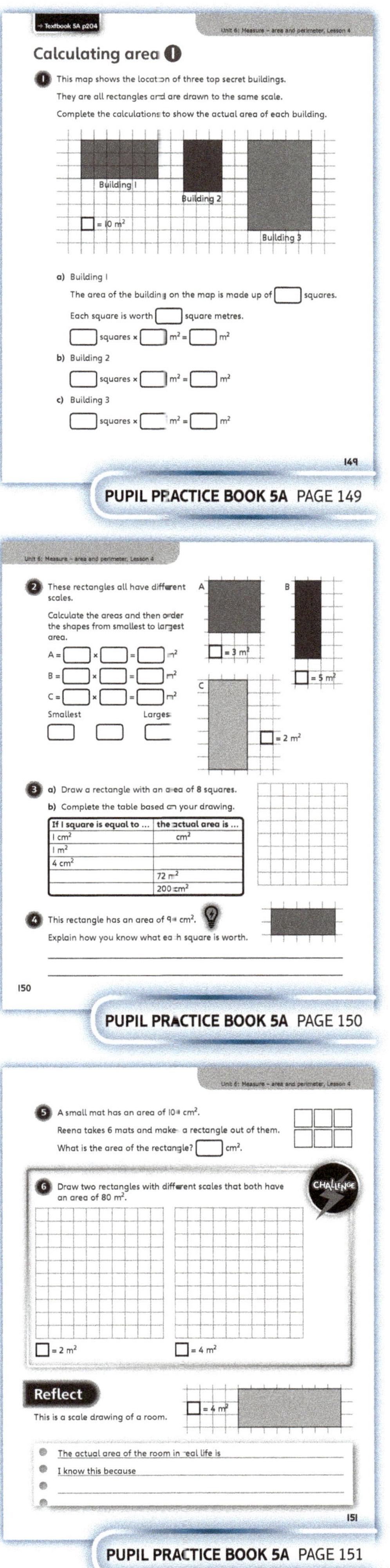

PUPIL PRACTICE BOOK 5A PAGE 149

PUPIL PRACTICE BOOK 5A PAGE 150

PUPIL PRACTICE BOOK 5A PAGE 151

Calculating area ❷

Learning focus

In this lesson, children will explore the relationship between a rectangle's length and width, and its area. They will link the number of squares to related arrays and use multiplication to derive the area.

Small steps

→ Previous step: Calculating area (1)
→ **This step: Calculating area (2)**
→ Next step: Comparing area

NATIONAL CURRICULUM LINKS

Year 5 Measurement

Calculate and compare the area of rectangles (including squares), including using standard units, square centimetres (cm^2) and square metres (m^2) and estimate the area of irregular shapes.

ASSESSING MASTERY

Children can confidently calculate the area of rectangles by multiplying their length by their width. They can use the *length × width = Area* formula correctly and can use their knowledge of arrays to explain why this can be used to calculate area.

COMMON MISCONCEPTIONS

Children may continue to rely on counting squares to find area despite the introduction of a more efficient method. They may apply the formula without understanding how it works. Ask:
• *What is the area of a shape? Why does multiplying a rectangle's length by its width result in its area?*

STRENGTHENING UNDERSTANDING

Help children to make the connection between a shape's length and width and its area by making arrays from objects or counters within squares. Ask children to find the total by counting up in multiples of the number in a row (length), then in the number of columns (width), finally multiplying the length by the width. Ask: *Which numbers do you need to multiply to get the area?* Turn the arrays through 90° to show that the width becomes the length but the area remains the same.

GOING DEEPER

Set an investigative challenge involving the length/width/area relationship. For example, ask: *Which area between 20 cm^2 and 25 cm^2 can be made with the most different rectangles?* (24 cm^2.) *How do you know?* Encourage children to apply their knowledge of factors and factor pairs as they explore the answer. Ask: *How can you organise your results in a clear way so you can see whether you have found all the possibilities?*

KEY LANGUAGE

In lesson: area, length, width, square centimetres (cm^2), square metres (m^2), array, rows, columns, table, factor, factor pair

Other language used by the teacher: measure, formula, efficient method, 90°

STRUCTURES AND REPRESENTATIONS

arrays, 2D shapes

RESOURCES

Optional: plastic counters, squared paper

 In the eTextbook of this lesson, you will find interactive links to a selection of teaching tools.

Before you teach ⏸

• Is children's knowledge of times tables secure?
• How might you scaffold questioning to help children reflect on their assumptions about area?
• Are children confident with the term 'square metres' from the last lesson?

Discover

WAYS OF WORKING Pair work

ASK

- Question ① a): *Can you show what one square metre looks like in real life?*
- Question ① a): *How is Amal trying to work out the area of his flower bed?*
- Question ① b): *What is the length and the width of the third flower bed? How do you know?*

IN FOCUS The focus in question ① a) is on Amal starting to use an inefficient method to find the area of his flower bed but realising there should be a quicker way. Ask for suggestions as to what that method might be. It is important that children understand that there is a more efficient way than counting each individual square.

PRACTICAL TIPS Give children the opportunity to visualise the problem using concrete representations. Mark out a grid of square metres on the ground (using chalk or masking tape) or use squared paper and counters. Ask children to place objects (for example, plastic cones) in each square metre, counting aloud. Ask: *Is there a quicker way to find out the total number of square metres?* Using a different coloured set of cones (counters) for each row will highlight that the grid is formed from several rows of the same amount (an array).

ANSWERS

Question ① a): The area of Amal's flower bed will be 20 m².

Question ① b): There are two possible rectangles with an
area of 8 m²:
Rectangle 1: length = 1 m, width = 8 m
(or vice versa)
Rectangle 2: length = 2 m, width = 4 m
(or vice versa)

Share

WAYS OF WORKING Whole class teacher led

ASK

- Question ① a): *Can you describe the flower bed using the words 'rows' and 'columns'?*
- Question ① a): *Name a different array that has an area of 20 m².*
- Question ① b): *1 × 8, 2 × 4, 4 × 2 and 8 × 1 all equal 8. Does this mean that four different rectangles can all have an area of 8 m²?*

IN FOCUS It is necessary that children make conceptual connections between rectangles split into squares and arrays that are used to represent multiplication facts. It may be worthwhile to ask children to represent the 5 × 4 array of squares in a different way (for example, using dots). Ask children to share their arrays before returning to the multiplication fact their arrays represent.

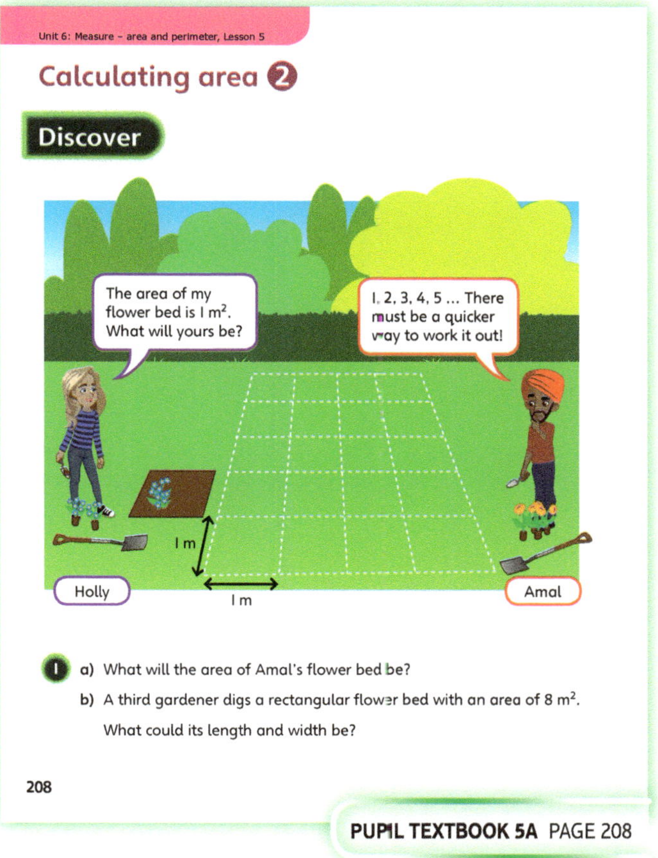

PUPIL TEXTBOOK 5A PAGE 208

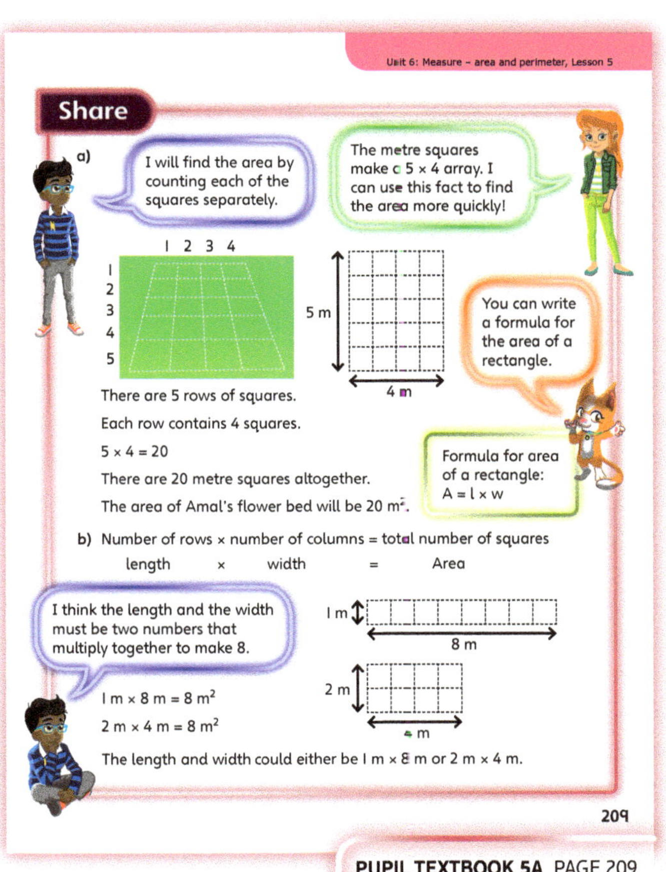

PUPIL TEXTBOOK 5A PAGE 209

Think together

WAYS OF WORKING Whole class teacher led (I do, We do, You do)

ASK

- Question **1**: *How does knowing the length and width help you to find the area?*
- Question **1**: *Where do the two numbers in the multiplication calculation come from?*
- Question **2**: *What units are you measuring in?*
- Question **2**: *Is it possible to calculate the area of the rectangle when not all the inner squares are shown?*
- Question **2**: *Does it matter which of the two numbers you call the length and which you call the width? Why or why not?*
- Question **3**: *What is a factor of a number? What is a factor pair of a number?*

IN FOCUS Questions **1** and **2** focus on using multiplication of length by width to find the area, with the level of scaffolding reduced in question **2**. Ask: *Is there a way to find the area of the rectangle in question **2** without drawing all the squares inside it? Would you still be able to find the answer if all you saw were the dimensions?*

STRENGTHEN If children are still counting individual squares, allow them to make arrays to match the rectangles shown in the questions. Encourage children to write the related multiplication facts and identify the factor pair of the array.

DEEPEN Use question **3** to explore the connection between the factors/factor pairs of 24 and the different rectangles with an area of 24 cm^2. Ask children to write down the dimensions of all their answers, then ask: *Can you arrange your rectangles so that they are in a clear order? What do you notice?* Note that a 4×6 rectangle is a 6×4 rectangle rotated 90°.

ASSESSMENT CHECKPOINT Are children confident in deriving the area of a rectangle by multiplying its length by its width? Can children explain this in terms of arrays of squares? Can some children see the connection between factor pairs and a given area?

ANSWERS

Question **1**: There are 3 rows of metre squares. Each row contains 6 squares.
3 × 6 = 18. There are 18 metre squares altogether. The area of the flower bed is 18 m^2.

Question **2**: 5 cm × 3 cm = 15 cm^2. The area of the rectangle is 15 cm^2.

Question **3**:

Length	Width	Area
24	1	24 cm^2
12	2	24 cm^2
6	4	24 cm^2
8	3	24 cm^2

Children's explanation should mention that finding the area is about looking for pairs of numbers that multiply to give 24 and so they do not need to draw each separate rectangle.

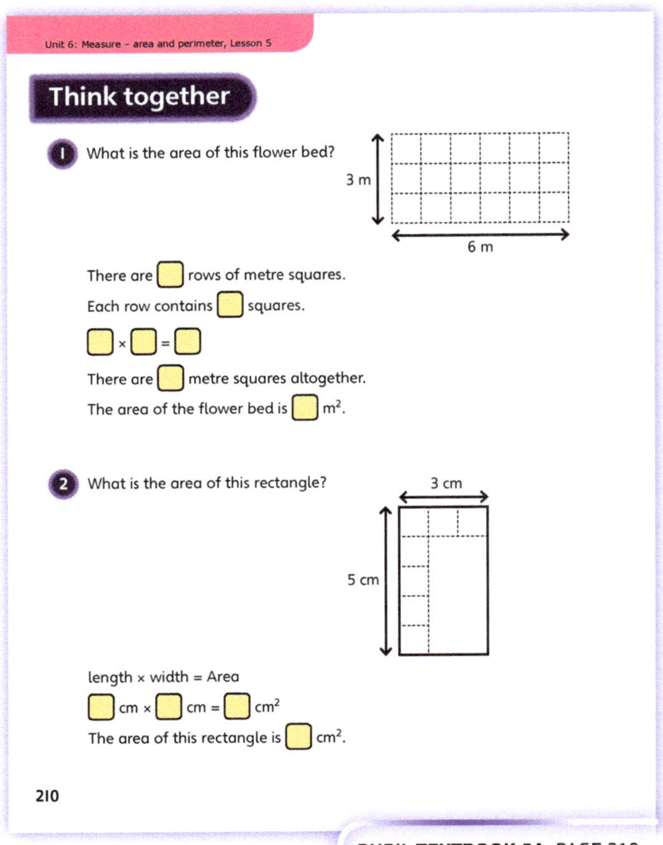

PUPIL TEXTBOOK 5A PAGE 210

PUPIL TEXTBOOK 5A PAGE 211

Practice

WAYS OF WORKING Independent thinking

IN FOCUS By question **3**, counting squares separately to find area has become an inefficient strategy as there are too many squares or all the squares are not drawn. The table emphasises the multiplicative relationship between each shape's length and width and its area. Ask: *How are the three numbers in each row connected?*

Question **4** requires children to use factor pairs of the given area as the length and width dimensions to make all the possible rectangles with that area.

STRENGTHEN To help children to make the connection between the dimensions of a shape and the multiplication facts needed to calculate its area, extend question **2**. Ask children to make the different arrays by placing plastic counters onto enlarged squared paper. Using different colours for each row may help. Ask children to draw around the array and remove the counters, leaving a rectangle. Encourage use of terminology such as a '2 by 8' rectangle, which draws children's attention to the two values that multiply together to calculate the area.

DEEPEN Using question **5**, encourage children to generalise. Ask: *If the dimensions of the card were different, is there a method you could use to predict the size of the square?* Extend the activity by giving children different paper rectangles. Ask them to predict the remaining area of the card once the largest square has been cut. Limit them to only measuring the length and width before they predict.

ASSESSMENT CHECKPOINT Children should be using the *length × width* formula confidently to find the area of rectangles. Can children use times-tables or factor pairs of a number to suggest different possible dimensions for rectangles with a given area?

ANSWERS Answers to the **Practice** part of the lesson appear in a separate **Practice and Reflect answer guide**

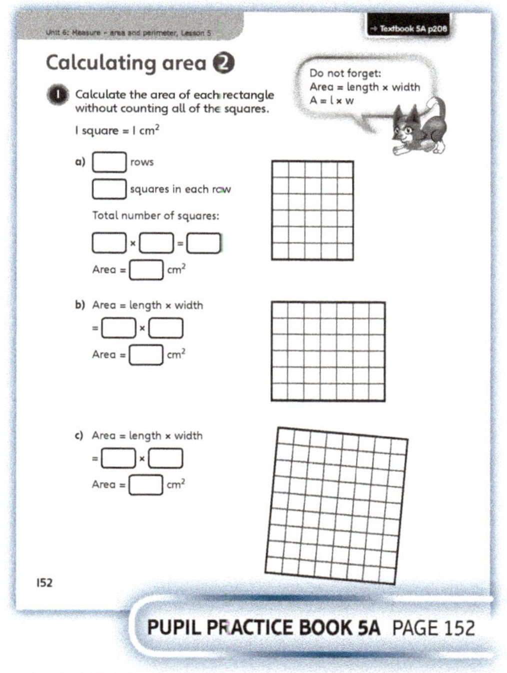

PUPIL PRACTICE BOOK 5A PAGE 152

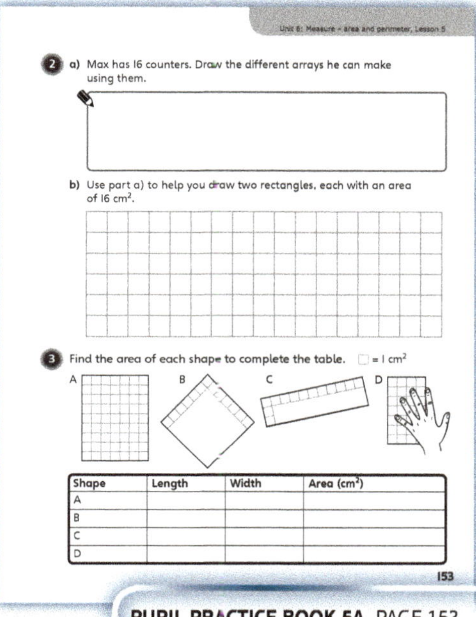

PUPIL PRACTICE BOOK 5A PAGE 153

Reflect

WAYS OF WORKING Independent thinking

IN FOCUS It is important that children apply their knowledge and understand that they simply need to multiply the board's length by its width. Note whether any children recognise that a chessboard has a square shape overall and, as such, they need only find the length of one of its sides to then be able to calculate the area.

ASSESSMENT CHECKPOINT Look for children who recognise how to multiply length by width to determine the area of the chessboard.

ANSWERS Answers to the **Reflect** part of the lesson appear in a separate **Practice and Reflect answer guide**

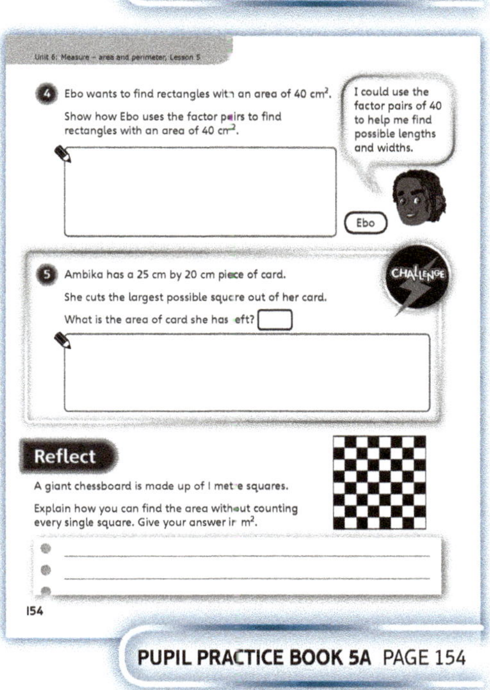

PUPIL PRACTICE BOOK 5A PAGE 154

After the lesson ⏸

- Have children mastered the formula for finding the area of a rectangle?
- Can children explain the difference between the area and the perimeter of rectangles and squares?
- Are children confident in the terminology of 'square centimetres' (cm^2) and 'square metres' (m^2)?

Comparing area

Learning focus

In this lesson, children will compare the area of rectangles (including squares). They will continue to apply their knowledge of multiplication to calculate areas.

Small steps

→ Previous step: Calculating area (2)
→ **This step: Comparing area**
→ Next step: Estimating area

NATIONAL CURRICULUM LINKS

Year 5 Measurement

Calculate and compare the area of rectangles (including squares), and including using standard units, square centimetres (cm^2) and square metres (m^2) and estimate the area of irregular shapes.

ASSESSING MASTERY

Children can confidently compare the areas of several rectangles. Children can calculate the area of a rectangle using the *length × width* relationship.

COMMON MISCONCEPTIONS

Children may compare two shapes visually (for example, thinking that the taller or wider of the shapes has the larger area), rather than by calculating their specific areas. Present two rectangles where one is significantly taller than the other yet has a smaller area. Ask:

• *Which shape has the larger area? How can you prove it?*

STRENGTHENING UNDERSTANDING

Challenge children to work backwards to design rectangles to match different comparisons. For example, ask children to draw a shape with an area of 20 cm^2. Ask: *What is the length and the width of your shape? Can you draw a shape with a greater or lesser area? What is the length and the width of your new shape?*

GOING DEEPER

Challenge children to explore some of the relationships between perimeter and area, particularly when comparing the different possibilities of shapes with given constraints. For example, give children a piece of string 30 cm long and ask them to use it to make rectangles that have a perimeter of 30 cm, but different dimensions (whole centimetres). Ask: *Which has the largest area? Is it always, sometimes or never true that, where several rectangles have the same perimeter, the 'squarer' a rectangle looks, the greater the area it has?*

KEY LANGUAGE

In lesson: area, largest/larger, smallest/smaller, square centimetres (cm^2), square metres (m^2), greater than (>), less than (<), equal to (=), compare, order, multiplication, units of measure, perimeter, length, width

Other language used by the teacher: prove, relationship, possibilities

STRUCTURES AND REPRESENTATIONS

2D shapes

RESOURCES

Mandatory: squared paper

Optional: rulers, paper squares, straws, plastic counters, string

 In the eTextbook of this lesson, you will find interactive links to a selection of teaching tools.

Before you teach 

• How confident are children with the *length × width* formula?
• What opportunities for reasoning about the area of rectangles and squares can you introduce into this lesson?

Discover

 Pair work

ASK

- Question ❶ a): *What do the labels tell you about the windows?*
- Question ❶ a): *Simply looking at the shape of each window, which looks the largest/smallest? Why?*
- Question ❶ b): *In what way are the diagrams of the windows the same? How are they different?*

IN FOCUS As in the previous lesson, the focus is on using the dimensions of width and length to find the area, but in this lesson two or more areas are compared. Ask: *Which window looks the larger/the largest? How could you prove, without counting squares, which window is larger?*

PRACTICAL TIPS Provide children with activities that get them thinking about area and how this might affect the speed that a job is completed. Mark out two simple rectangles made from metre sticks on the floor and challenge children to cover the area with newspaper. Ask: *Which rectangle will need most newspaper to cover it? How do you know?*

ANSWERS

Question ❶ a): 25 > 24, so window A has the larger area.

Question ❶ b): The area of window C is 20 m^2.

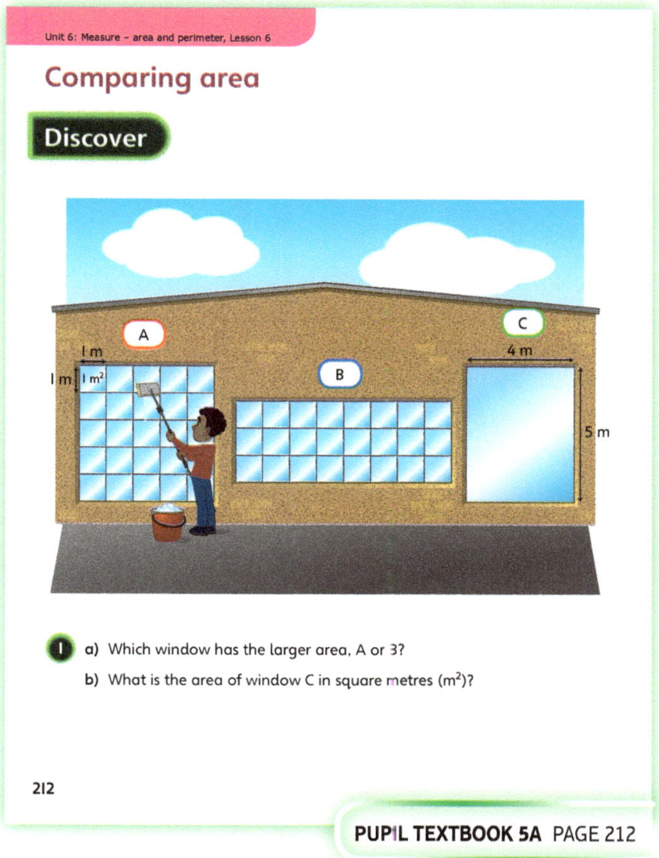

PUPIL TEXTBOOK 5A PAGE 212

Share

 Whole class teacher led

ASK

- Question ❶ a): *What information do you need to know to be able to compare the area of each window?*
- Question ❶ a): *What is the relationship between the number of rows and the number of panes of glass in each row, and a window's area?*
- Question ❶ b): *What is the most efficient method of finding the area?*

IN FOCUS Discuss the three diagrams and ensure that children understand why they have been labelled in this way. Ask: *How could you describe A and B in terms of rows and columns? How would you describe the shapes in terms of length and width?* Check that children are able to explain why area is found by multiplying length and width together.

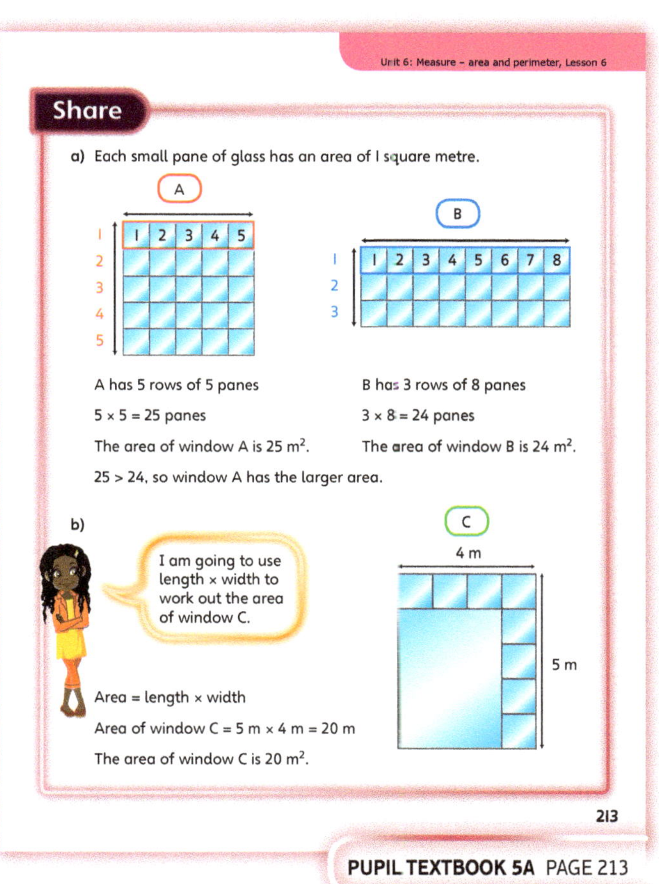

PUPIL TEXTBOOK 5A PAGE 213

Think together

WAYS OF WORKING Whole class teacher led (I do, We do, You do)

ASK

- Question **1**: *Which shape is wider/taller? Does this help you to tell which has the larger area? Why/why not? Turn the page 90° and work out the answer. What do you notice? How do you know which shape has the largest area?*
- Question **2**: *There are no squares to count, so how can you find each area? Explain how you know which two numbers to multiply to find the area of each shape. Is it always, sometimes or never true that the shape with the smallest area is the one with the smallest length or width? Explain your answer.*
- Question **3**: *What information is important in deciding which shape has the larger area? Is it just the numbers or is something else important?*

IN FOCUS It is assumed in this lesson that children will use the formula of *Area = length × width* introduced in Lesson 5. Discuss the width of Shape Z in question **2** and the fact that it is a decimal. What would this look like on squared paper? Ensure that children understand that the only way they can calculate the area of this shape is through multiplication. Revise multiplying decimal numbers by 10 if necessary.

STRENGTHEN Allow children to make arrays using counters to show the shapes in questions **1** and **2** or to draw the shapes on squared paper. Discuss the number of counters in each row and in each column, making the connection with the width and length and recording each area as a multiplication. Turn the array 90° and repeat to show that the width becomes the length but the area remains the same.

DEEPEN In question **3** ask children to explain how they know that Aki's rectangle is larger. Ask: *Are the rectangles drawn to scale? Can you use what they look like to decide which is larger? Can you think of a rectangle measured in centimetres that would be larger than one measured in metres?* (It would have to involve hundreds of cm: 1 m × 1 m < 101 cm × 100 cm, for example.)

ASSESSMENT CHECKPOINT Children should be confident using the *length × width* formula to calculate the areas of rectangles, using this information to compare and/or order rectangles according to their areas.

ANSWERS

Question **1** Shape A has 4 rows of 7. 4 × 7 = 28 cm²
Shape B has 6 rows of 5. 6 × 5 = 30 cm²
30 cm² > 28 cm²
Shape B has the larger area.

Question **2**: a) Area of X: 9 × 4 = 36 m²
Area of Y: 5 × 7 = 35 m²
Area of Z: 10 × 4 = 40 m²

Question **2** b): Z X Y

Question **3**: Aki is correct. Although 2 × 12 and 4 × 6 both equal 24, it is important to look at the units of measurement. 24 m² is larger than 24 cm².

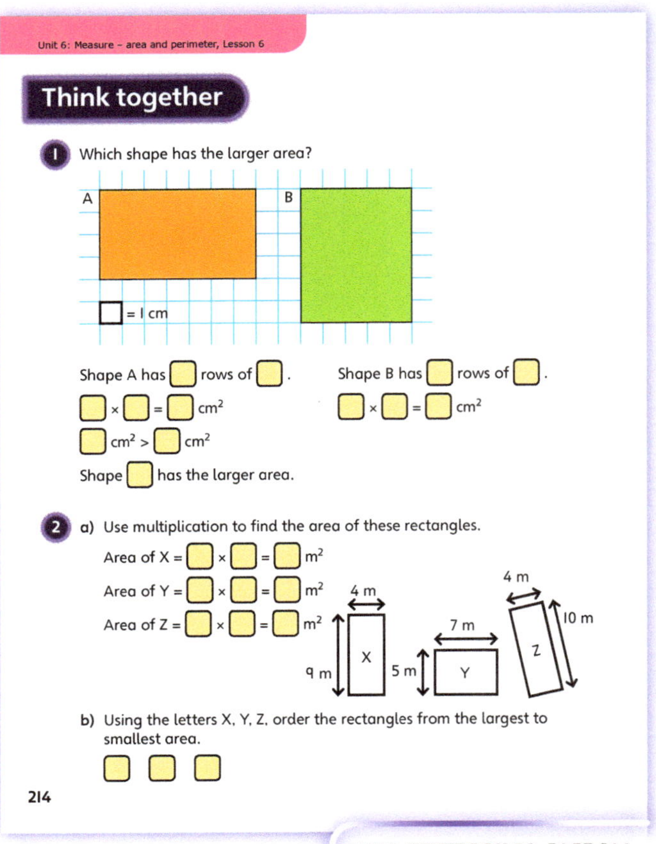

PUPIL TEXTBOOK 5A PAGE 214

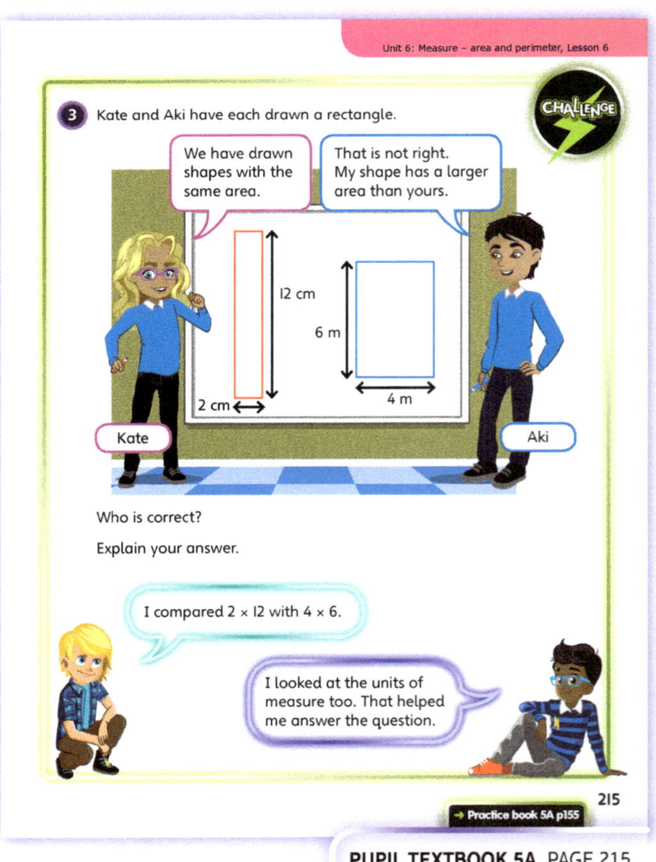

PUPIL TEXTBOOK 5A PAGE 215

Practice

WAYS OF WORKING Independent thinking

IN FOCUS Question ❹ encourages children to think carefully about the relationship between one side length and the area of squares and rectangles. The square with the longer side length will have the greater area (x^2 increases as the value of x increases). However the area of a rectangle depends on the values of both the length and the width (a rectangle with a shorter width will not necessarily have a smaller area). Ask children to give examples to prove that the statement is false, for example: $10 \times 2 > 4 \times 3$ (length × width).

STRENGTHEN If children need support with multiplication, provide multiplication grids, as the important concept here is using the formula *length × width* to work out area and understanding that you cannot always tell which is the larger shape by looking; you need to work out each area. For question ❹, provide squared paper for children to explore the areas of squares and of rectangles with varying widths and lengths.

DEEPEN For questions ❸, ❹ and ❺ encourage children to discuss in small groups the strategies they used to answer the questions. Ask: *How did you make sure you found all six rectangles/squares (question ❸)? What did you need to work out first (question ❺)? Whose strategy was easiest/ quickest?* Ask children to make up similar problems and to try using a different strategy to solve them.

THINK DIFFERENTLY For question ❸, children need to think systematically to ensure they find all the possible squares and rectangles. Ask: *What lengths of straws do you need to choose to make a square? What about a rectangle? How many different squares/rectangles could you make?* It may be useful to support children by giving them paper straws cut to the three different lengths and getting them to model their answer concretely.

ASSESSMENT CHECKPOINT Children should be calculating the areas of rectangles using multiplication in order to compare or order them according to their areas.

ANSWERS Answers for the **Practice** part of the lesson appear in the separate **Practice and Reflect answer guide**.

Reflect

WAYS OF WORKING Independent thinking

IN FOCUS Use this question to check children's reasoning. Encourage children to write two steps: multiplying the length by the width to find the areas and then comparing the two values to determine which has the larger area and therefore is the larger shape.

ASSESSMENT CHECKPOINT Can children describe the two steps needed to compare the areas of two rectangles? (Calculating the length × width then comparing values.)

ANSWERS Answers for the **Reflect** part of the lesson appear in the separate **Practice and Reflect answer guide**.

After the lesson ⏸

- Can children confidently compare the areas of two or more rectangles?
- Are there any ways that you can consolidate children's comparison skills in the following lesson, when they are estimating the area of irregular shapes?

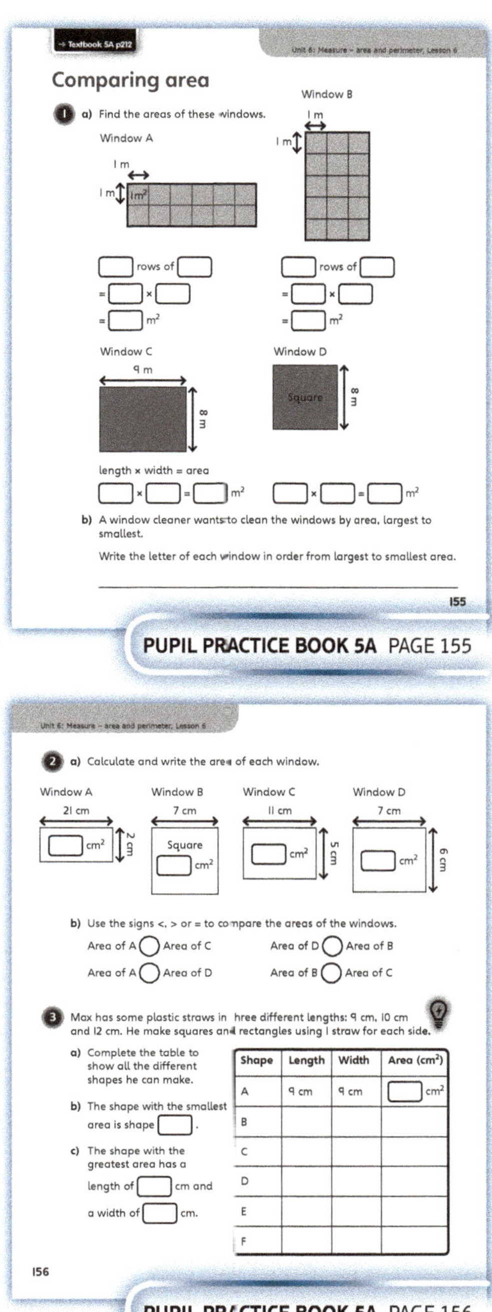

PUPIL PRACTICE BOOK 5A PAGE 155

PUPIL PRACTICE BOOK 5A PAGE 156

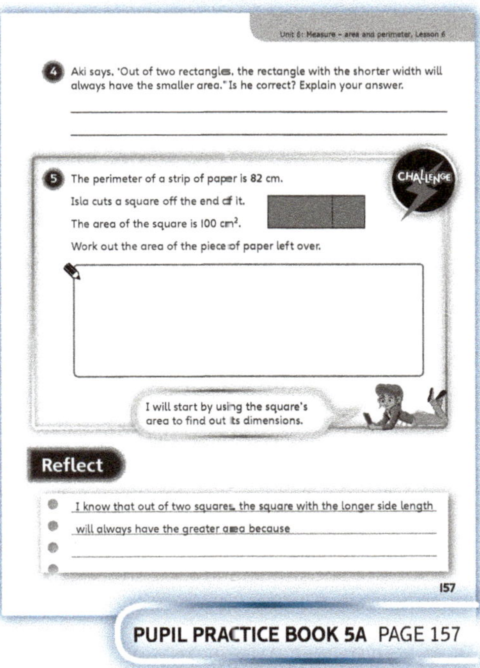

PUPIL PRACTICE BOOK 5A PAGE 157

Estimating area

Learning focus

In this lesson, children will apply their knowledge of area to estimate the area of irregular shapes.

Small steps

→ Previous step: Comparing area
→ **This step: Estimating area**
→ Next step: Multiplying numbers up to 4 digits by a 1-digit number

NATIONAL CURRICULUM LINKS

Year 5 Measurement

Calculate and compare the area of rectangles (including squares), and including using standard units, square centimetres (cm^2) and square metres (m^2) and estimate the area of irregular shapes.

ASSESSING MASTERY

Children can confidently estimate the area of irregular shapes by counting the numbers of whole squares, almost-whole squares, half squares and less-than-half squares. Children are able to recognise that this is an estimate of area and can explain why this is.

COMMON MISCONCEPTIONS

Children may disregard any squares that are not whole (or, conversely, count every part of a square as one whole, regardless of its size) and so misjudge their estimation. Ask:
• *Which squares should you count in this shape? Which ones should you ignore? Why?*

Children may confuse estimation with calculating the actual area, because it involves counting. Ask:
• *Can you describe whether you have estimated or calculated the actual area?*

STRENGTHENING UNDERSTANDING

Ensure children understand that they cannot know for definite the area of irregular shapes, but they will learn a method to estimate this. Provide children with shapes drawn on enlarged grids and plastic counters in four different colours. Use the counters to colour code the diagram, placing differently coloured counters inside whole squares, almost-whole squares, half squares and less-than-half squares. Ask: *Can you compare your shape with a partner's? Have you estimated in the same way?*

GOING DEEPER

Challenge children to set their own estimation problems for a partner to solve. Encourage children to give a visual estimation before counting. For example, they draw three irregular shapes on squared paper and ask their partner: *Without counting any squares, can you put these shapes in order from smallest to largest area?* They should then ask their partner to count the different types of squares to give a more accurate estimation.

KEY LANGUAGE

In lesson: area, square centimetres (cm^2), estimate, whole, part, almost-whole, half, less than (<), more than (>)

Other language used by the teacher: irregular

STRUCTURES AND REPRESENTATIONS

2D shapes

RESOURCES

Mandatory: coloured pencils

Optional: transparent squared overlay, counters in four colours

 In the eTextbook of this lesson, you will find interactive links to a selection of teaching tools.

Before you teach 

• How confident are children with the concept of estimation?
• What practical activities could children do to practise estimation?

Discover

WAYS OF WORKING Pair work

ASK

- Question ❶ a): *In what way are the shapes different than in previous lessons? In what way are they the same?*
- Question ❶ a): *Why do you think you are being asked to estimate the area of these shapes and not calculate them?*
- Question ❶ b): *How will the squared grid be useful when estimating the areas of the fish?*

IN FOCUS Discuss how these diagrams are different to the rectilinear shapes in previous lessons. Children should notice that the shapes are made up of a mixture of full and partial squares. Ask: *When might someone need to estimate the area of a shape that is not rectilinear?* (For example, when buying carpet for a room that is an unusual shape.) Ask: *Which fish looks like it will be the easiest/most difficult to estimate the area of? Why?*

PRACTICAL TIPS Play estimation games with children split into two teams. Display two irregular shapes drawn on squared grids and, without counting squares, ask each team to predict which shape has a larger area. Begin with simple shapes drawn from whole and half squares, gradually including almost-whole and less-than-half squares too. Talk about the usefulness of a 'thoughtful guess' (a useful way to describe an estimate).

ANSWERS

Question ❶ a): One way is to count the whole squares (8), the almost-whole squares (6), the half squares ($\frac{1}{2}$) and to ignore any squares that are less than half. $8 + 6 + \frac{1}{2} = 14\frac{1}{2}$, so the area of fish A is about $14\frac{1}{2}$ cm².

Question ❶ b): Fish C has the largest area.

Share

WAYS OF WORKING Whole class teacher led

ASK

- Question ❶ a): *Why do you think it is important to find all the whole squares first, then all the almost-whole squares and so on? Why not just look at each square one by one and decide which category it fits in?*
- Question ❶ a): *How is this method estimation? Is it not just a way to calculate area?*
- Question ❶ a): *How does this method use rounding?*

IN FOCUS Prior to discussing each of the illustrated steps with children, ask them to count each type of square on the fish for themselves. Ask: *How many half squares can you see? Compare your answer with the purple squares on the image. How many whole squares does this equal?* Ensure children understand that they need to combine two half squares into one whole square.

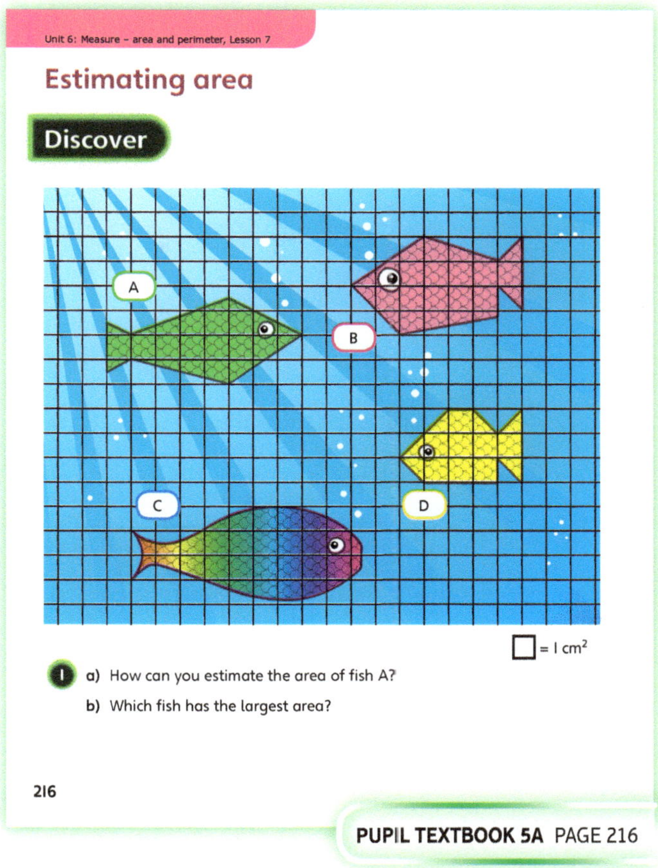

PUPIL TEXTBOOK 5A PAGE 216

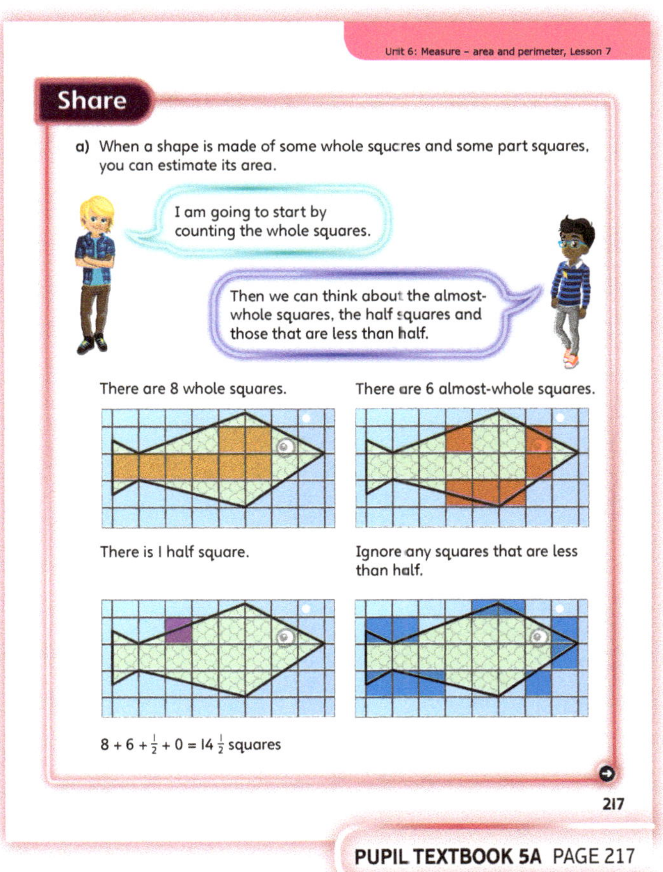

PUPIL TEXTBOOK 5A PAGE 217

Think together

WAYS OF WORKING Whole class teacher led (I do, We do, You do)

ASK

- Question ❶: *Estimate the area of the crab. What are you basing your estimate on?*
- Question ❶: *What makes a shape like this difficult to estimate?*
- Question ❶: *How can you make sure that you have counted all of the squares?*
- Question ❷: *Before you start counting the different types of squares, which sail looks largest? Why?*
- Question ❷: *How does using a table help with this activity?*

IN FOCUS Although questions ❶ and ❷ both require children to estimate areas by classifying the different types of squares that are covered by the shape, it may be worthwhile to begin each activity by asking them to make a more informal quick estimation. Ask: *Without counting anything, what do you estimate the area of this shape is? Explain your answer.* Challenging children to do this first will help them build visual connections.

STRENGTHEN To support children initially when dealing with part squares, display an image drawn from whole and half squares. Choose different children to come up and cross out a whole square each time. Ask: *How many whole squares are there?* Repeat, but this time cross out two half squares each time. Ask: *How many half squares were crossed out? How many wholes is this?* Ensure children are confident when working with wholes and halves before moving on to less familiar parts of squares.

DEEPEN Set children the task of drawing their own shapes made up of whole squares, half squares and less-than-half squares to swap with a partner. Drawing these out will help increase understanding of the different type of squares.

ASSESSMENT CHECKPOINT At this point, children should be working with greater confidence, using the method they have been taught to estimate the area of irregular shapes. Children should be able to count and record the different types of squares using tables or similar.

ANSWERS

Question ❶: 17 whole squares, 6 almost-whole squares, 6 half squares = 3 squares.
The area of the crab is about 26 squares. Note that children will use their judgement and answers may vary slightly.

Question ❷:

Sail	Whole squares	Almost-whole squares	Half squares	Less-than-half squares	Estimated area (squares)
A	13	4	1 = $\frac{1}{2}$ whole squares	4	17 $\frac{1}{2}$
B	15	0	6 = 3 whole squares	0	18

Question ❸: Various answers possible. Answers should mention the fact that as the almost-whole squares are being rounded up to 1 the less-than-half squares can be rounded down to zero.

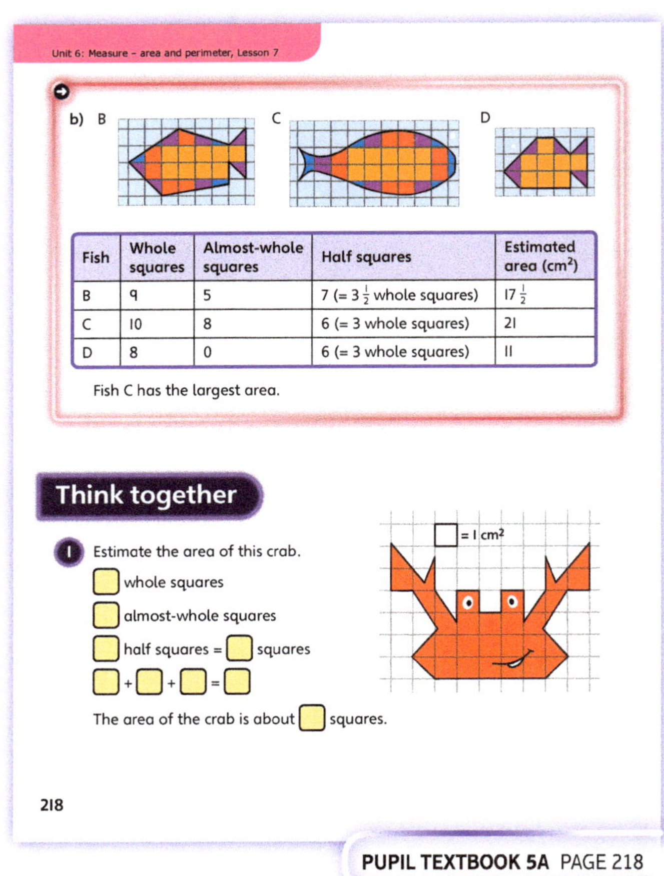

Fish	Whole squares	Almost-whole squares	Half squares	Estimated area (cm²)
B	9	5	7 (= 3 $\frac{1}{2}$ whole squares)	17 $\frac{1}{2}$
C	10	8	6 (= 3 whole squares)	21
D	8	0	6 (= 3 whole squares)	11

PUPIL TEXTBOOK 5A PAGE 218

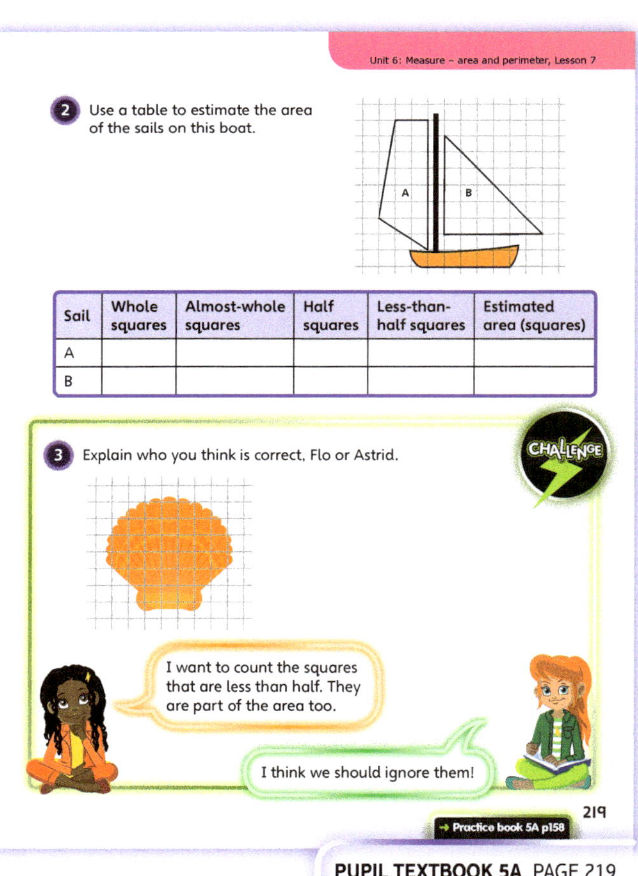

Sail	Whole squares	Almost-whole squares	Half squares	Less-than-half squares	Estimated area (squares)
A					
B					

PUPIL TEXTBOOK 5A PAGE 219

Practice

WAYS OF WORKING Independent thinking

IN FOCUS Children are led to consider half-squares in detail in question **2**. As well as the familiar representations (half shaded horizontally, vertically or diagonally), encourage children to think about less obvious ways (perhaps by using curves or splitting up the shading). This should develop children's ability to spot pairs of squares that can be combined as part of their estimation to make one whole.

STRENGTHEN Provide children with enlarged copies of each of the footprints in question **1**. Ask them to cut out each footprint and sort square or part-square into groups (whole squares, almost-whole squares, half squares and less-than-half squares). Ask: *Can you match up your squares to make as many whole squares as possible?* (For example, by placing two half-squares together to make a whole or by placing an almost-whole square with a less-than-half square to complete it.) This will strengthen children's understanding of how estimation of area involves a degree of rounding.

DEEPEN Question **5** deepens children's understanding by challenging them to estimate the area of their hand. Ask: *Where will you place your fingers to make it easier to estimate an area?* Ask children to predict their answer and then count to find a more accurate estimate.

THINK DIFFERENTLY In question **4**, children are challenged to draw shapes with an estimated area of 15 squares. Ask: *Do you think you could just draw a 3 × 5 and a 1 × 15 rectangle to answer the question? Why/why not?* Discuss with children what method they might use to create an irregular, less-obvious shape (for example, drawing a starting rectangle of 2 × 4 and then adding almost-whole, half and less-than-half squares around it to make a total of 15).

ASSESSMENT CHECKPOINT Children should be confident when estimating the area of irregular shapes. They should display a secure knowledge of a strategy to help them do this (by counting the numbers of whole, almost-whole, half and less-than-half squares). They should be able to recognise that this is an estimate of area and explain why this is.

ANSWERS Answers for the **Practice** part of the lesson appear in the separate **Practice and Reflect answer guide**.

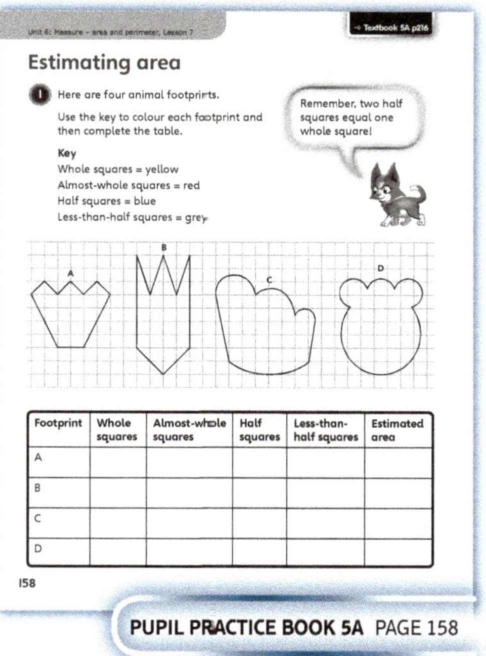

PUPIL PRACTICE BOOK 5A PAGE 158

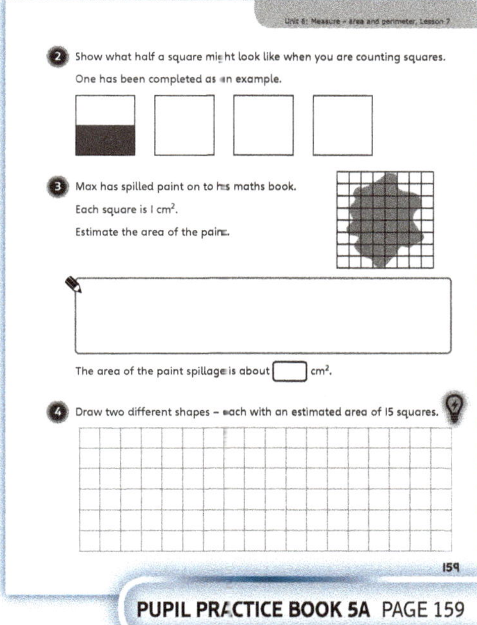

PUPIL PRACTICE BOOK 5A PAGE 159

Reflect

WAYS OF WORKING Independent thinking

IN FOCUS This question is used to help children look back on the strategy they have used to estimate the area of irregular shapes. Ask: *How would you describe an irregular shape? Is it easier to find the area of a circular or a square coffee mat? Why? Why do you think you have been asked to estimate the area of irregular shapes and not calculate them?*

ASSESSMENT CHECKPOINT Look for children who are able to describe how to count the different types of squares covered by an irregular shape.

ANSWERS Answers for the **Reflect** part of the lesson appear in the separate **Practice and Reflect answer guide**.

After the lesson ⏸

- Did children understand how to identify whole squares, half squares, almost-whole squares and less-than-half squares?
- Are children confident estimating the area of irregular shapes?

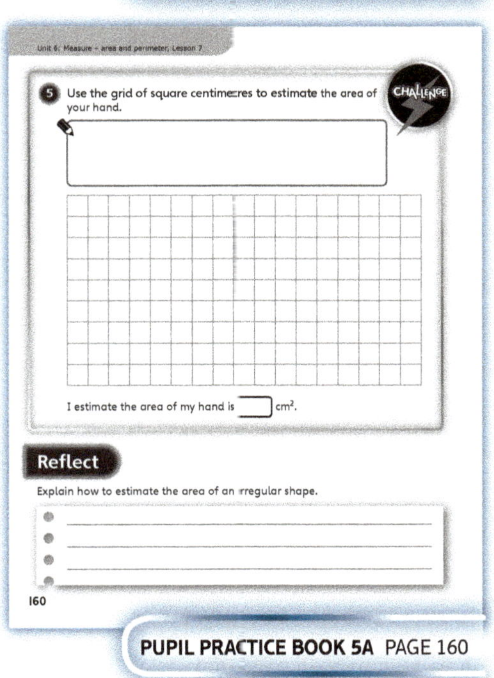

PUPIL PRACTICE BOOK 5A PAGE 160

End of unit check

Don't forget the *Power Maths* unit assessment grid on p26.

WAYS OF WORKING Group work adult led

IN FOCUS

- Question **1** assesses whether children can find the perimeter of a rectangle by measuring it.
- Question **3** assesses whether children can calculate the perimeter of a rectilinear shape where not all the side lengths are provided.
- Question **5** assesses whether children can calculate the area of a composite rectilinear shape.
- Question **6** is a SATs-style question where knowledge of how to calculate area is necessary to solve the problem.

ANSWERS AND COMMENTARY

Children who have mastered this unit will be able to find the perimeter of rectilinear shapes by accurately measuring a shape's sides and by using the given dimensions, including rectilinear shapes where not all the dimensions are supplied and the unknowns need to be derived. They can apply their knowledge inversely to suggest the dimensions of a shape given its perimeter. Children can calculate the area of rectangles by using the *length × width = Area* formula correctly. They can apply their knowledge of area inversely to suggest the dimensions of a shape given its area. Children can estimate the areas of irregular shapes, explaining which squares are to be counted.

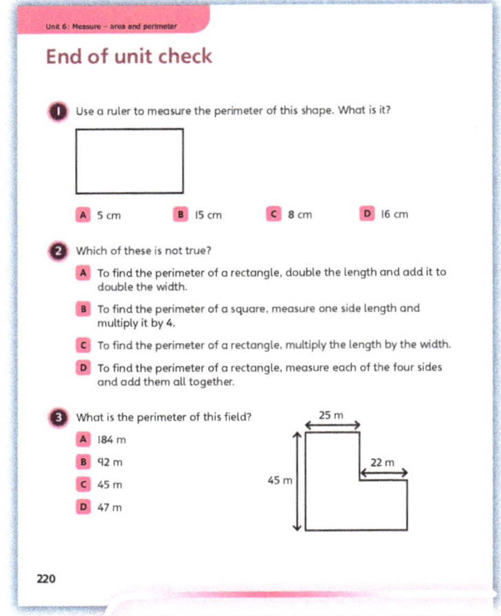

PUPIL PRACTICE BOOK 5A PAGE 220

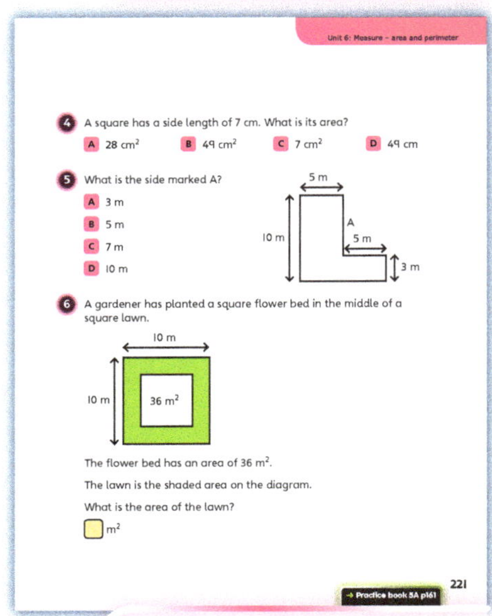

PUPIL PRACTICE BOOK 5A PAGE 221

Q	A	WRONG ANSWERS AND MISCONCEPTIONS	STRENGTHENING UNDERSTANDING
1	D	A suggests the child has found the length of the shape. B suggests they have confused area and perimeter. C suggests they have added one length and one width but forgotten to double them.	Provide different representations, both pictorial and concrete, to enable children to build conceptual connections. These may include the following: • Making shapes out of paperclips, paper straws or similar, then moving the paperclips to model the perimeter and to show combinations of equal sides. • Making arrays out of manipulatives (for example, plastic counters) and relating these concrete representations to the number of squares found in rectangles drawn on squared paper. Encourage children to work backwards too, starting with abstract representations of area and perimeter and then modelling these concretely and pictorially.
2	C	A, B and D each suggest that the child does not recognise all the different methods to calculate perimeter.	
3	A	B suggests the child simply added the given measurements. C and D both suggest that the child has calculated one dimension (the length and width, respectively).	
4	B	A shows a confusion between area and perimeter. D suggests that the child knows how to calculate area, but has forgotten to use the correct unit of measure.	
5	C	A, B and D suggest the child has not realised that the answer requires a simple subtraction.	
6	64 m²	Ensure the unit of measure is given, not just the number.	

My journal

WAYS OF WORKING Independent thinking

ANSWERS AND COMMENTARY

Question **1** a): I know that the perimeter of the shape is 58 cm because the perimeter is the total of double the length (which is 40 cm) plus double the width (which is 18 cm).

Question **1** b): I know that the area of this shape is 63 m^2 because the area of a rectangle is found by multiplying the length by the width. $9 \times 7 = 63$

If children need support answering either question, ask:
- *What does the word perimeter/area mean?*
- *What information are you told about the rectangle? How can you use it to help?*

Power check

WAYS OF WORKING Independent thinking

ASK
- *What did you know about area and perimeter before you began this unit?*
- *What do you know now?*
- *How has what you have learnt made it easier to calculate the area and perimeter of shapes?*

Power puzzle

WAYS OF WORKING Pair work

IN FOCUS The purpose of this puzzle is for children to explore the different possible dimensions of rectangles in order to maximise the given perimeter and area. In order to solve the puzzle, children will need to apply their knowledge of perimeter and area and of number bonds and factor pairs.

ANSWERS AND COMMENTARY

Question **1** : The possible rectangles are: 24 cm × 1 cm (perimeter: 50 cm), 12 cm × 2 cm (perimeter: 28 cm), 8 cm × 3 cm (perimeter: 22 cm) and 6 cm × 4 cm (perimeter: 20 cm).
The rectangle that maximises perimeter, therefore, is the one with the longest length (24 cm × 1 cm).

Question **2** : The possible rectangles are: 11 cm × 1 cm (area: 11 cm^2), 10 cm × 2 cm (area: 20 cm^2), 9 cm × 3 cm (area: 27 cm^2), 8 cm × 4 cm (area: 32 cm^2), 7 cm × 5 cm (area: 35 cm^2) and 6 cm × 6 cm (area: 36 cm^2).
The rectangle that maximises area, therefore, is the one that is square (6 cm × 6 cm).

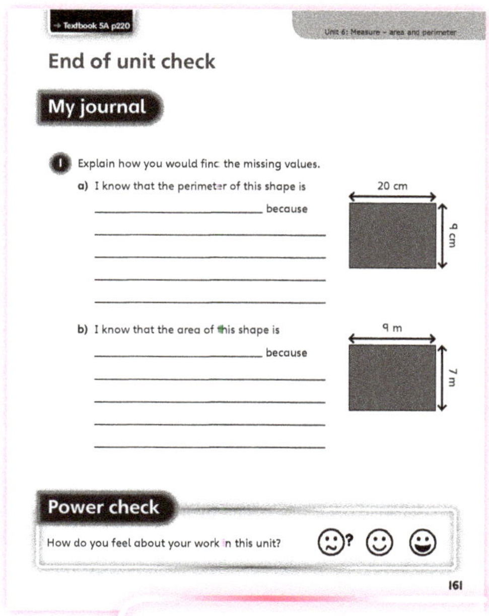

PUPIL PRACTICE BOOK 5A PAGE 161

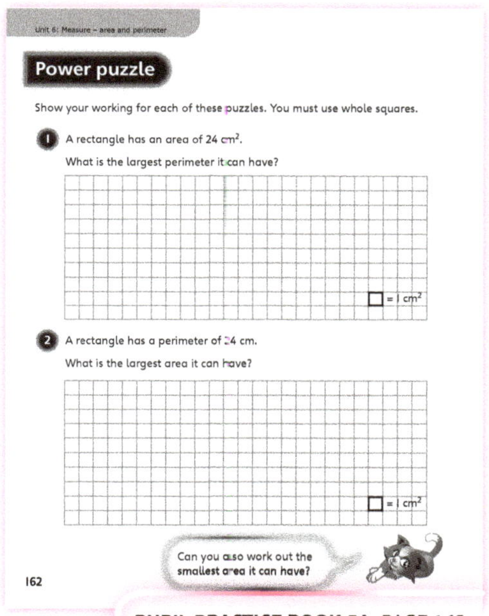

PUPIL PRACTICE BOOK 5A PAGE 162

After the unit ⏸

- Can you think of opportunities where you could use the concepts of area and perimeter in other curriculum areas to keep children's knowledge simmering?

Strengthen and **Deepen** activities for this unit can be found in the *Power Maths* online subscription.

Published by Pearson Education Limited, 80 Strand, London, WC2R 0RL.

www.pearsonschools.co.uk

Text © Pearson Education Limited 2018
Edited by Pearson, Little Grey Cells Publishing Services and Haremi Ltd
Designed and typeset by Kamae Design
Original illustrations © Pearson Education Limited 2018
Illustrated by Diago Diaz and Nadene Naude at Beehive Illustration; Emily Skinner at Graham-Cameron Illustration; and Kamae.
Cover design by Pearson Education Ltd
Back cover illustration © Diago Diaz and Nadene Naude at Beehive Illustration.

Series Editor: Tony Staneff
Consultants: Professor Liu Jian and Professor Zhang Dan

The rights of Tony Staneff, Liu Jian, Josh Lury, Zhou Da, Zhang Dan, Zhu Dejiang, Kate Henshall, Wei Huinv, Hou Huiying, Zhang Jing, Steph King, Stephanie Kirk, Huang Lihua, Yin Lili, Liu Qimeng, Timothy Weal, Paul Wrangles and Zhu Yuhong to be identified as authors of this work have been asserted by them in accordance with the Copyright, Designs and Patents Act 1988.

First published 2018

22 21 20 19
10 9 8 7 6 5 4 3

British Library Cataloguing in Publication Data
A catalogue record for this book is available from the British Library

ISBN 978 0 435 19040 8

Copyright notice
Pearson Education Ltd 2018

Printed in Great Britain by Ashford Colour Press Ltd.

www.activelearnprimary.co.uk

Note from the publisher
Pearson has robust editorial processes, including answer and fact checks, to ensure the accuracy of the content in this publication, and every effort is made to ensure this publication is free of errors. We are, however, only human, and occasionally errors do occur. Pearson is not liable for any misunderstandings that arise as a result of errors in this publication, but it is our priority to ensure that the content is accurate. If you spot an error, please do contact us at resourcescorrections@pearson.com so we can make sure it is corrected.